AMBIGUITIES IN DECISION-ORIENTED LIFE CYCLE INVENTORIES

ECO-EFFICIENCY IN INDUSTRY AND SCIENCE

VOLUME 17

The titles published in this series are listed at the end of this volume.

Ambiguities in Decision-oriented Life Cycle Inventories

The Role of Mental Models and Values

by

Frank Werner
Environment and Development,
Zürich, Switzerland

Springer

A C.I.P. Catalogue record for this book is available from the Library of Congress.

ISBN-10 1-4020-3253-6 (HB)
ISBN-10 1-4020-3254-4 (e-book)
ISBN-13 978-1-4020-3253-0 (HB)
ISBN-13 978-1-4020-3254-7 (e-book)

Published by Springer,
P.O. Box 17, 3300 AA Dordrecht, The Netherlands.

www.springeronline.com

Cover image of the world: *NASA, http://visibleearth.nasa.gov/data/ev2/ev246_indoex2_a.mov*

Printed on acid-free paper

Printed in the Netherlands.

Contents

CASE STUDY I: DEVELOPMENT OF THE VALUE-CORRECTED SUBSTITUTION FOR ALUMINIUM WINDOWS

Preface

This book is the result of the research work I have been carrying out during my stay at the Swiss Federal Laboratories for Materials Testing and Research (EMPA) in Duebendorf[1]. From 1996, I have conducted various comparative life cycle assessments (LCA) of building materials and products, often in close collaboration with companies. Over the years, I have become quite disillusioned about the usefulness of LCA-results for the companies involved. This concern was less related to the often discussed assessment step within LCA methodology, but with fundamental accounting principles, concerning the question of which material and energy flows are considered causally linked to a product.

'Descriptiveness of the product system' has become a key term and guiding principle for my inquiries and methodological research. I am convinced now that the LCA community will have to abandon a much loved conception of LCA being able to objectively record the material and energy flows related to one product. Instead – and as a consequence of fundamental epistemological considerations – I am very much committed to a subjectivistic conception of LCA as a decision support tool.

There is a difference between doing the right things and doing things right. Conducting an LCA according to somehow established rules does not necessarily provide consistent, 'descriptive' models that are useful as a decision support tool. How life cycle models of a product should be set up in the

[1] This text was accepted as a PhD-thesis at the Swiss Federal Institute of Technology (2002) under the title "Interdependencies between LC-modelling and the use of LCA in product design-related decision situations; with special emphasis on the influence of cognitive models and values on the modelling of reuse & recycling and other end-of-life options"

life cycle inventory analysis and which criteria they should fulfil are primary research questions of this book. Furthermore, I have become more and more interested in the question of which subjective elements finally guide the setting up of a product system and how they should be handled to provide a 'descriptive' and reliable model as basis for rational decision-making.

I have carried out the studies on which this book is based hoping to adapt LCA-methodology – particularly the life cycle inventory step – to the decision-maker's needs and questions. Saying 'good-bye' to the fiction of recording the environmental relevance of products in an objective way and accepting the influence of subjective elements throughout the life cycle inventory analysis might be disillusioning. On the other hand, it allows one to understand LCA in a different, 'functional' way: as a decision-support tool that supports efficient and effective decisions in a case-specific context. I am very much convinced that this perception of LCA uncovers potential for LCA that has largely lied fallow.

Taking into account the decision-maker's (subjective) conceptualisation of the decision situation and his/her values is an indispensable condition for the 'descriptiveness' of LCA and thus for its usefulness in a case-specific context. With this different perception of LCA this method can hopefully gain back the trust and acceptance it has been losing during the last years.

This book would not have been possible without input and support from a large number of people. I am particularly grateful to:

- Dr. Jürgen Sell and Dr. Klaus Richter at EMPA for their moral support and for leaving me the freedom to carry out the studies on which this book is based;
- Prof. Dr. Roland W. Scholz for encouraging me and accepting the reference of my PhD-thesis on which this book is based as well as for his valuable guidance;
- Prof. Dr. Stefan Schaltegger for accepting the co-reference and for his critical well-founded inputs;
- Dr. Susan Kythzia, Dr. Irmi Seidl and Dr. Carmenza Robledo for encouraging me to undertake the Ph.D-thesis;
- Dr. K. Buxmann for introducing me to the world of aluminium and aluminium recycling;
- Inge Reichart, Hans-Jörg Althaus and Dr. Klaus Richter for their critical inputs and for improving clarity of various chapters of this book;
- The Swiss Federal Institute of Technology Zurich for funding the drawing up of this book;
- Dominic Corbett for his careful lectorate.

Furthermore, I express my gratitude to several people who have facilitated data collection or with whom I have shared fruitful discussions, among them Dr. Olaf Tietje; Dr. Gjalt Huppes; Dr. Erwin Lindejier; Dr. Bo Weidema; Dr. Johannes Gediga; Patrick Corbat; Dr. Jörg Schweinle; Hans Udry; Dr. Rolf Frischknecht; Andreas Keel; Hans-Peter Graf and the members of working group WG3 of COST E9, especially Gerfried Jungmeier (Graz) and Anna Jarnehammer (Stockholm). Special thanks also go to the members of the Wood Department at EMPA, who with their warm and care-taking attitude have created an optimal milieu for my work.

Frank Werner, Zurich 2004

The functionalistic perspective of LCA modelling and application - a foreword

Life Cycle Analysis or Life Cycle Assessment (LCA) has become a widespread evaluation approach that has almost attained the status of a branch of environmental sciences. The current state of the art of this approach has been formed to a large extend by a Dutch group around the Centre of Environmental Science – Leiden (CML). The current state of the art knowledge is presented in the powerful 'Handbook on Life Cycle Assessment' (Guinée *et al.* 2002).

In the last decade, LCA development and application build up a meeting point of natural scientists, engineers, public health researchers and some social scientists. From the social science side, in particular economists working at the fringes of environmental sciences and economical thinking came up to this field. However, also some other social sciences such as cultural theory, cognitive psychology, decision sciences have touched the field. In my opinion of specific importance for LCA, however, is the theory of science.

Naturally, it seems to be in the nature of any theory development that the reflections on the '*epistemological status*' of the basic assumptions or on the status of the outcomes or the theory itself have to be postponed until a certain level of maturity and/or practical success stories are at hand. Epistemology is particularly important for applied sciences as – casually formulated – epistemology tells us what we can know and how good certain knowledge is. Statements on the 'epistemological status' should thus clarify what the results or basic statements of a theory embody. In this context, a very fundamental question concerning LCA is, whether the components of the methodological framework (*i.e.* goal and scope definition, inventory analysis, impact assessment, interpretation; according to ISO 14040) are conceived as objective natural science knowledge or whether LCA is conceived as a subjective, interest related construction in such a way as it has been, postulated in the theory of probabilistic functionalism for a perceptional (Brunswik 1950) or evaluative human activity (Scholz and Tietje 2002).

LCA was initially developed by natural scientists and engineers. Thus criteria from the traditional natural sciences were dominating theory construction from the beginning. The pure, objectivistic natural science thinking of the LCA community was pronouncedly expressed by the following statement: LCA should be:

> "objective like mathematics where people all over the planet can do the same work and get the same answer" (Elkington and Hailes 1993:16).

This 'pure objectivity' assumption got 'relativised' later by the suggestion that we should distinguish between objective and subjective elements in LCA. Proponents of this opinion, however, always stressed that in particular the Inventory analysis was still considered as purely objectivistic natural science work.

Thus a counter position got developed, which was elaborated in the work of Hofstetter (1998, 1999). Thus the core statement of a 'social science and epistemology' perspective was that "subjective elements should be integrated into all phases of LCA rather than the subjective elements" (Hofstetter *et al.* 2000:161) are ubiquitous in all steps. The present book of Frank Werner on 'Ambiguities in decision-oriented Life Cycle Inventories' elaborates this complementary functionalistic, social science perspective for the Inventory analysis. As it is documented in the Introduction, Werner is not only referring to theoretical arguments, but developed this perspective from reflections on practical experiences gained in several case studies on recycling studies, which were conducted at the Swiss Institute of Material Testing in the 1990s (Werner and Richter 2000). Werner and colleagues obviously noticed that the outcomes of an LCA strongly depend on subjective perspectives and model assumptions of the user.

Frank Werner's seminal analysis reveals that LCA is both a *theoretical construct* and a *tool to model environmental impacts* of products or processes. As he discusses, the results of a LCA depend on the perspective and contexts of the modeler or user. On the background of two thorough case studies on aluminum window frames and railway sleepers from beech, he carefully discloses fundamental assumption and limitations of LCA such as the ceteris paribus assumption, which ignores context specific interactivity, the ubiquitous linearity assumption of LCA approaches, the consequences that mass- and energy flows are completely abstracted from their temporal and spatial context, the context dependence of setting system boundaries or allocations and last but not least the value choices that are made in particular on the product system that underlies the inventory stage. Werner concludes that LCA is a highly abstracted, simplified 'unit-world, input-output model'. In particular he infers that any inventory model is relatively bound to the available knowledge, the specific model assumptions made and on the context of modelling and/or application.

In line with a functionalistic perspective, Werner considers LCA as a decision support tool. In practice, it is selected in a specific decision situation if it is assumed suitable for providing 'right' answers to a problem in a given context. This managerial perspective stresses the context under which LCA is performed. Crucial context factors for instance are the material and market characteristics of the main materials involved in an LCA as well as their desireable material-specific 'sustainable' use patterns. As Werner points out, a meaningful or appropriate application of LCA – in particular the coping with the allocation problem – is only possible if the modeller has the respective structural knowledge, is familiar with the model and modelling characteristics of LCA and is aware of the epistomological preconditions of LCA.

As Jungbluth (2000) and Jungbluth *et al.* (2000) showed, LCA must be tailored to the information and to the needs of the user or to express this in other terms:

> "LCA only makes sense, if it both appropriately represents the subject/issue to be balanced and evaluated including the relevant environmental impacts *and* if it meets the values and utility functions of the user" (Scholz and Weidenhaupt 1998:41; translated from German).

If LCA is considered a decision support tool, the user thus does not only have to be capable to master the algorithmic modelling framework but especially the 'qualitative' aspects of decision making. As Scholz & Zimmer (1997:9) pinpointed, the mastering of the qualitative side of decision making depends on having the appropriate *concepts* at your disposal to reasonably dealing with *complexity* and *contexts.* Those who want to gain mastership in this sense in the context of LCA or Life Cycle Inventories are strongly recommended to the book at hand.

Prof. Dr. Roland W. Scholz

Swiss Federal Institute of Technology (ETH) Zurich
Chair of Environmental Sciences, Natural
and Social Science Interface

I.

INTRODUCTION

Chapter 1

INTRODUCTION

If there is no unambiguous way of modelling the life cycle of a product, then there is a subjectively best way.

This book is about modelling in life cycle assessment (LCA). Life cycle assessment is a decision support tool within environmental management, which has been developed to compile and assess the environmental implications of products.

It is investigated how the decision-maker's mental models, knowledge representations of how real world things work and interdepend, as well as how his/her values influence the attribution of material and energy flows to a product under study.

Emphasis is laid on modelling reuse & recycling and other end-of-life options as methodological discussion on these issues is still going on. Furthermore, the decisions related to the modelling of end-of-life options can have decisive impact on the result. Thirdly, the modelling of reuse and recycling is the most complex methodological topic related to attribution and allocation in LCA. This implies that, when treating this issue, the decision-maker's mental models and values and their impact on the modelling within LCA emerge in a very clear way.

The book concentrates on the product-related use of LCA, *e.g.,* the use of LCA in product-design related decision-making. This restriction is made to limit the scope of the book and to leave aside other applications of LCA such as the environmental assessment of systems (*e.g.,* regional or national waste treatment systems) or of companies (*i.e.* the environmental performance evaluation).

Focus lies on the modelling aspect of LCA, and in the life cycle inventory analysis in particular. The role of LCA in planning processes and investment decisions of companies from the business management point of view is not of primary interest. Also, it is not the aim of this book to adapt or improve LCA within different concepts of strategic decision-making.

Instead, reference will be made to the series of standards ISO 14040 and to the concept of LCA outlined therein[2]. The practice perspective is assured by discussing experiences made with LCA as a decision support tool in product design-related decision-making.

1.1 Rising of the disquiet: a retrospective

The context of discovery of a problem often provides valuable insights into the structure and the logic of a problem solution; the motivation for the investigation and the context of the problem solution become clear: the rising of the disquiet, a retrospective.

It all began in 1996. The life cycle assessment (LCA) of different window frames, which was carried out and published by EMPA in 1996,[3] was widely received with great interest. As one of the first studies within the building sector, various constructions of window frames out of steel, wood, wood-aluminium, aluminium, and PVC were looked at. The study included all processes from raw material extraction (or collection and recycling activities), frame production, compensation of heat losses during the use phase, and finally demolition of the different frames. Further on, some construction options were assessed, such as different coatings, higher recycling rates and different shapes of the profiles searching for the best environmental improvement option.

A project group consisting of representatives of each frame material accompanied the study.

In contrast to general expectations, the main results were not clear. As a multi-objective impact assessment method[4] had been used, a clear-cut ranking of the different materials could not be established. However, for all window constructions some general recommendations and improvement options were outlined in the interpretation of the study. Among them:

> "For all the recyclable materials, the scenarios with the maximal recycling rates show the lowest environmental implications, as expected. A higher recycling rate (in fact a higher recycled metal content because of the specific modelling in the study; WF) shows to be much more relevant

[2] The series of standards ISO 14040 provides guidance for the descriptive type of LCA and not for marginal LCA, although not explicitly mentioned.

[3] Richter *et al.* 1996

[4] the CML-classification method after Heijungs *et al.* 1992

> for environmental improvement than all other measures examined" (after Richter *et al.* 1996:115; translated from German by WF).

This outcome was in fact *expected* and thus not very surprising. In public opinion, recycling was (and still is) considered one of the most important strategies for lowering the material intensity and environmental impacts of the socio-economic system[5]. Recycling was and is perceived to be a reduction of raw material consumption, of the environmental impacts related to primary material production and of waste going to landfill. Indeed, for many materials, the collection and recycling of secondary material is much less environmentally intense than primary material production. Thus, this outcome of the study intuitively seemed right.

Even actors within the aluminium industry followed this line of reasoning. As an aim stated in the environmental objectives of his newly developed Environmental Management System, an aluminium window producer declared to raise the content of recycled aluminium in his products as an environmental target. He should have known better.

Another point was highlighted in the study:

> "Due to the fact that no material clearly scores best, it may be concluded that a boycott of certain materials or the ban of their application cannot be factually legitimated".[6]

Indeed, some of the German Bundeslaender had claimed bans and restrictions on several window materials in public submissions. In Berlin for instance, the use of PVC in window frames was prohibited as part of the chlorine discussion, and the use of aluminium for windows was made dependent on the proof of a separate 'closed-loop' re-collection system of old aluminium windows ('windows to windows'). The authorities legitimised this restriction as being part of the implementation of the German 'Kreislaufwirtschaftsgesetz'.

There is no evidence that the restriction on aluminium windows in Berlin was based on LCA-results. Nevertheless, the restriction was very much in line with the recommendations of the EMPA-study: in a scenario analysis, the increase of the recycled metal content from an average of 35% to an optimistically estimated future 85% had lowered the environmental burden of the aluminium frame material of more than 50%! In this study, it was furthermore assumed that a recycling rate of 85% would be reached as a steady state after the closing and stabilisation of the material cycle of aluminium window frames ('windows to windows'). This last assumption was made

[5] see, *e.g.*, Anonymous 1994d

[6] after Richter *et al.* 1996:115; translated from German by WF

implicitly as a consequence of the characteristics of the cut-off allocation procedure[7] combined with the closed-loop assumption: whereas a recycled metal content of 35% represented an assumed[8] average ratio between primary aluminium production and secondary material production of the Swiss aluminium market, a recycled metal content of 85% could only be assumed if the window fraction was separated from the common aluminium recycling flows.

This favourable closed-loop scenario had to be understood as a recommendation for the environmental authorities (and for the aluminium industry) to intervene on the existing recycling logistics and to realise such closed material loops within one and the same application.

Aluminium window producers in Germany were not at all willing to accept the legislative restriction of their market. Instead, they set out to put the closing of the recycling loop for aluminium windows into practice. An initiative 'Aluminium und Umwelt im Fenster- und Fassadenbau' (A/U/F) was launched with considerable PR-effort to implement a collection and recycling system for aluminium window scrap separate from the already existing collection and recycling routines for aluminium. Success was granted to A/U/F and cumulated in broad media presence[9]. This time, the ban could be avoided.

All is well that ends well?

In 1998, representatives of the aluminium window industry came back to EMPA and said: "Look, the separate collection and recycling of the aluminium window scrap is against common sense! It requires more transportation and all you achieve is that the previous buyers of (high-quality) window scrap are now forced to buy primary material instead to upgrading their scrap[10]. Considering the socio-economic system as whole, the environmental impacts must have risen because of the setting up of separate collection systems. This cannot be all that LCA is able to recommend?!"

To fully understand this argument one should take some general characteristics of aluminium and its market into account[11]:

- Aluminium is a metal. It can be remelted (recycled) without changes to its inherent material properties;

[7] for details see Chap. 11.1

[8] As remelted recycled aluminium consists of a mixture of various scrap types, ranging from industrial, direct scrap to post-consumer scrap and sometimes even primary aluminium to achieve a certain alloy composition, this data has in fact little meaning (see also Chap. 9.3.).

[9] see, *e.g.*, Anonymous 1997c; Anonymous 1997a

[10] see, *e.g.*, Buxmann 1998a:27

[11] see first case study (Chap. 8 to Chap. 13) for a complete reasoning and references

- Adding a variety of alloy elements modifies the inherent properties of aluminium influencing its processability (*e.g.*, wrought aluminium and casting aluminium) and its functional material characteristics. For the inherent material characteristics it is of no relevance if an alloy is composed of primary aluminium having the alloy elements added or of a combination of recycled aluminium scraps already containing the alloy elements;
- Some of the most common alloy elements – especially iron and zinc – lead to an economic devaluation of aluminium. These alloy elements restrict the possible processing and applications of aluminium, although they are required in certain quantities for achieving specified processing characteristics[12]. Removal technologies are theoretically available but far from being applicable due to high costs and due to the expanding aluminium market with corresponding primary material extraction;
- Secondary (recycled) aluminium alloys are composed by remelters on demand out of varying types of scrap (and primary aluminium) in varying quantities;
- The amount of recycled aluminium available depends on various factors such as the quantity of primary material put into technosphere, the service life of aluminium products, the recycling rate and the yield of the recycling processes;
- Aluminium scrap often has a positive monetary value, especially if collected in the building, printing and transport sector or as industrial scrap. This leads to an almost complete recollection in these sectors, as quantities to be recollected are high, and their composition is homogeneous;
- As a relatively young material[13] much more primary material is actually entering technosphere than is actually being recycled. The annual average increase of the aluminium pool is about 3.6%. One of the reasons, especially in the building and transport sectors, is aluminium is used in applications with long service life such as windows and facade elements. From a mass flow perspective, these applications with long service life can be considered temporal sinks in a system not in a steady state. As a consequence and taking into account the penultimate point the worldwide demand of aluminium in all its applications has to be considered the cause for the amount of primary material entering the technosphere.

[12] Among the representatives of the aluminium industry, this even lead to some arguing if it was justified to address a higher iron content (or a higher alloy content in general) as 'devaluation'.

[13] compared to iron or brass; processing on an industrial scale started around 1890 with the invention of the Hall-Héroult-processing based on electrolysis (Regitz and Falbe 1995, keyword: aluminium).

Coming back to the example of the aluminium window, the overall consumption of primary aluminium is not affected by separating single aluminium streams[14]. Thus, taking into account the additional collection efforts, the separate collection system leads to an increase of the overall environmental impacts.

Overall, the representatives of the aluminium industry felt that if the separate closing of the material cycle for aluminium frames and a higher recycled metal content were the main improvement options deducible from LCA, this decision support tool was not suitable for them. Its main deficiencies were perceived to be the following:

- It recommended improvement options at the input side of the product system – a higher recycled metal content – which were against the logic and the characteristics of the aluminium market;
- It recommended improvement options at the output side of the product system – a closed-loop recycling system for windows – leading to a higher environmental overall impact if reassessed from a meta-perspective; again, the logic and characteristics of the aluminium market were not adequately addressed;
- It didn't depict environmental improvement options influencing the output side of the product system that are currently followed in the design of aluminium products. Such improvement strategies are, *e.g.*, uni-grade constructions or an improved separability of different materials aiming at a 'value preservation of aluminium'. Neither did efforts in improving recycling technology show up in LCA results.

The arguments they presented were not very flattering for LCA in general, and for the results of the windows study in particular. But what were the reasons?

One of the reasons for the misleading outcome certainly was that the window study was planed and conducted in the years of birth of LCA-methodology and with few practical world-wide experiences at the time. The series of standards ISO 14040 giving some guidelines for the conduct of LCA were not then available; the methodology applied was based on earlier recommendations of SETAC and CML[15], at that time being state-of-the-art.

But there were more reasons, much more fundamental for the methodological discussion:

- The part cut out of the worldwide aluminium flows as product system and as such attributed to the aluminium window was considered inade-

[14] see also Chap. 9.3

[15] see, *e.g.*, Consoli *et al.* 1993; Fava and Denison 1991; Fava *et al.* 1991; Heijungs *et al.* 1992

quate. It seemed that the product system set up according to wide spread 'accounting rules' was inconsistent with their view of the characteristics of the aluminium market. As a consequence, optimisation strategies derived from the LCA had to lead to environmentally inefficient solutions on a real world level;

- Furthermore, the 'accounting rules' applied did not allow for addressing the values the representatives of the aluminium industry stated in their general ideas and strategies on how to improve the environmental profile of aluminium products: the 'value preservation of the material' as the prevention of downcycling;
- As a consequence of the 'inadequate' system definition, the action space of the decision-makers could not properly be depicted, as recycling characteristics had not been taken into account.

It was obviously the life cycle inventory analysis of the study that needed closer consideration, especially the attribution principles underlying the definition of the product system. Setting system boundaries and allocation related to recycling of aluminium window frames[16] were of special interest: during these steps within the inventory analysis, the interconnection of the product system with the representatives' mental models of real world material and energy flows was defined[17].

A follow-up project was launched to find a way in which a more 'appropriate' product system of an aluminium window could be developed. It should allow one to depict the improvement strategies stated by the representatives of aluminium industry, and thus should be in line with their mental models and values of how to manage aluminium flows in a more sustainable way. Among other requirements, 'value preservation over the life cycle' or 'prevention of downcycling' required taking the attention from the input side (the recycled metal content) and redirecting it to the recycling characteristics and recycling improvement options of the window.

For this purpose:

- A corresponding paradigmatic description of their values could be found in the cascade model of Sirkin and ten Houten (1994);
- The effective recycling flows of an aluminium window were gathered during a series of visits to recycling sites;

[16] In LCA-terminology 'allocation' means a partial attribution of the material- and energy flows of a shared or joint process to the product under study according to specified allocation rules (see also Chap. 5.6).

[17] These mental models of the characteristics of the aluminium market can be backed up by literature from economics, material sciences, etc. For the question of whose mental models and values are to be depicted, see Chap. 3.2.4.

- An allocation procedure 'value-corrected substitution' was developed for aluminium products taking into account the material and market characteristics of aluminium. The value-corrected substitution is based on the cascade model of Sirkin and ten Houten (1994) and complies with the series of standards ISO 14040[18].

As a result of this study, 'adequate' recommendations could be given for the construction of environmentally improved aluminium windows. Such windows should (amongst others) not contain other non-ferrous metals, especially no zinc-diecasted corner cleats or brass parts because the separation of aluminium from other non-ferrous metals requires additional treatment leading to aluminium losses and poor separation efficiency. Furthermore, iron fittings should be constructed and integrated into the window profile in a way that they could easily be removed during stripping[19] or could easily be separated during mechanical separation (*e.g.,* shredding). Such measures would allow for an almost complete and pure separation of the aluminium fraction, which in turn would lead to an economic and environmental value preservation of the material.

Indeed, during the last session of the project a representative of an aluminium window producer proudly presented technical drawings of a pure aluminium construction in which the zinc-diecasted corner cleats had been replaced with aluminium parts made from the same alloy as the window profiles. He had finally convinced his technical director of the need to modify the window construction itself as well as the production processes. The environmental argument had outweighted the economic costs of the modifications.

In contrast to the positive perception within the aluminium industry, a fierce controversy arose during the peer review with some leading exponents of LCA community. How could it be that system modelling was case-specific and that there was no generic way of conducting either an LCA in general or in particular of modelling the life cycle of a product? How about objectiveness and credibility of LCA results if everybody did whatever he/she believed to be adequate on a case-specific basis? Wasn't there an objective step by step procedure for the setting up of the product system derived from system analysis: first the definition of the unit processes, then the setting of system boundaries, and finally the allocation of multifunctional

[18] see Werner and Richter 2000a; Werner and Richter 2000b; Huppes 2000a

[19] The cutting out of zinc-containing corner cleats currently made with larger aluminium windows to allow for the collection of 'pure' profile aluminium, separated from a smaller fraction of zinc-containing aluminium

processes? It was argued that subjective and objective parts of LCA should be kept separate!

The debate has raised several issues that need clarification. They are related to the interdependence between a specific decision situation and the modelling of the life cycle of a product in LCA. Among the issues raised are the needs to:

- Clarify the interconnection of the modelling characteristics of LCA and the actual and desired sustainable material flows of the whole socio-economic system;
- Examine the claimed objectiveness of the setting up of the product system to check if a generic methodology for definition of a product system is tenable;
- Develop a set of requirements a product system and its respective life cycle inventory (LCI) has to fulfil to be 'appropriate';
- Distinguish the decision situations that lead to the definition of the product system to bring out the influence of mental models and values;
- Look at differing attribution principles for the definition of the product system and to evaluate their appropriateness for different decision situations in LCA, *e.g.,* their suitability for different materials.

For the clarification of these questions, LCA is looked at from the standpoint of normative decision analysis. The decision analysis framework provides a formal way of describing the decision-making process taking into account the encoding of subjective judgements. As such it helps to understand the conduct of an LCA as a decision situation itself.

As a glimpse on the things to come in this book, the following will be assessed: the influence of (subjective) mental models and values can be traced throughout the conduct of an LCA, also in the life cycle inventory analysis. There is no objective way of defining a product system as an artefact out of the complex system that is our world. As a consequence of the specific modelling characteristics of LCA, consistency of the product system with the real world situation can only be achieved by considering the attribution rules:

- The material and market characteristics of the materials involved on a case-specific basis; and
- The management rules for the sustainable use of the materials involved on a case-specific basis.

Only in this way will decision-makers find their mental models of reality and their values incorporated in the product system and their action space properly depicted. Otherwise, the implementation might lead to environmentally inefficient solutions. Or even worse, LCA-results will be ignored and will have no influence on the decision-making process at all. There is no escape from this.

The claimed objectivity of the life cycle inventory analysis is at stake. Welcome to controversy.

1.2 Objectivity and subjective elements in LCA

The above-mentioned research questions *per se* contradict frequently heard opinions and often stated characteristic of LCA, namely that the attribution of material and energy flows *has to be* made in an objective way in accordance with some accounting rules and (subjective) values are the basis of the assessment step only.

Objectivity and the influence of subjective elements within LCA have been discussed since the broader use of this tool. As further introduction, embracing subjectivity within the whole framework of LCA during the last decade is described in a certainly more anecdotal than complete way.

Originally derive from energy accounting, LCA has been developed as a tool to 'measure' the environmental implications of products. This idea of 'measurability' of environmental implications is present in the following quotations:

According to the SETAC definition of 1993, life cycle assessment

> "...is an objective process to evaluate the environmental burdens associated with a product, process or activity by identifying and quantifying energy and materials used and wastes released to the environment, to assess the impact of those energy and material uses and releases to the environment, and to evaluate and implement opportunities to affect environmental improvements" (Consoli *et al.* 1993:7).

In much the same direction tend the following authors:

> "LCA ought to be like mathematics, where people all over the planet can do the same work and get the same answer" (Elkington and Hailes 1993:16).

A slightly different position on objective and subjective elements within LCA was held by the research group at the Centre of Environmental Science in Leiden (CML), which in 1992 proposed a groundbreaking methodology for the assessment step of LCA. This group spotted the subjective parts in the valuation step. 'Objective' and 'subjective' elements were distinguished within the framework of LCA as follows:

> "In the goal definition, discussions take place between different participants such as commissioners, consumers and LCA scientists and technological information is needed about product alternatives that can be significantly compared with each other in relation to the goal of the study.

The inventory is pre-eminently a subject of systems analysis theories and process technology. The classification is based on environmental sciences, while the valuation is a subject of social sciences (*e.g.*, decision theory). The improvement analysis is based on applied mathematics and knowledge about process technology" (Guinée *et al.* 1993: 3).

In view of the 'subjective' parts in the valuation step, this group switched from the 'term life cycle analysis' to the term 'life cycle assessment' in 1991[20].

Over the last few years, doubts have arisen concerning the claim of objectivity of LCA, even in its life cycle inventory analysis – this made Klöpffer state in 1998 as a summery of the current state of affairs:

"Furthermore subjective items were identified even in the Inventory Analysis: allocation in the case of coproducts and open-loop recycling" (Klöpffer 1998:61).

Indeed, individual authors have commented on some subjective parts in the case of allocation, *e.g.,* in the:

- Distinction between products, co-/byproducts and waste[21];
- Choice of an allocation factor[22];
- Choice of a procedure for the allocations related to open-loop recycling[23].

The considerable and growing body of scientific literature[24] on allocation and setting system boundaries over the last years could anyhow be interpreted as conclusive evidence for the existence of subjective elements in the life cycle inventory analysis.

However, the continued belief in a generic LCA-methodology (with some subjective choices) is unbroken and is manifested in various attempts in separating 'objective' and 'subjective' elements within LCA-methodology. Miettinen and Hämäläinen (1997) are quoted deputising for various authors and representatives of the LCA community[25]:

"...to capture the subjective elements inherent in any decision-making, the LCA process needs to be clarified... A distinction between the objec-

[20] Gabathuler 1997:188

[21] Frischknecht 1994:123; also Heiskanen 2002:428, referring to multi-output allocation in general

[22] Frischknecht 1994:123; Hofstetter 1998:314

[23] Ekvall and Tillman 1997; Hofstetter 1998:314

[24] Throughout this book, reference is made to a broad set of literature on LCA, especially in Chap. 5 and Chap. 6.

[25] see, *e.g.*, Consoli *et al.* 1993; Heijungs *et al.* 1992; ISO/EN 14042; ISO/EN 14040

tive and subjective elements of LCA would clarify the structure of the method" (Miettinen and Hämäläinen 1997).

Some years later, Hofstetter *et al.* (2000) argue that

> "...subjective elements should be integrated into all phases of LCA rather than treated separately, but that the subjective elements have to be organised in a sophisticated manner that acknowledges social science knowledge in particular" (Hofstetter *et al.* 2000:161).

They suggest explicitly complementing the models for the technosphere and the ecosphere with a model for the valuesphere as the decision-maker's disposition based on his/her value system

> "...as this framework provides a consistent treatment of value judgements and is able to deal with distinct worldviews ...The proposed framework thus allows ...an incorporation of the decision-maker's value systems into goal and scope definition, inventory analysis, impact assessment and valuation" (Hofstetter *et al.* 2000:161ff).

Obviously, Hofstetter *et al.* (2000) propose a different perception of LCA. It consists in a radical departure from the general consensus, namely to describe ecosphere and technosphere by pure 'objective' elements only and to allow 'subjective' elements solely in the assessment step. In this alternative approach,

> "…it is the subjective elements that determine the view of the eco- and of the technosphere and shape the models representing them" (Hofstetter *et al.* 2000:162).

This book very much supports the line of reasoning by Hofstetter (1998) and Hofstetter *et al.* (2000)[26]. Whereas Hofstetter (1998) and Hofstetter *et al.* (2000) concentrate on the impact assessment step within LCA, this book focuses on the life cycle inventory analysis and emphasises on modelling open-loop recycling and other end-of-life options.

It will be shown that the implicit or explicit application of subjective elements – mental models and values – is unavoidable for the 'causal' attribution of material and energy flows to a product under study.

Two exemplary methodological decisions outline the problem:

- Should the material and energy flows of the keeping of a cow (a joint coproduction process from a methodological point of view) be partially allocated to milk and cream based on fat content or on market prices?

[26] see also Keeney and Raiffa 1976:12

- Under which circumstances *may*

  "...reuse and recycling ... imply that the input and outputs associated with unit processes for extraction and processing of raw material and final disposal of products are to be shared by more than one product" (ISO/EN 14041, chap. 6.5.4, point a)?

A sensitivity analysis is often recommended for such ambiguous decisions[27]. But a sensitivity analysis can only provide insight into the impact of a methodological decision on the overall result. It has often been analysed that methodological decisions in LCA can have a significant impact on the overall result. In such cases, sensitivity analysis cannot provide guidance for methodological choices. Nonetheless, methodological decisions have to be made to provide an unambiguous model as a useful basis for decision support.

So, the questions remain:

- Which are the methodological decisions necessary for the setting up of the product system as life cycle model within the life cycle inventory analysis?
- Which mental models and value choices influence these methodological decisions?
- How should these methodological decisions be made to obtain a representative, consistent[28] product system and a consistent LCA, from which 'useful' improvement options can be derived?

1.3 From real world to models and back

The main purpose of LCA is to generate information on the environmental implications of products[29]. This information should support the process of efficient and effective decision-making[30] in terms of providing the environmentally soundest action alternative[31]. LCA generates this information by modelling.

[27] see, *e.g.*, ISO/EN 14041, chap. 6.5.2, point 3

[28] in the sense of substantiatable, intersubjectively comprehensible

[29] including services

[30] It is assumed here that LCA-results are normally consulted in the analytic mode of thought. For different modes of thought in product design, see Ehrlenspiel 1995:63ff; for models of the analytic and intuitive mode of thought, see Scholz 1987:170ff.

[31] LCA also serve other purposes, such as environmental hot spot identification. Nonetheless, also an product system for hot spot identification must be set up in a way that it would fit the modelling-evaluation-realisation-(re-evaluation) circle presented below. If

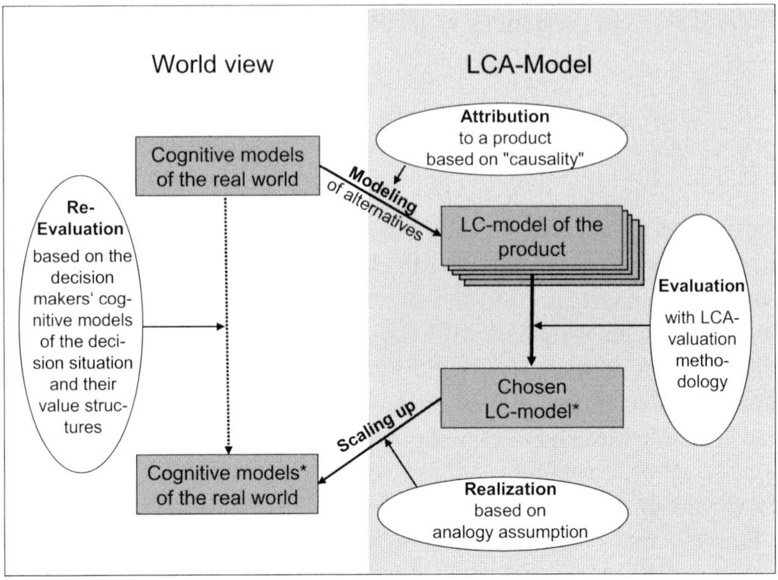

Figure 1-1. The modelling-evaluation-realisation-(re-evaluation) circle of LCA

Figure 1-1 illustrates the conduct of LCA as a modelling technique and decision support tool and its relation to the decision-maker's mental models of the real world. Mental models (also called cognitive models) are mental conceptualisations of partial aspects of our world, *e.g.,* knowledge representations of how our socio-economic system works[32].

A decision-maker uses LCA for generating information on the environmental implications of products. For this purpose a model is set up covering the material and energy flows attributed to a product and their evaluation in view of their environmental impact. The decision-maker expects LCA to depict the environmental implications of products 'adequately' and to provide recommendations that are in line with his/her understanding of the real world (his/her mental models on the real world).

Modelling in the life cycle inventory analysis consists in the 'causal' attribution of material and energy flows to the product under study (see Chap. 2.3.1). LCA-methodology can be understood as a set of creative rules for the conduct of an LCA, particularly for the setting up of the product sys-

---

not, there is a good chance that the hot spot identified turns out to be none... For the clarity of the reasoning, this use of LCA is not discussed specifically.

32 see, *e.g.*, Newell and Simon 1972; Johnson-Laird 1983; Pennington & Hastie 1993; Jungermann *et al.* 1998; also called cognitive models by Anderson 1985:73; Zimbardo 1992:304ff

tem in the life cycle inventory analysis. The resulting model is the life cycle model of the product, also called the product system. The product system is a partial, theoretical enclosure of real world material and energy flows.

In the transdisciplinary context of LCA, the choice of creative rules is ambiguous for fundamental epistemological reasons[33]. Consequently, a variety of methodological propositions have been made for the implementation of the 'causality' principle, which guides the attribution of material and energy flows to the product out of the complexity of the socioe-economic system.

In decision situations where LCAs are conducted, at least two alternatives are defined and evaluated with impact assessment methods. The corresponding model of the alternative selected ('Chosen product system*' in Figure 1-1) must be understood as a creative model for real world interventions (realisation), assuming analogy between the LCA-model and the decision-maker's mental models of the real world. In LCA, this intervention is, *e.g.*, the choice of an environmentally preferable product or the implementation of an environmentally preferable design option. With this intervention, the mental models of the real world are transferred to the environmentally preferable situation ('Mental models* of the real world' in Figure 1-1). This means that the action, for which the information was generated, is executed.

Ideally, the interventions deduced from LCA-models for realisation reflect the decision-maker's mental models of the decision situation from a meta-perspective and his/her value system, *e.g.,* the management rules for the environmental dimension of sustainable development (see Chap. 3). This means that they lead to 'real' environmental improvement if re-evaluated from a meta-perspective. The 'descriptive' power of LCA – and in fact its usefulness as a decision support tool – is given by the degree of how well the real world interventions deduced from LCA-models reflect the decision-maker's mental model of the decision situation and his/her value system if re-evaluated from a meta-perspective.

The aluminium example outlined in Chapter 1.1 has shown that the 'descriptive' power of LCA depends on the product system as a model of the life cycle of a product developed in the life cycle inventory analysis.

The 'descriptive' power of LCA in product design-related decision situations and its dependency on mental models and value choices for the definition of the product system is at the centre of interest in this book.

33 as different scientific disciplines or even 'schools' within a discipline can provide differing creative models because of differing creative rules. A entertaining example is provided by Cerutti 2000. He takes the shaking of a garden table to illustrate how different scientific disciplines develop different problem descriptions. These problem descriptions lead to totally different creative models for the solution of the problem.

1.4 Impossibility to validate LCA-models

In a decision situation, several action alternatives are compared. The selected alternative can be understood as creative model for real world interventions. The implementation of the creative model puts the selected alternative into practice, creating 'reality'.

What of the appropriateness of LCA-models and their creative rules? Can LCA-models be validated and its results verified?

In fact, there is no empirical approach for validating any of the models used or set up during the conduct of an LCA[34]. Several characteristics of LCA are responsible for this:

- The life cycle of a product is a theoretical construct, which has no clearly definable temporal or spatial boundaries in a complex context. The resulting model is a-temporal and spatially undifferentiated as an answer to the complexity of real social, economic and environmental systems (Chap. 2.3.2). Or as Heiskanen (2002) puts it:

  "Product life cycles do not exist as distinct entities 'out there in nature', but they have to be modelled on the basis of what we assume to be the purpose of the activity, and where we place responsibility for environmental interventions" (Heiskanen 2002:428);

- The '*ceteris paribus*' assumption made to reduce the complexity of real social, economic and environmental systems makes the validation of such models impossible[35];
- Mental models are applied and value choices are made during the setting up of the product system and its respective LCI as well as during the inventory assessment. These subjective elements are necessary for the choice of creative rules in a transdisciplinary context[36] (see Chap. 1.3).

Consequently, no empirical approach based on measurements is feasible to validate the models set up or used in an LCA. The environmental cause-effect relations can still be approached partially by empirical methods for the assessment step, whereas the measurement of total material and energy flows caused by a product system lies beyond this option.

LCA takes its justification from the suitability to address a specific problem in a practical context: the assessment of the environmental implications of products. The validity of an LCA (or of a methodological proposition for LCA-methodology) is reflected in the appropriateness and internal

[34] see also Scholz and Tietje 2000:262
[35] see, *e.g.*, Ulrich 1981:9
[36] see, *e.g.*, Keeney and Raiffa 1976:12

consistency ('descriptiveness') of the models and of the values applied in a specific decision situation in the eyes of the modeller, the decision-maker and possibly other stakeholders of the study[37]. Validation can also be seen as a negotiation process, in which criteria for validity are negotiated corresponding to the uncertainties, stakes and interests involved (see Haag and Kaupenjohann 2001:56)[38].

However, models developed in the context of decision-making can be quasi-validated: if the outcomes of an assessment tool – and thus its underlying models – become relevant for the decision-maker in a way that they influence his/her action[39]. Indeed, this is the real challenge for further methodological development of LCA.

1.5 Scientific approach: theses

Life cycle assessment as part of applied sciences cannot be attributed to the theory of one single scientific discipline. Due to its transdisciplinary nature and due to the impossibility of an empirical approach to LCA, a hypothesis on methodological matters cannot be formulated nor can a hypothetical hypothesis be disproved. Thus, this book must choose another than the 'classical'[40] scientific approach.

Instead of trying to disprove a hypothesis, a constructivistic approach is chosen[41]. Several theses are formulated based on the introducing retrospective (Chap. 1.1) as initial position. The scientific approach relies on combining knowledge of different sciences to a substanciatable, intersubjectively comprehensible line of reasoning.

These theses also allow the reader a 'guided' reading:

Thesis I: The definition of a product system as attribution of material and energy flows to a product out of a complex reality cannot be made in an unambiguous way. Modelling the product system contains subjective elements, *i.e.* the use of mental models and value choices. An objectively right,

37 see, *e.g.*, Hertwich *et al.* 2000

38 The peer review process according to ISO/EN 14040, which involves scientists from a disciplinary point of view, can also be seen as a validation process of the LCI-model by externals (see Scholz and Tietje 2000:309). This type of validation was applied in the two case studies with peer reviews of the scientific report in the aluminium example (Chap. 8ff.) and with peer reviews of a related scientific paper in the railway sleeper example.

39 'face validation' by the decision-maker, according to Scholz and Tietje 2000:309

40 critical-rationalistic

41 see, *e.g.*, Glaserfeld 1984 and 1985; Holzkamp 1972; Berger and Luckmann 1980; Maturana and Varela 1987; Foerster 1984; Watzlawick 1984; see also Vaassen 1996

generic way of modelling a product system in the life cycle inventory analysis is not definable.

Thesis II: Modelling characteristics of LCA address the environmental dimension of sustainable development only partially. Recommendations derived from the LCA-model could be perceived sub-optimal from a meta-perspective if attribution of material and energy flows to the product(s) under study is not made considering the specific characteristics of the decision situation, *i.e.* the material and market characteristics of the materials involved, and the 'descriptive' power of LCA in view of the material-specific management rules for their sustainable use.

Thesis III: Decisions related to modelling the product system in the life cycle inventory analysis such as module definition, setting system boundaries, allocation and the implementation of the guiding principle of 'causal relationship' for the attribution of material and energy flows to the product and the definition of the functional unit are interdependent. They cannot be treated separately and in a subsequent order. They have to be addressed in a consistent way, reflecting the decision-maker's mental models and values in the decision situation.

Thesis IV: Structural uncertainty ('ignorance') related to changes within the socio-economic system over time is a predominant aspect of modelling and allocation related to reuse & recycling in the life cycle inventory analysis. Attitude towards risk is an essential determinant of modelling and allocation related to reuse & recycling within LCA.

1.6 Structure of the book

The following Figure 1-2 summarises the structure of the book; it consists of six parts. The numbers in brackets indicate the respective chapters.

Part I, the introduction, contains a retrospective as the context of discovery, the positioning of LCA as modelling tool within science and the scientific approach chosen in this book, the theses and an overview of the structure of the book (Chap. 1).

Part II investigates the requirements of product systems and their LCIs in product design-related decision-making.

In Chapter 2, life cycle assessment as an environmental management tool is introduced. Structure, scope and general modelling characteristics of LCA are outlined. Due to the scope of this book, emphasis is put on:

a) How LCA reduces complexity of the real world socio-economic system and its interconnection with the ecosphere during the attribution of material and energy flows to the product under study as an enclosure of socio-economic reality;

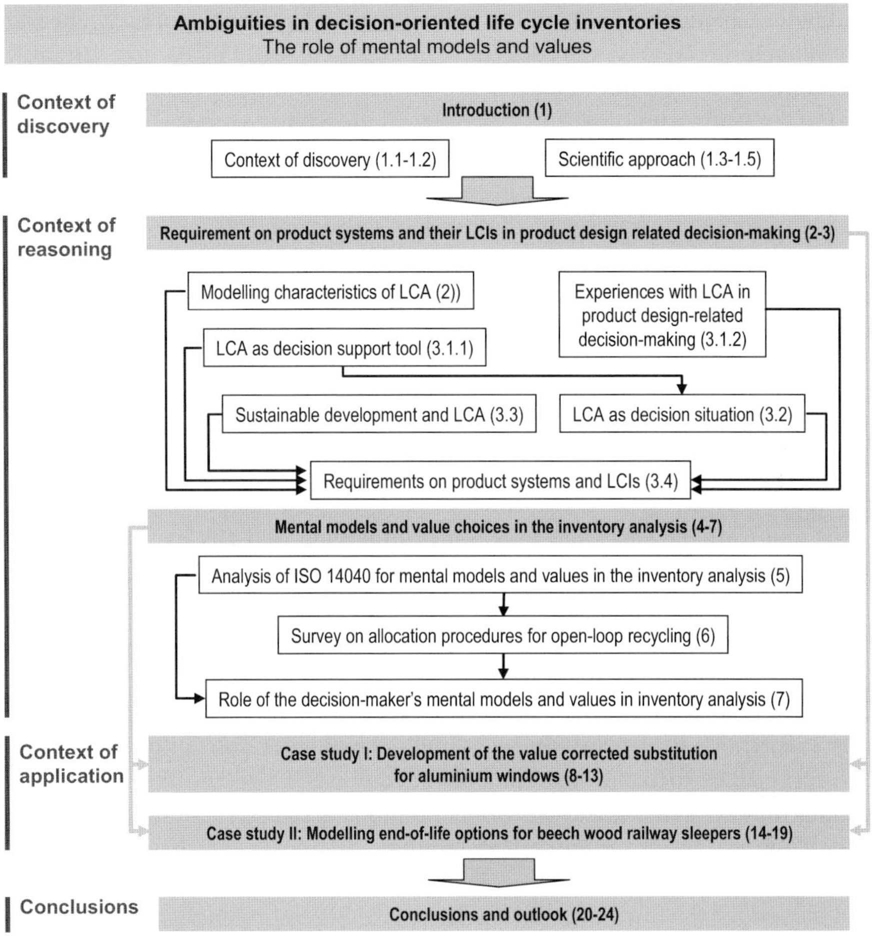

Figure 1-2. Structure of the book (the number of the respective chapters in brackets)

b) The characteristics of the resulting model of the life cycle of a product.

The setting of system boundaries (including allocation) during the definition of the product system is identified as possible source of systematic error

Chapter 3 treats the question how decisions requiring the use of mental models and value choices should be handled to provide the 'best' product system and LCI.

A set of requirements is deduced from:

- Practical experience of LCA in product design (Chap. 3.1);
- Decision theory considering LCA a decision support tool (Chap. 3.2);
- Comparison of the 'descriptive' power of current impact assessment methods in view of the environmental dimension of sustainable development underlying LCA as a normative concept (Chap. 3.3).

Under Chapter 3.2, a model of normative decision analysis by Sieben and Schildbach (1994) is introduced. The elements of LCA are structured according to this model to clarify the influence of mental models and values during the setting up of an LCA as a decision-making process (Chap. 3.2.3).

As a result of this chapter, conclusions are drawn concerning the requirements of LCA-modelling. In line with the scope of this book, emphasis is put on the definition of the product system and its LCI as a model of the life cycle of a product (Chap. 3.4).

Part III investigates, which mental models and values guide the life cycle inventory analysis.

After an introducing Chapter 4, the standard ISO/EN 14041 is analysed in-depth for attribution rules concerning the setting up of the product system in the life cycle inventory analysis (Chapter 5). The series of standards ISO 14040 is taken as minimum consensus of the actual discussion of LCA-methodology. The attribution rules of ISO/EN 14041 are screened for decisions requiring the (implicit) use of mental models and values.

LCA-literature is consulted to illustrate the influence of mental models and values on decisions made during the life cycle inventory analysis. Due to the scope of this book, special emphasis is put on decisions made in situations related to reuse & recycling.

In Chapter 6, a literature review provides an overview of current allocation procedures for open-loop recycling. The chapter concentrates on different approaches to handling reuse & recycling in LCA and the influence of mental models and values. Recommendations are made for the choice of an allocation procedure for reuse & recycling.

Chapter 7 summarises the role of the decision-maker's mental models and values for the attribution in the life cycle inventory analysis.

Part IV contains the first case study, an LCA of aluminium window frames where the value-corrected substitution is developed (introduced with Chapter 8). The case study starts with the question how recycling of aluminium in building applications should be integrated into product systems to meet the requirements of product systems and their LCIs of Chapter 3.4 in an optimal way.

Taking into account market and material characteristics as well as the respective management rules for a sustainable use of the material has been claimed essential for meeting 'causality' in the attribution of material and energy flows to the product under study (Chap. 3.2.3). Chapter 9 treats the characteristics of aluminium as a material and its primary and secondary material markets. Sustainable management rules for aluminium are postulated. The value-corrected substitution (VCS) is outlined as allocation procedure for aluminium products in the building sector that meets the above-stated requirements best (Chap. 10).

The VCS is compared to traditional allocation procedures like the 'cut-off method' (applied in Richter *et al.* 1996) or the closed loop procedure based on a material pool model described in ISO/TR 14049: 2000:27ff. A further procedure, the strict coproduct allocation according to Huppes (1994:84f) and Lindeijer and Huppes (2001:309), is discussed but not included into calculations, as it is not yet considered fully developed (Chap. 11).

These allocation procedures are applied in an LCA of a fictitious aluminium window in two variants, one containing zinc-parts and the second being free of zinc and brass (Chap. 12). As different allocation procedures rely on different accounting principles, so the system boundaries differ for each of the allocation procedures.

The suitability of the allocation procedures is discussed on an argumentative basis referring to the set of requirements deduced in Chapter 3.4 and to the market and material characteristics of aluminium and on the strategies for its sustainable use (Chap. 9.6). A sensitivity analysis is conducted on the inclusion and exclusion of dross treatment for each of the allocation procedures (Chap. 13).

Part V contains the second case study, which covers modelling of end-of-life options for beech wood railway sleepers.

Chapter 15 and Chapter 16 provide an overview of the current market characteristics of post-consumer wood in Switzerland; the characteristics and use options of post-consumer wood and on the wood flow management politics in the light of sustainable development.

In Chapter 17, the attribution and allocation problems related to wood are outlined and two wood-specific guidelines for LCA of wood products – the User's guide by the American Forest & and Paper Association[42] and the methodological appendix of the European research project Life-Sys Wood[43] – are presented and discussed.

Chapter 18 contains the example itself. As the modelling end-of-life options in the life cycle inventory analysis can depend on methodological decision made further up the process chain, several allocation problems throughout the wood processing chain are discussed. They cover the allocation of forestry processes to the different products from forestry, the allocation of transport from the forest to the production site, the allocation of the production processes to the products from sawmill and sleeper production, and the allocation related to the two end-of-life options for railway sleepers, *i.e.* the incineration with co-generation of thermal energy and electricity or

42 Fava *et al.* 1996

43 Esser and Robson 1999a

their use as a construction element in landscape architecture. Finally, (partial) life cycle models are built using best-suited allocation procedures for each of the allocation problems aiming at optimally consistent models for the end-of-life allocation procedures selected.

General methodological conclusions and wood-specific conclusions and recommendations are drawn in Chapter 19.

The book concludes with Part VI, with contains a review of the theses where conclusions are drawn from the methodological part as well as from the case studies (Chap. 20). Furthermore, the consequences for LCA as a decision support tool are outlined (Chap. 21); the consequences for a future revision of ISO/EN 14041 are drawn (Chap. 22) and a new modelling approach to LCA as group-model-building process is outlined (Chap. 23). In Chapter 24, need for further research is detected.

II.

REQUIREMENTS OF PRODUCT SYSTEMS AND THEIR LIFE CYCLE INVENTORIES IN PRODUCT-RELATED DECISION-MAKING

Chapter 2

LCA AS METHOD AND ITS MODELLING CHARACTERISTICS

Product-oriented environmental policies have increasingly been incorporated by governmental and non-governmental institutions as well as by industry and its interest groups. The need for instruments and tools to support product-oriented decision-making has increased correspondingly. Life cycle assessment (LCA) is one of the environmental instruments considered to be important in the context of product-related decision-making[44].

LCA can be used for the development and optimisation in environmentally conscious product design; environmentally conscious procurement; strategic planning; policymaking; marketing and so forth[45]. This book focuses on the application of LCA within the decision process in development and optimisation of environmentally conscious product design.

As a relatively young tool, the position of LCA to other decision support tools such as risk assessment, environmental performance evaluation or site-related environmental auditing is still provisional[46] and methodological development is ongoing.

[44] see, *e.g.*, ISO/EN 14040; see also Chap. 3.1.1

[45] ISO/EN 14040, chap. 4.2

[46] see, *e.g.*, Beck and Bosshart 1995, Hofstetter 1998:18ff or Hopfenbeck and Jasch 1995:98ff for the description of different environmental management techniques and decision tools

2.1 Methodological structure of LCA

In this chapter, the methodological structure of LCA is outlined. Reference is made to the series of standards ISO 14040 covering the methodology of the tool as one of the most up-to-date documents on LCA-methodology.

ISO/EN 14040 defines LCA as a

> "compilation and evaluation of the inputs, outputs and the potential environmental impacts of a products system throughout its life cycle" (ISO/EN 14040, chap. 3.9).

Normally, four different steps of an LCA are distinguished[47]. The methodological structure of LCA is shown in Figure 2-1.

Step 1: Goal and Scope Definition. A preliminary definition of the technical framework of an LCA study is given. Step one implies the statement of the intended application of the study and the specification of the products (the 'functional units'[48]) and their respective life cycle to be considered; a first provisional setting system boundaries and choice of the allocation procedures; the impact assessment method to be applied; the data requirements; the type of critical review and the format of the report.

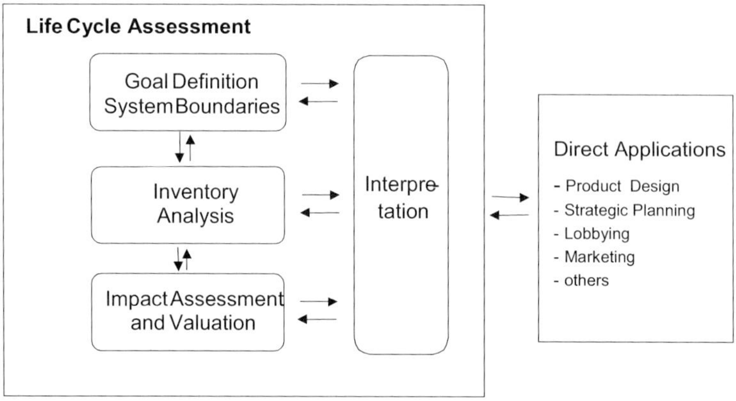

Figure 2-1. Methodological structure of LCA (ISO/EN 14040:10; reprinted with permission from the European Committee for Normalisation, Brussels)

Step 2: Life Cycle Inventory Analysis (*Inventory Analysis*) involves data collection and calculation procedures to quantify relevant environmental

[47] see ISO/EN 14040, Lindfors *et al.* 1995a:21ff

[48] According to ISO/EN 14040 the functional unit is *the quantified performance of a product system for use as a reference unit in an LCA study (*ISO/EN 14040, chap. 3.5).

interventions of a product in a life cycle model[49], called the 'product system' according to ISO/EN 14040[50]. These in- and outputs should include the use of resources and releases to air, water and land associated with the system. The summary of the environmental interventions of a product over its life cycle is called the life cycle inventory (LCI).

Step 3: Life Cycle Impact Assessment evaluates the inventoried in- and outputs with regard to their environmental implication. Several steps can be distinguished within the assessment: the classification (the assigning of inventory data to impact categories), the characterisation (the weighting of the assigned inventory data within impact categories) and the valuation (the aggregation of the impact categories by normalisation and addition).

Step 4: Interpretation where the findings of the inventory analysis and the impact assessment are interpreted. Conclusions and recommendations are drawn with regard to goal and scope of the study.

The above-mentioned succession of steps gives only a rough idea how LCA is conducted. In practice, establishing an LCA is an iterative process, as adjustments of decisions from preceding phases have to be made during the course of the study. This may be, *e.g.*, due to the lack of data or data quality available, due to needs for the avoidance of allocation or because of inequalities encountered in the definition of different functional units to be compared.

2.2 Some characteristics of current LCA-methodology

The consideration of environmental impacts caused over the entire life cycle of a product from raw material extraction, production, use and disposal is a constitutive characteristic of LCA. But there are different ways of applying life cycle thinking on specific questions. This has basically been the motivation and driving force behind all the methodological discussion during the last ten years, when LCA was mainly developed. The following list summarises the current state of development in line with the reasoning in the SETAC-working group[51] or the series of standards ISO 14040.

LCA *sensu stricto* can be described to be an LCA which[52]:

[49] According to ISO/EN 14040 the life cycle of a product is "the consecutive and interlinked stages of a product system, from raw material acquisition or generation of natural resources to the final disposal" (ISO/EN 14040, chap. 3.8).

[50] According to ISO/EN 14040 the product system is the "collection of materially and energetically connected unit processes which performs one or more defined functions" (ISO/EN 14040, chap. 3.15); the life cycle model of a product (or functional unit).

[51] see, *e.g.*, Consoli *et al.* 1993; Fava and Denison 1991; Fava *et al.* 1991

[52] originally taken from Hofstetter 1998:5ff, modified and added

- Analyses functional units that are small enough to cause only marginal changes[53];
- Respects *ceteris paribus*, *i.e.* assumes that all other processes will stay as they are;
- Relies on fully quantitative information on a ratio scale[54];
- Is based on simple linear or linearised models of the socio-economic system (technosphere), environmental processes and the environmental impact chain (both considered to be the ecosphere)[55];
- Follows the 'less is better' approach[56];
- Includes no detailed information on time pattern of releases and it is assumed consequently that all environmental interventions occur in an unspecified point in time [57];
- Includes no detailed information on the site of releases. It is assumed that all environmental interventions occur in one place with meteorological conditions, ecosystem sensitivities and densities of receptors that are typical for, *e.g.*, Western Europe;
- Allows for trade-offs between different kinds of environmental impacts, *i.e.* they have a compensatory character (additive disutility function)[58];
- Can be made based on two approaches:
  - Static, descriptive LCA aiming at attributing all environmental impacts caused by one product on the basis of average data;

[53] Frischknecht 1998 has developed a framework for addressing multifunctionality situations, requiring the allocation of multi-output processes based on decision situations differing in the scale of impact on the system under study (among other influences).

[54] although qualitative aspects can be discussed in the interpretation phase

[55] The linearity assumption is fundamental for the matrix calculation procedure as proposed in Heijungs 1997:33ff, Frischknecht 1998:32ff or Heijungs and Frischknecht 1998; see also Frischknecht 1998:28ff on the linearity assumption of the production function. See footnote 131 for the matrix calculation procedure

[56] thus, not taking into account the complex environmental damage functions with non-effect levels and thresholds such as carrying capacity above which self-regulation of an ecosystem is exceeded (see, *e.g.*, Messner 1999:148ff; for damage function and LCA, see, *e.g.*, Müller-Wenk 1996a; Müller-Wenk 1996b; Hofstetter 1998:87ff)

[57] although several attempts have been made to use McKay-modelling for fate modelling in the inventory phase and adjusted characterisation factors for the assessment step (see, *e.g.*, Guinée *et al.* 1996; Hillier *et al.* 1997; Beck *et al.* 2000; Hauschild and Potting 2000). This topic has also been the starting point for an extensive discussion of the relationship of LCA and risk assessment and their possible combination (see, *e.g.*, Scheringer *et al.* 1999)

[58] This is especially obvious for single-score impact assessment methods such as the ecoscarcity method (Brand *et al.* 1998) or the Eco-indicator 95 (Goedkoop 1995) and Eco-indicator 99 (Goedkoop and Spriensma 2000).

- Change-oriented LCA (also called marginal LCA) aiming at mapping the environmental impacts caused or avoided by marginal changes in the life cycle of a product[59,60];
- Focuses on one of the three dimensions of sustainability: the dimension 'environment', *i.e.* the dimensions 'society' and 'economy' are not analysed. Occupational health is also disregarded.

As methodological development of LCA is continuous, the positioning of LCA as a decision support tool can only be provisional. The main characteristics of LCA in comparison to other tools are the following[61]:

- LCA is a tool to model the environmental dimension;
- LCA is used as a tool within the decision support process and does not in itself encompass the whole decision-making procedure;
- LCA is designed to support decisions on a micro level, *i.e.* products including services (and plants) are objects of analysis;
- LCA assesses changes in single human activities or average human activities and cannot describe the state of the environment or societal responses to environmental stress;
- LCA assesses potential global environmental interventions and damages by assuming a unit-world with average meteorological and ecological conditions;
- LCA compresses the time dimension in the modelling phase as well as in the impact assessment phase and does not discount the future[62];
- LCA relies on input-output accounting.

2.3 Complexity as challenge for the inventory analysis

In the following sub-chapters, the modelling characteristics of product LCA in view of the complexity of the real world are addressed. Emphasis is laid on the modelling characteristics of the life cycle inventory analysis. The life cycle inventory analysis consists in the definition of the product system as model of the life cycle of a product.

59 see, *e.g.*, Frischknecht 1998:47ff; Heijungs 1997:177f; Hofstetter 1998:10; Weidema *et al.* 1999; Weidema 1998a. Heijungs 1997:178 proposes to use descriptive LCA for the identification of key issues (dominant aspects, hot spots) and change-oriented, marginal LCA for investigating change induced by switching to one alternative.

60 see Ekvall 1999:33 for different terms used to address the descriptive and the change-oriented type of LCA

61 originally taken from Hofstetter 1998:24, modified and added

62 Apart from this methodology-inherent reasoning, there is dissent among economists on the moral justification of a positive discount factor of future utility in the intergenerational context (see, *e.g.*, Messner 1999:78ff/141f/204ff; Pearce 1993:54ff; Siebert 1983:277ff).

The modelling of the ecosphere as part of the assessment step of LCA is considered only in so far as it determines the way in which the product system as part of the technosphere is modelled.

2.3.1 The attribution problem in defining the life cycle model of the product

During the life cycle inventory analysis, the life cycle of a product is modelled 'from cradle to grave'. The product system is usually modelled in a modular way, with modules describing the life cycle stages of the product, *e.g.,* 'primary material extraction', 'production', 'use phase', 'disposal', etc. The modules are further refined and completed with up-stream and down-stream processes. Ideally, the material and energy flows are tracked upstream and downstream until they become elementary flows. Elementary flows, also called 'environmental interventions' or 'environmental releases', are considered a link between the socio-economic system (the technosphere) and the environment. They are assumed to be taken out of the ecosphere as resources or to be released into the ecosphere as emissions or waste[63].

The 'life cycle of a product' is a theoretical construct. There is no locatable, 'tangible', physical and thus limited object that can be described as 'the life cycle of a product', in contrast to, *e.g.*, a production site. Therefore, rules have to be established on what is accounted for as the life cycle of a product.

The essence of the definition of the product system in the life cycle inventory analysis is to demarcate, structure and describe a small part of the real world as a model, dependent on the product to be assessed[64]. The definition of the product system requires dealing with the socio-economic system – the technosphere – and its interconnection with the ecosphere as a complex whole-system[65] (see Chap. 2.3.2).

The definition of the product system as a partial enclosure of socio-economic reality is part of a problem that has been called the 'attribution problem' by Heijungs (1997):

[63] if waste treatment processes are not modelled to its very end as environmental interventions, *e.g.*, in the case of land filling for the lack of data on behaviour and emissions of the land filled waste

[64] see, *e.g.*, Luhmann 1994:73 or Spencer-Brown 1972 for distinction and designation as precondition for observations and descriptions

[65] Complex structures consist of elements and transitions. Their main characteristics are feedback loops with information flows counter-current to the effect ('control loop'). Complex structures are characteristically inter-linked as cause-effect nets and show a high variability; their behaviour is commonly non-linear, unpredictable and elude intuition (after Ashby 1971:10ff; Forrester 1971:15; Vester 1980:24ff/49ff/59f).

"The attribution problem is the question which environmental problems are to be attributed to which economic activities" (Heijungs 1997:4).

The attribution problem is a key issue in LCA. The attribution problem consists in the question concerning which material and energy flows and which environmental impacts a product is held responsible for (Figure 2-2).

"Responsibility must be understood here in a primarily causal sense, without necessarily having any moral connotation" (Heijungs 1994:8).

The attribution problem in LCA is twofold:

Attribution I: the material and energy flows and its related environmental interventions (emissions, resource consumption and wastes) 'caused' by the product have to be determined. In LCA-methodology this is done in the life cycle inventory analysis by defining the product system as an enclosure of socio-economic reality.

Attribution II: types and amounts of environmental impacts 'caused' by these environmental interventions have to be determined. Applying impact assessment methods addresses this issue.

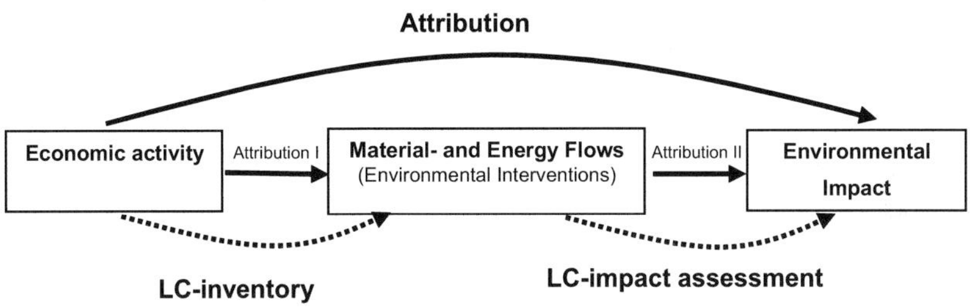

Figure 2-2. The twofold attribution problem in LCA

The primary interest in this book lies on the first attribution, the definition of the product system as a partial enclosure of socio-economic reality.

Some attribution principles for the definition of a product system in LCA are explicitly stated in the standard ISO/EN 14041. However, in view of the complexity of the real world, the attribution of material and energy flows to the construct 'life cycle of a product' cannot be unambiguous.

Ambiguity arises, *e.g.*, in the setting of system boundaries between nature as ecosphere and the socio-economic system as technosphere.

Another source of ambiguity in the attribution of processes to the life cycle of the product lies within the socio-economic system: if a process fulfils more than one function. Multifunctionality of processes requires the application of so called 'allocation rules'[66] that allow the partial attribution of such processes to the life cycle of the product under study. Examples of such multifunctional processes are:

- Coproduction processes providing several products;
- Waste treatment processes treating wastes from different sources simultaneously;
- Reuse & recycling.

In the context of LCA-methodology, the term 'allocation' is used in a specific meaning. Allocation in LCA-terminology refers to the partitioning of environmental interventions of multifunctional processes (see Chap. 5.6). The allocation problem in LCA is a sub-problem of the attribution problem. The term allocation, however, is also used for other purposes, especially in economic theory with respect to the allocation of scarce resources.

Ambiguity in the setting up of the product system requires the implicit or explicit application of mental models or value choices[67]. The influence of mental models and value choices on the attribution is a central aspect of this book, which is intensely discussed in Part III.

2.3.2 Complexity reduction during the setting up of the life cycle model of a product

Modelling is reductionistic by definition. Models reflect reality in a symbolic way[68]. Developing models of part of the socio-economic system requires dealing with three difficulties related to the complexity of the socio-economic system:

- Definition and reduction of the number of elements and their interconnections (or 'transitions'), *i.e.* the definition of processes and flows representing the life cycle of the product;
- Setting system boundaries as the demarcation of the system, *i.e.* the life cycle of the product;
- Assumption of the determinants of the behaviour of the system, *i.e.* on the organisational principle of the socio-economic system.

The definition and reduction of the number of elements and their interconnections as well as the setting of system boundaries are related to the question of which processes and flows should be considered in the product system (the first part of the attribution problem, see previous chapter).

[66] ISO/EN 14041, chap. 6.5
[67] This topic will intensively be treated in Chap. 4 to Chap. 7
[68] see, *e.g.*, Vester 1980:104

The description of processes and flows is not sufficient for the characterisation of the behaviour of a system. Assumptions must be made on the organisational principle of the socio-economic system if changes should be represented in a model.

In marginal LCA, the necessity of considering the organisational principle of the material and energy flows in the socio-economic system are obvious, as the changes induced to the socio-economic system by an additional (or avoided) functional unit is the basic accounting principle.

In the average (descriptive) approach, the consideration of the organisational principle of these flows is less obvious. It is hidden in the guiding principle for setting system boundaries and allocation: the 'causal' relationship[69] between the product to be assessed and the material and energy flows attributed to it. The need to make assumptions on the organisational principle in descriptive LCA becomes visible when, for instance, joint coproduction processes have to be allocated[70]; furthermore, if improvement options determined by descriptive LCA have to be considered against material and market characteristics, if re-evaluated from a meta-perspective (see Chap. 3.2.3).

The assumption concerning the organisational principle as determinant of the behaviour of the socio-economic system made implicitly or explicitly during the development of the product system must also be seen as an element of complexity reduction. Different mental models of the organisational principle and thus on 'causal relationship' are extensively discussed in Chapter 5.7.

Complexity reduction and the setting of system boundaries are as characteristic as crucial for dealing with practical problems[71]. Many of the above-stated characteristics of LCA (Chap. 2.2) can be interpreted as response to the complexity of the socio-economic system and its interrelationship with the ecosphere. These characteristics must be seen as the intent to reduce this complexity to generate a depictable, limited, simple and thus 'manageable' model as a basis for efficient and effective decision-making[72]:

- The *ceteris paribus* assumption to limit the extension of the system under study;
- The linearity assumption to reduce the complexity (in fact the non-linearity[73]) of economic, social and environmental cause-effect relations;

[69] ISO/EN 14041, chap. 6.5.3, point b
[70] for details, see Chap. 5.6
[71] see Ulrich 1981:8f; Keeney and Raiffa 1976:11
[72] see von Winterfeldt and Edwards 1986 and Keeney and Raiffa 1976:5 in Chap. 3.2.1 for requirements on decision support tools
[73] see, *e.g.*, Vester 1980:44/71ff

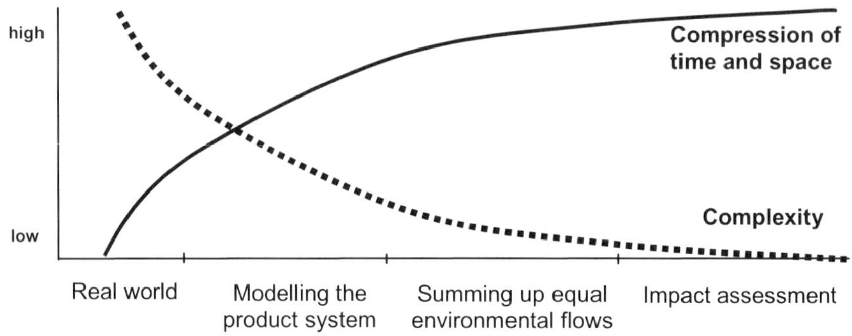

Figure 2-3. Compression of time and space as a strategy to reduce complexity over the inventory and assessment step of LCA

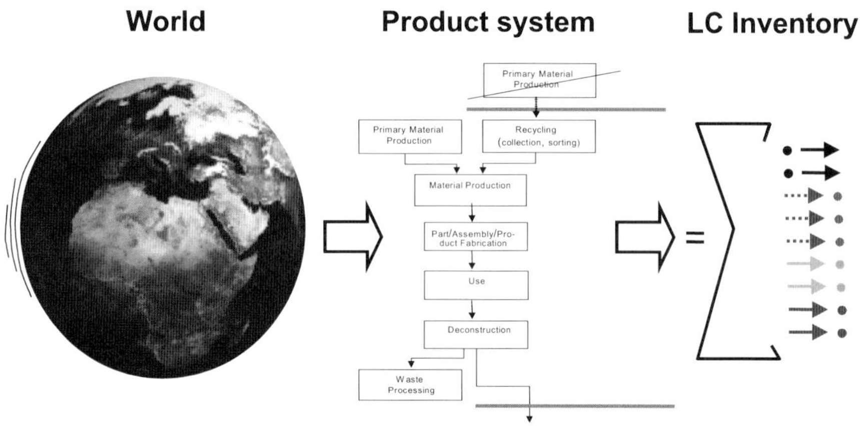

Figure 2-4. Compression of time and space as modelling characteristics of LCA as a strategy to deal with the complexity of the socio-economic system and its interconnection with the natural environment. Note that the dots represent current accounting categories in a LCI: resources, airborne emissions, emissions to water, emissions to soil (see also Figure 2-5)

- The static modelling as a compression of time (see Figure 2-3);
- The compression of space by assuming a socio-economic and environmental unit-world (see Figure 2-4);
- The assessment of potential damages as a consequence of the compression of time and space.

Finally, a remark on the practical side of LCA and its implementation in computer tools from a mathematical perspective: it is the above characteristics that allow matrix calculation procedures for LCA[74].

2.3.3 System boundaries as source of systematic errors in input-output models

As a consequence of the socio-economic and environmental unit-world assumption mentioned previously, equal environmental interventions can be summed up for the impact assessment.

Using this model, the mass and energy flows are completely abstracted from their temporal and spatial context.

The resulting model is an:

- Unit-world, input-output model (see Figure 2-5).

Spatial and temporal information is, in principle, not completely lost during the construction of the product system and its related material and energy flows. Tracking emissions back to their causing processes is still possible. The interpretation step within LCA-methodology still allows one to address temporal and spatial aspects of the inventory or assessment step on an argumentative basis. Nonetheless, spatial and temporal information is not integrated into the calculation routines of the inventory step nor in the assessment step.

This leads to:

- An a-historical, site-independent input-output model representing the life cycle of a product, with no linkage to local, regional, global, historic, actual or desired sustainable material and energy flows;
- An assessment of a-historic and spatially undefined environmental interventions and their potential environmental impacts.

The static modelling and the assumption of a unit-world have to be considered a rather simplistic approach for dealing with the complexity of the real world[75].

The question arises about the representativeness and adequacy (validity) of the static, input-output model of the life cycle of a product in view of the complexity of the real world material and energy flows (and their environmental impact).

[74] see Frischknecht 1998:31ff; Heijungs 1997; Heijungs and Frischknecht 1998 and footnote 131. Examples where the matrix procedures have been applied are the ESU-inventories for energy systems by Frischknecht *et al.* 1996.

[75] The static modelling and the assumption of a unit-world limits the huge data requirements of dynamic and site-specific modelling approaches to a still considerable quantity (see, *e.g.*, Hofstetter 1998:23/88ff).

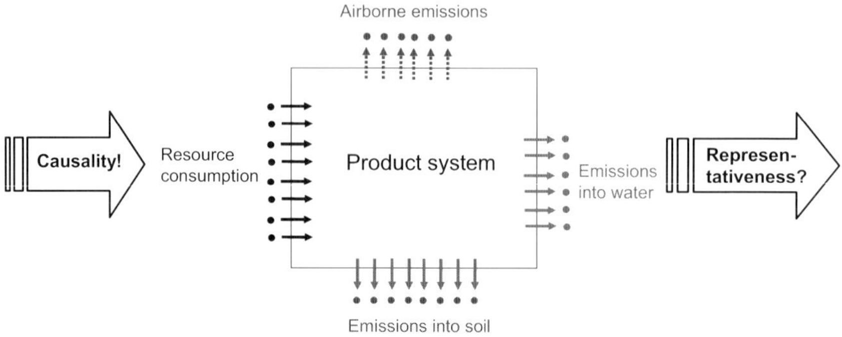

Figure 2-5. The model of the life cycle of a product as unit-world input-output model; the importance of system boundary setting in achieving causality in the attribution and the question on the representativeness and adequacy (validity) of the model in view of real world complexity

Referring to Figure 1-1, the question can be reformulated: how can 'causality' be assured in the construction of the product system so recommendations derived from LCA-models can be considered 'suitable' for interventions on the real world material and energy flows, given the constraints of LCA-modelling characteristics?

Complexity reduction and system boundary setting is crucial for dealing with practical problems[76]. In view of the complexity of the depicted object, the static input-output model resulting from LCA-methodology has to be considered rather simplistic. In view of the input-output modelling characteristic of the product system, special attention must thus be paid to the adequate setting of system boundaries to obtain a 'causal' relationship between the object to be depicted and the model.

Pohl *et al.* (1996) have the setting of the system boundaries as a possible source of systematic errors:

> "Systematic errors occur …when the structure of a system is not included completely in the calculation model ...This kind of systematic error in LCA occurs, *e.g.*, when defining boundaries of a system.." (Pohl *et al.* 1996:54).

They refer to a first type of systematic error in the life cycle inventory analysis when they write:

> "Systematic errors occur when relevant processes are excluded and the model structure is therefore incomplete. For example the question of

[76] see Ulrich 1981:8f; Keeney and Raiffa 1976:11

whether one-way bottles or deposit bottles are less polluting cannot be answered without considering transport" (Pohl *et al.* 1996:56).

This kind of systematic error is concerned with if processes held relevant for the environmental impacts of a product are *considered at all.*

This book postulates a second source of systematic error related to the setting of system boundaries (as final defining element of an input-output model) in the life cycle inventory analysis: the reductionistic way of modelling in the life cycle inventory analysis resulting in an input-output model and the representativeness and adequacy (validity) of the model in view of the characteristics of the real material and energy flows of the socio-economic system. If system boundaries are not set representing the 'causalities' of the specific decision situation, improvement options derived from the model can lead to sub-optimal solutions if re-evaluating them from a meta-perspective.

This kind of systematic error is related to the point if processes held relevant for the environmental impacts of a product are *considered in the 'right' way*.

Chapter summary

- The definition of the product system and its LCI requires the attribution of material and energy flows as a partial enclosure of the socio-economic system. In view of the complexity of the real world, this attribution cannot be unambiguous and thus only 'objective' under a set of pre-set assumptions. It requires assumptions on the structure and the behaviour of the socio-economic system;
- LCA reduces the complexity of the socio-economic system and its interconnection with the natural environment by compressing time and space. The resulting model – the product system – is an input-output model, a-temporal and site-unspecific in terms of spatial differentiation;
- These modelling characteristics of LCA have to be understood as an answer to the most complex and thus highly data dependent character of the object to be modelled;
- Setting system boundaries as decisive elements of an input-output model determines the interrelationship of the product system as model within LCA-methodology with the real world situation. System boundaries correspondingly determine the 'descriptive' power of a specific LCA. Therefore, the setting of system boundaries (including allocation of multifunctional processes) is of crucial importance for the representativeness (validity) of an LCA-model. Inadequate setting of system boundaries can lead to systematic errors depending on *if* and *how* processes are considered.

Chapter 3

REQUIREMENTS OF LCIS IN PRODUCT DESIGN-RELATED DECISION-MAKING

It has been claimed in previous chapters (and will be shown in Chap. 3.2.3) that the definition of the life cycle of a product as causal attribution of material and energy flows requires the implicit or explicit use of mental models and value choices. The following chapters treat the question of how decisions requiring the use of mental models and value choices should be handled to provide the 'best' product system and thus the 'best' life cycle inventory (LCI). Requirements are deduced from:

- Practical experience of LCA in product design (Chap. 3.1);
- Decision theory considering LCA a decision support tool (Chap. 3.2);
- Comparison of the 'descriptive' power of current impact assessment methods in view of the environmental dimension of sustainable development underlying LCA as the normative concept (Chap. 3.3).

3.1 Requirements from the use of LCA in product-related decision-making

A first set of requirements of LCA and the product system is derived from several surveys on practical experiences with LCA in product design-related decision situations.

3.1.1 Use of LCA as a decision support tool

LCA is a decision support tool. It is selected in a decision situation if it is assumed suitable for providing 'right' answers to a problem in a given context. According to ISO/EN 14040, LCA can be used in different contexts, such as for the development and optimisation in (environmentally conscious) product design, strategic planning, policy-making, marketing, etc.[77]. In decision situations supporting environmentally conscious product design, LCA can be used to structure the environmental dimension (or at least part of it, see Chap. 3.3.4)[78].

The purpose of an LCA should dictate the method to be used, the quality of data to be collected and the time and resources to be allocated to the project. Lewis (1996) distinguishes four different methods depending on the data quantity and quality required[79]:

- Life cycle review or 'life cycle thinking', a simple flow chart or process tree, which shows all of the main components of the product life cycle. This can help to identify the system boundaries and provides useful hints how to structure a more in-depth LCA;
- A simple qualitative matrix approach, which is used to identify areas for product improvement. A matrix is drawn up with environmental concerns along one dimension and with the life cycle stages along the other. This method relies on professional judgement rather than the collection of data for raw materials use, emissions, etc.;
- Streamlined or bottleneck LCA, which is a quantitative LCA which does not necessarily collect data on all inputs and outputs at every stage of the life cycle. Instead, streamlined LCA focuses, *e.g.*, on particular issues that are known to be particularly important in the life cycle of a product; or it collects primary data only for processes of particular interest, *e.g.*, the production processes, and uses average data for other processes;
- Comprehensive ('full') LCAs are purely quantitative and rely on measured, calculated and estimated data, whereas abbreviated approaches rely more on professional judgement.

The considerations in this book focus on quantitative LCA, streamlined and comprehensive ('full') LCAs. Nonetheless, the reasoning is also valid for more qualitative approaches.

According to international surveys[80], LCA-results are most often used for several purposes. In most cases, the driving force to conduct an LCA is the benchmarking of different products as well as the evaluation of environmental improvement options on a product-specific basis. Thus, two different levels of comparison can be distinguished:

[77] ISO/EN 14040, chap. 4.2
[78] see also Hofstetter 1998:18ff
[79] see also Wenzel 1998:281ff
[80] Anonymous 1996e; Anonymous 1999e:11ff/27ff

- *Inter-product comparison* comparing the environmental implications of different products (benchmarking), differing, *e.g.*, in different materials or different suppliers;
- *Intra-product comparison* comparing:
  - Different life cycle stages of a product, *e.g.*, for an environmental hot spot identification;
  - Different design options for one product, *e.g.*, the material selection or the selection of a production process.

Stakeholder information[81], *i.e.* marketing and lobbying, are secondary but nonetheless frequent uses of LCA-results (see also Chap. 3.1.2). Furthermore, LCA-results can be used as a basis for environmental labelling[82].

LCA results are consulted mainly in multi-objective decision situations[83], *e.g.*, the selection of a product, the selection of a supplier for the same product as procurement decisions, or the evaluation of different design options. In all these decision situations, the environmental criterion is only one among many. The financial leeway is one of the further criteria common to all decision situations mentioned[84].

From a management science perspective however, two characteristics are necessary for an efficient and effective use of LCA:

- *Balance between benefits and costs of LCA*: the marginal costs of collecting and computing more and better information about environmental impacts must not exceed the marginal benefits of information for the natural environment[85];
- *Eco-efficiency*: the ratio of environmental benefits to the economic costs of the tool must stand in an efficient relationship compared to other investment options[86].

In the following considerations, experiences made with LCA in comparative uses in product-related decision-making stand in the centre of interest for the derivation of a first set of requirements of the modelling within the life cycle inventory analysis.

81 for the distinction of different stakeholders of an LCA, see Lundie 1999:105ff

82 see ISO/TR 14025

83 requiring trade-offs between and within the financial, environmental and social dimensions, between internal and external constraints and expectations, and so forth (see, *e.g.*, Hopfenbeck and Jasch 1995:53)

84 see, *e.g.*, Ehrlenspiel 1995:121 for determining factors during product development; see, *e.g.*, Clark *et al.* 1997 for techno-economic issues in materials selection; see, *e.g.*, Ghodsypour and O'Brien 1998 for the multi-objectiveness of supplier selection

85 Schaltegger 1997b:95

86 Schaltegger 1997a:2f; Schaltegger and Burritt 2000:49ff

3.1.2 Experiences with LCA in product-related decision-making

Several surveys have been conducted around Europe to investigate the current and future use of LCA[87]. It has been shown that in practice, the link between LCA and a decision situation is not particularly direct:

> "However, today, LCA is by far not yet applied as a routine tool for product innovation, as it is still used more in a retrospective than in a prospective way. This is suggested by the common trend in all countries (Germany, Sweden, Italy, and Switzerland; WF) indicating that LCA is mostly used for some existing products, and is clearly not used for green products only. In general, it is more frequently applied to existing products than to new products. Moreover, LCA is usually applied to some products and not to all products..." (Frankl and Rubik 1998:54).

This statement is in line with an observation of Bhamra *et al.* (1999):

> "Tools such as Life Cycle Assessment (LCA) are often used at the later stages to assess a product's environmental performance in relation to a previous generation of the product or a competitor's. This tool therefore assists designers when they are at the stage of selecting materials and other components for the product" (Bhamra *et al.* 1999:332).

The indirect impact of LCA on product design is also visible in the different actors involved in product design and LCA. Frankl and Rubik (1998) state:

> "most of these studies (LCA-studies; WF) have been carried out within environmental departments, which are 'far away' from accounting and production departments" (Frankl and Rubik 1998:35).

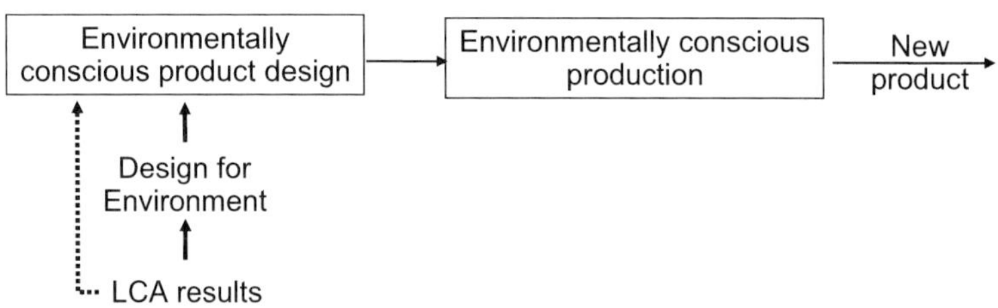

Figure 3-1. The indirect integration of LCA-results in the design process (after Gungor and Gupta 1999:818; reprinted with permission from Elsevier)

[87] see, *e.g.,* Bhamra *et al.* 1999; Brezet *et al.* 1999; Frankl and Rubik 1998; Hannsen 1995; Keller *et al.* 1999; Poole *et al.* 1999; Ritzén and Norell 1999; Wenzel and Alting 1999

Instead of directly supporting decisions in product design, LCA-results are more likely integrated – among other sources – into environmental design strategies and their tools such as check lists, recommendations, guidelines, standards etc. (Figure 3-1)[88]. Depending on the focus of these environmental design strategies, they are called 'Design for Environment'[89], 'Design for Recycling'[90], 'Design for Remanufacturing[91]' or 'Life Cycle Design'[92].

These design strategies are applied in later phases of product design concentrating on a small part of what is left to be influenced. The early phases of product design – such as the planning and conceptual phases – are not covered by these strategies or their tools. As a consequence, the fulfilment of the 'function' of the product – its environmental function efficiency[93] – is not evaluated in its integrity; as a result, only partial more environmentally efficient solutions can be provided[94].

Wenzel and Alting (1999), in a somewhat provocative outcome of their Dutch survey, state:

"Eco-design is not for the designers" (Wenzel and Alting 1999:378).

They conclude that generally the most effective environmental improvement options require decisions that typically lie in the hands of the company management or of someone else higher in the decision hierarchy than the designer[95].

It is in fact during the early phases, *i.e.* during product planning and conceptual design, where knowledge gained from LCA should be used if a major influence on the specification of a product is expected. However, during

88 after Ritzén and Norell 1999:347; Gungor and Gupta 1999:812

89 see, *e.g.*, van Weenen 1995; Gungor and Gupta 1999:820ff

90 see, *e.g.*, Frech and Schemme 1993

91 see, *e.g.*, van Weenen 1995; Kimura 1999; Kimura and Suzuki 1996

92 see, *e.g.*, Hannsen 1999

93 see Schaltegger and Burritt 2000:50f: environmental ('ecological') function efficiency measures how much environmental impact is associated with the provision of a specific function in each period of time.

94 after Frei 1998:38f; Frei and Züst 1998:124ff; also Bhamra *et al.* 1999:331; Wenzel and Alting 1999; Ritzén and Norell 1999:347; Schaltegger and Burritt 2000

95 In accordance with the above observations, Wenzel *et al.* 1999:378 have detected three different stakeholders having each their own kind of responsibility in eco-design[95]:

- The environmental specialist, responsible for the environmental assessment of products and for the development of environmental priorities;
- The company management, for assessing business opportunities and setting company strategies; and
- The designer, being responsible for fulfilling targets and following guidelines.

these phases, the data needed to conduct an LCA is not clear and probably not decided until the detail design phase[96]. These reverse trends of increasing knowledge about the product and decreasing influence on product design are illustrated in Figure 3-2.

These reverse trends of increasing knowledge about the product and decreasing influence on the product design are not only characteristics of environmentally conscious design. Rather, they are a constant factor in cost conscious design processes in general[97].

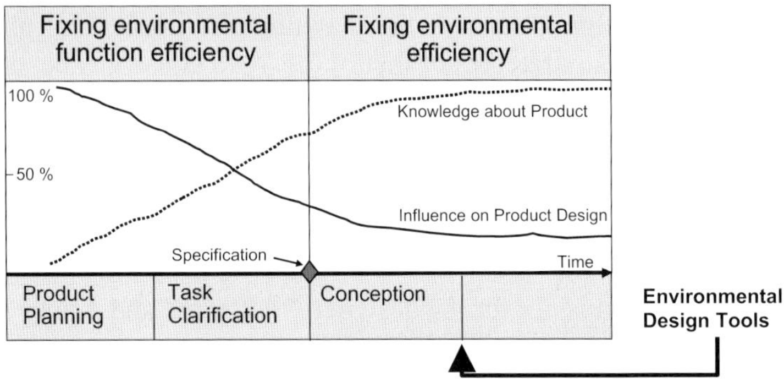

Figure 3-2. Strategic versus operational decisions. Possible influence on product design and knowledge about the product over the design process (after Frei and Züst 1998:125; modified according to Schaltegger and Burritt 2000:49ff reprinted with permission from Eco-Performance Verl.)

On the other hand, once the most effective environmental improvement options are detected, *e.g.,* with the help of an LCA, they are valid for a whole set of similar decision situations[98] and could *a priori* be taken into account, especially in early stages of product design.

Current experiences with LCA in environmental product design are summarised from various surveys. The strengths of LCA are perceived to be as follows:

- LCA considers the whole life cycle of a product for material and process selection[99];
- LCA makes the connection between product features and environmental impacts understandable[100];
- LCA allows one to understand environmental trade-offs[101];

[96] see, *e.g.*, Bhamra *et al.* 1999:332
[97] see, *e.g.*, Binder 1998:47ff
[98] see, *e.g.*, Wenzel and Alting 1999:378; Brezet *et al.* 1999:37
[99] Bhamra *et al.* 1999:332; Ritzén and Norell 1999:347; Brezet *et al.* 1999:37
[100] Bhamra *et al.* 1999:332; Ritzén and Norell 1999:347
[101] Bhamra *et al.* 1999; Ritzén and Norell 1999:347

- LCA produces a learning effect on environmental matters[102];
- "LCA is not simply a method for calculation, but, potentially, a completely new framework for business thinking" (Portisch 1997:100).

On the other hand, LCA is burdened with some weaknesses impeding its wider use:

- LCA-results are disputable[103];
- LCA still is burdened with general methodological difficulties[104];
- LCA has high data demand in the early stages of product development, but at the early stages of design data is available in low volume and of low quality[105];
- LCA generally has large data requirements[106];
- LCA is a very cumbersome tool, but at all stages of design designers often only want a tool which will allow a 'quick alternatives analysis', enabling them to make decisions about which material or option to select[107];
- LCA is costly; however, costs are not always regarded as a handicap for the use of LCA. This aspect is stressed more in Switzerland[108] and Germany than in Sweden or Italy[109].

Despite these weaknesses, an increased direct application of LCA in research & development and design is expected. This can be interpreted as a reorientation of LCA applications from a retrospective, analytical and descriptive perspective towards a forecasting and planning perspective.

This will imply an integral use of (a probably modified) LCA in product development and design and an integration of LCA into design software:

> "In all countries (Germany, Sweden, Italy and Switzerland; WF), and particularly in Italy and Germany, the use of LCA for radical changes in the product life cycle is expected to increase. This might suggest a (very modest) hint at a future relationship between LCA and product innovation" (Frankl and Rubik 1998:54).

Portisch (1997) puts it as follows:

[102] Ritzén and Norell 1999:350
[103] Frankl and Rubik 1998:34
[104] Frankl and Rubik 1998:34; Portisch 1997:99; Brezet *et al.* 1999:36
[105] Bhamra *et al.* 1999:332; Ritzén and Norell 1999:347
[106] Portisch 1997:100
[107] Bhamra *et al.* 1999:332; Wenzel and Alting 1999:378
[108] see Schaltegger 1996; Schaltegger 1997a. Apart from private costs of conducting an LCA, Schaltegger stresses the meta-economic costs of inventory data collection if financed by public funds.
[109] Frankl and Rubik 1998:34; also Brezet *et al.* 1999:37

> "The creation and development of environmental management systems, including extended producer responsibility and product stewardship responsibility, pollution prevention strategies, 'green' procurement guidelines and eco-labeling programs are evidence of the growing importance of life-cycle concerns" (Portisch 1997:102).

Two more experiences with the use of LCA in the product development context should be highlighted:

- The more specific an LCA for a decision situation the better it supports decision-making. This affects the type and quality of data required for setting up an LCA[110]. Site-specific data should be used as far as possible and available to obtain the most representative data for a decision situation.

  "If the producer (or consumer/waste treater; WF) is known, the firm's environmental situation and behaviour (*e.g.*, eco-audit) should be taken into account. The choice of a 'cleaner' production site may be environmentally more relevant than the choice of a specific product" (Pohl *et al.* 1996:67).

  The terms 'foreground' and 'background system' have been introduced to distinguish specific data and general data in product systems[111]. Foreground processes consist of processes that are under control or freedom of choice of the decision-maker. The background system is largely under control of others and cannot be shifted into a state that might be desired by the decision-maker;
- The more actor-based an LCA the better it supports decision-making and fosters practical consequences. The following passage illustrates the divergence between the needs of industry for practicable and suitable tools and the academic claim of 'scientific truth':

  "Evolution at the leading industries has been from defensive to proactive, from necessity to opportunity, from the standalone to full integration into the business. …The proactive industry approach is actor based (there are owners) and is therefore very effective in its implementation. LCA has a useful but not a dominating or subordinate role. The academic approach generally is holistic (there are no specific owners) and centred around LCA. With respect to business there is generally a self-chosen apartheid,

[110] Portisch 1997:102; Schaltegger 1996; Wenzel 1998:284

[111] see, *e.g.*, Frischknecht 1998:57. Although the use of site-specific process data related to in-house processes of the commissioner or of parties involved in an LCA is common practice, site-specific impact assessment for the foreground system has not been made so far (see also Hofstetter 1998:10f).

which in our opinion seriously hampers practical implementation" (Brezet *et al.* 1999:38).

In the next chapter, a first set of requirements can be deduced from current experiences gained during the use of LCA in product-related decision-making.

3.1.3 Requirements of LCA in product-related decision-making

The requirements from the use of LCA product-oriented decision-making can be summarised as follows:

- Comparable;
- As simple, manageable, transparent, cheap and quick but still as 'adequate' as possible under the given economic constraints depending on the goal and scope of the study and allows instant re-calculation (for direct applications);
- Actor-based, *i.e.* reflects the action space and the decision-maker's mental model of his/her range of responsibility in view of the management rules of sustainable development;
- As site- and case-specific as possible, *i.e.* uses as much site-specific information as possible.

Additionally, the marginal costs of collecting and computing more and better information about environmental impacts must not exceed the marginal benefits of information for the natural environment; the ratio of environmental benefits to the economic costs of the tool must stand in an efficient relationship compared to other investment options.

The importance of site-specificity and of an actor-based perspective supports the claim derived from decision theory that tools applied in decision-making should be "*...descriptive of the problem environment"* (see Chap. 3.2.2). This means they should reflect the decision-maker's mental model of the decision situation as much as possible (see also Chap. 3.2.4).

3.2 Requirements derived from decision theory

A second set of requirements of LCA as a decision support tool is derived from decision theory. In the first sub-chapter, requirements are listed that every decision support tool has to fulfil according to decision theory. In the following sub-chapters, the process of setting up a model within a decision-making process is looked at more closely. For this purpose, the theory of normative decision analysis is outlined and a model of the decision-making process by Sieben and Schildbach (1994) is introduced. LCA-methodology is structured according to the elements of the model of the decision-making process. This allows one to clarify the interrelationship of the object to be depicted (the life cycle of a product and its environmental implications) and the resulting LCA model, as well as the influence of the decision-maker's mental models and values on the setting up of the LCA model.

This comparison allows one to refine the requirements of decision support tools derived from decision theory (Chap. 3.1.3) for the specific case of LCA.

3.2.1 Requirements of decision support tools derived from decision theory

Decision theory treats the question how problems are solved, or should be solved, under given constraints[112]. The author shall abstain at this point from going into the details of decision theory and its different orientations. A more profound introduction to decision theory, particularly to normative decision analysis, will be provided in Chapter 3.2.2. In this chapter, a set of general requirements of models built within a decision support tool is presented.

During the conduct of an LCA, models are set up that describe the life cycle of a product and its environmental implications as a problem description. These models have to comply with some requirements common to all models developed within decision support tools.

According to von Winterfeldt and Edwards (1986), problem descriptions of decision support tools should be[113]:

- *Simple*: the decision-maker should be able to understand inputs, processes and outputs of the problem description as a precondition to affect their decisions;
- '*Descriptive' of the problem environment*:

  "A problem description must, of course, capture the analyst's (and with some luck, the decision-maker's) intuitions about the important aspects of the problem, including values, structures and other features of the organization and processes, entities and phenomena that specify its environment" (von Winterfeldt and Edwards 1986:35f);

- *Manageable*: value choices that are too numerous or too difficult for the decision-maker and excessively tedious or expensive computations which do not serve the decision-making process.

[112] Keeney 1992; Sieben and Schildbach 1994; von Winterfeldt and Edwards 1986; Kleindorfer *et al.* 1993

[113] after von Winterfeldt and Edwards 1986:35f

Keeney and Raiffa (1976:50) mention the following as desirable properties of a problem description:

- *Complete*: it should cover all the important aspects of the problem;
- *Operational*: it can be meaningfully used in the analysis;
- *Decomposable*: aspects of the evaluation process can be simplified by breaking it down into parts;
- *Non-redundant*: double counting of impacts can be avoided;
- *Minimal*: the problem dimension is kept as small as possible.

This is, of course, also true for LCA as Scholz and Weidenhaupt (1998) state, especially related to the second of the above-mentioned points of von Winterfeldt and Edwards (1986):

> "An LCA ...only makes sense if the object to be inventoried and assessed as well as the related impacts are recorded appropriately and the value structures of the user are taken suitably into account" (Scholz and Weidenhaupt 1998:39; translated from German, WF).

It is at the very least doubtful whether results provided by a decision support tool will be considered in the decision-making process if the models developed do not fulfil all the points mentioned by von Winterfeldt and Edwards (1986) and by Keeney and Raiffa (1976).

The second of the points mentioned by von Winterfeldt and Edwards (1986) requires further clarification:

- Which are the elements that make an LCA, and the product system in particular, *'descriptive'* of the problem environment?
- How do mental models and values influence the way an LCA model, and especially the product system, is set up?

Alternatively, referring to the above quotation by Scholz and Weidenhaupt (1998):

- When are environmental interventions related to an object recorded *appropriately*?
- When are the value structures of the user taken *suitably* into account?

In the next chapters, decision theory is addressed to clarify the above questions. The structure of LCA is viewed from the viewpoint of normative decision analysis to investigate the interdependency of the decision-maker's mental models and values as well as the choice of elements for the problem description in the LCA-model. For this purpose, a model of rational decision-making after Sieben and Schildbach (1994) is introduced to which the elements of LCA-methodology are attributed.

3.2.2 Introduction to normative decision analysis

Decision theory is committed to problem solving. Problems arise when an unwanted initial state should be transferred into a wanted target state. The main characteristic of problems is that the means for achieving the target state are unknown[114].

The issue of the theory of normative decision analysis is the rational choice of means. It treats the question of how one alternative action can be determined out of a set of alternatives that best achieves a given (or assumed) target state. The structured search for the best possible alternative to reach a given target under given constraints is considered a rational choice[115]. Or put the other way round: the selection of an alternative is rational because the choice is in accordance with the pre-set values of the decision-maker. This rationality of the choice of means has to be clearly distinguished from the rationality of the choice of ends[116].

Figure 3-3 provides a simple model of the process of decision-making by (Sieben and Schildbach 1994:16) that will be further explained.

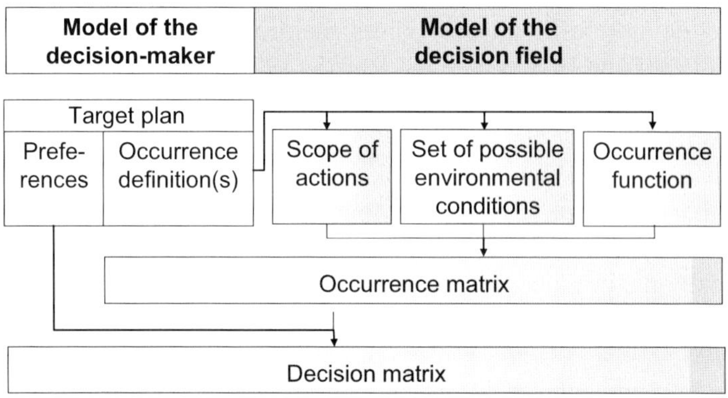

Figure 3-3. The basic elements of a model of normative decision-making (after Sieben and Schildbach 1994:16; reprinted with permission from Lucius & Lucius Verl.)

[114] This distinguishes problems from tasks, as the means for the task solving are known by definition. In the case of problems, even the target state sometimes cannot reasonably be defined. This kind of problems can be called 'ill-defined' problems (see Chap. 3.3.1; also Ehrlenspiel 1995:47f; Scholz and Tietje 2001:20).

[115] see, *e.g.*, Sieben and Schildbach 1994:1; von Winterfeldt and Edwards 1986:18. Transferred to management sciences, Sieben and Schildbach state: "A special form of the principle of rationality is pronounced in the striving for highest possible satisfaction of needs with a given stock of means" (Sieben and Schildbach 1994:1; translated from German by WF).

[116] von Winterfeldt and Edwards 1986:2

Practical normative decision analysis distinguishes two categories of information that describe the decision model (see also Figure 3-1)[117]:

- Information on the value structure of the decision-maker appearing in the 'target plan' as a *model of the decision-maker*. The target plan covers factual matters the decision-maker strives for and the intensity, he/she strives for them. The target plan is of purely subjective nature and has axiomatic character. The target plan regulates the setting up of the model of the decision field. Furthermore, the target plan serves as selection criterion for the choice of one action alternative that supports the striving of the decision-maker in the best way;
- Information on the decision situation as perceived by the decision-maker as a *model of the decision field*: this information includes the possible alternatives of action, the restrictions of the decision situation that cannot be influenced by the decision-maker and those of the consequences of the actions that are linked to the decision-maker's value system. The decision field is the result of conceptualising the decision situation. The decision field fixes the possible actions and their consequences in an inventory of the decision-maker's reality. The decision information is generated by the intent of an objective inventory of the alternatives for actions and their occurrences under the given environmental conditions; however, subjective influences are inevitable[118].

The information concerning the target plan is a model of the decision-maker; the information over the decision field is a model of his/her decision environment. Under everyday conditions, these models are generated and combined by simplifying them simultaneously within the decision-making process. No well-defined programme guides the procedure of decision-making itself; rather, it is guided by an undefined urge[119].

For more demanding decision situations, normative decision analysis provides a framework for the setting up of the two models in a structured way. In the following part, the elements of the decision model by Sieben and Schildbach (1994) are looked at in greater detail.

As already stated above, decision-making starts with the perception of a problem as a discrepancy of the actual state and a desired state, 'the target state'. A structured approach to problem solving requires a specification by the decision-maker from his/her value system. He/she has to define which occurrences, level, temporal course and certainty he/she wants to strive for

[117] after Sieben and Schildbach 1994:15ff

[118] after Sieben and Schildbach 1994:15

[119] after Feyerabend 1983:25 referring to Kirkegaard

or prevent in the future. The decision-maker has to define desired or undesired occurrences (*the occurrence definition*).

Given that various occurrences differing in kind, scale, temporal course or certainty seem desirable or undesirable to the decision-maker, the relative contributions to his/her subjective target state have to be determined. Weighting the different occurrences based on stated and unstated *preferences* does this.

The occurrence definitions weighted by the preferences are called *the target plan*, which depicts the value system of the decision-maker relevant to the specific problem. Metaphorically speaking, the target plan is the viewfinder through which the problem is perceived. It is through which the scope of possible actions, the possible external influences and the combination of actions and external influences are highlighted and through which the resulting occurrences are evaluated.

This close proximity between subjective elements of problem perception, target plan and description of the decision field is expressed with the two colours blending in Figure 3-3.

So, not only is it necessary to clarify the target plan (or clean the 'viewfinder'), but alternative scenarios of action under the constraints of the environment have to be formulated in consistence with the target plan: the *decision field.*

The components of the decision field are:

The *scope for actions*: a rational choice can only be made if a decision-maker has more than one action (including the action of omission) to choose from. The (limited) quantity of actions open to the decision-maker is called the scope for actions. Each action can consist of various action parameters that exclusively define the action in detail.

Decision analysis assumes that the decision-maker *a priori* is indifferent towards different actions. The different actions are attributed a value only by their consequences.

The *set of possible environmental conditions*: the decision-maker and his/her actions are imbedded in an environment that is beyond the influence and control of the decision-maker, which influence the consequences of his/her actions. In the situation of security, the environment will only adapt one condition; in the case of risk and uncertainty, various environmental conditions are possible. In a risk situation, the underlying probability distribution is known. In the case of uncertainty, the probability distribution is not known; the handling of an uncertain environment reveals the decision-maker's temporal preferences in his/her attitude towards risk[120].

[120] see, *e.g.*, Keeney and Raiffa 1976; Keeney 1992:7

Again, it is assumed that the decision-maker is indifferent towards different environmental conditions. They are covered by the value system only because of the influence on the consequences of the actions.

The *occurrence function*: each action alternative must be assigned occurrences under the given environmental conditions in an unambiguous way. These assignment rules are called the occurrence function. They consist of the decision-maker's mental models of the elements and organisational principles of reality.

The *occurrence matrix* summarises the three components of the decision field stated above. The head row contains the possible action alternatives; the columns contain the occurrences covered by the decision-maker's value system.

The action alternatives to be assessed still stand incomparably side-by-side in the occurrence matrix. Their occurrences have different features (such as type, scale, certainty and temporal course) so the occurrences first need to be weighted before the consequences of the actions can become comparable. For this purpose, the occurrences are submitted to the decision-maker's preference, which transposes them to the utility derivable from each action alternative. The depiction of the weighted occurrences on the utility level for each action alternative is found in the *decision matrix*. The decision matrix allows one to choose the action with the highest utility as the rational choice.

Chapter Summary

The target plan and decision field are mutually dependent. Any tool applied within the structuring process of the decision field has to be consistent with the target plan to be relevant in the decision situation

Mental models and temporal preferences related to the decision-maker's attitude towards risk are applied when action alternatives have to be assigned occurrences under a set of (uncertain) environmental conditions

3.2.3 Mental models and values in LCA

> "Accounting frameworks are more than just a set of bookkeeping rule and conventions. They represent a particular conceptualisation or worldview of how the economy and ecological systems operate"
> (Patterson 1998:108)[121]

[121] The quotation stems from an article on theories of value in environmental economics and is not restricted to environmental accounting of companies in a strict sense.

LCA cannot be looked at only as a decision support tool, but conducting an LCA – and in this case the definition of a product system – can also be understood as a decision problem itself. A 'best way' of conducting an LCA must be definable if the setting up of the product system cannot be made in an unambiguous way (see Chap. 2.3).

Focusing again on the inventory step, one can ask:

- Which are the elements necessary to make the model of the life cycle of a product 'descriptive'?
- How do the decision-maker's mental models and values influence the setting up of the model?

To answer these questions, LCA-methodology approached from the normative decision analysis perspective. LCA-methodology and in particular the life cycle inventory analysis are structured according to the model of normative decision analysis by Sieben and Schildbach (1994) (Figure 3-4).

Note that in a deviation of the model by Sieben and Schildbach (1994), a valuation function is introduced to explicitly depict the impact assessment methods applied in LCA (see below).

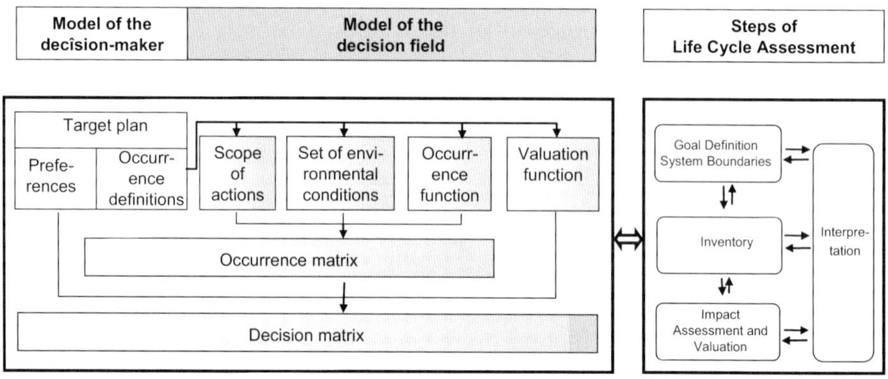

Figure 3-4. Merging LCA into the model of normative decision making after Sieben and Schildbach (1994). Note that in deviation of the model after Sieben and Schildbach (1994) a valuation function is introduced to explicitly depict the impact assessment methods applied in LCA

In an even broader sense, Feyerabend 1983 states that even "sciences do not know any 'facts' at all, but that all 'facts' that enter into our scientific findings, already are seen in a specific way and therefore are essentially ideal" (Feyerabend 1983:15f; translated from German WF). All statements on observations, results gained from experiments, statements on 'facts' contain theoretical assumptions or make use of them by the way of their application (after Feyerabend 1983:36).

Figure 3-5 shows the assignment of the explicit and implicit elements of LCA-methodology to the elements of the model of normative decision analysis.

The target plan depicts the preferences and the definition of occurrences depending on the conception of the decision situation:

Preferences are stated in the:

- Choice of the temporal and financial scope of the study in terms of the assignment of financial and human resources to conduct an LCA in dependency of the goal of the study;
- Definition of 'equivalency' of the functional units to be compared. The claim of equivalent functionality of two different products requires insight into the utility derived from these products. The determination of utility always has a subjective component, as it is a social construct[122]. Utility also depends on the uncertain service life of products on which assumptions have to be made;
- Structural uncertainty[123] ('ignorance') on future behaviour of the technosphere requires assumptions on future processes, market conditions, etc. during the setting up of the product system. Dealing with temporal aspects in LCA reveals the decision-maker's attitude towards risk and thus also the influence of his/her temporal preferences;
- Value choice inherent to the choice of an environmental impact assessment method stating environmental preferences expressed in its structure. Ecosphere cannot be described by one single disutility function because of its multidimensional character. The aggregation of environmental impact categories requires weighting the disutilities derived from each impact category[124];
- Overall target direction of 'sustainable development', on which LCA is based, also reveals preferences in a very general way.

The *occurrence definition* specifies the occurrences that are covered by the decision-maker's value system. For LCA, the occurrence definition contains:

[122] see, *e.g.*, Hofstetter *et al.* 2000:162; Sieben and Schildbach 1994:7f; Varian 1995:50ff

[123] Sources of uncertainty can be, *e.g.*, the variability of a system over time and imprecise measurement (both addressed as technical uncertainty), linguistic imprecision, subjectivity, disagreement, and approximations as a cause of ignorance on past, actual or future events (called structural uncertainty) (after Hofstetter 1998:313f referring to Pohl *et al.* 1996; Schwarz and Thompson 1990).

[124] A very good example is given by Hofstetter 1998, who distinguishes three different typologies of human preferences leading to different weighting factors for the safeguard subjects 'resources', 'ecosystems health' and 'human health' within Eco-indicator 99-methodology (Goedkoop and Spriensma 2000).

- Definition of environmental interventions (emissions, resources, wastes) as parameters of the LCI as linkage of the product system and of its LCI with the model of ecosphere;
- Definition of environmental implications expressed in the definition of 'environmental impact categories', to which the environmental interventions are attributed. The occurrence definition can also be interpreted in a broader sense than is actually done in the model by Sieben and Schildbach (1994). It is not only the wanted and unwanted occurrences of the decision situation, but occurrences can also be defined for the decision tool itself[125]. These occurrences define the requirements of the decision support tool, which the decision-maker must see fulfilled in the sense of desired occurrences.
  These requirements contain (among others) the representativeness of the:
- Mental model of the 'life cycle of a product'. It has been argued in Chapter 2.3.1 that the 'life cycle of a product' is a theoretical construct that has to be depicted properly in an LCA;
- Mental model of the organisational principle (determination) of the material and energy flows within the socio-economic system; this is relevant for solving the first part of the attribution problem related to the question which material and energy flows a product is held responsible for (see Chap. 2.3.1);
- Mental model of environmental implications expressed in the definition of adverse environmental impact categories and in the boundary setting between the technosphere and the ecosphere. The definitions of the 'ecosphere' as well as 'adverse environmental impacts' are human and are thus mental models in analogy to the conceptualisation of the structure and behaviour of the technosphere.
  It is now commonly agreed that the impact assessment implies value choices in the definition of the impact categories and in their further aggregation to single-score results[126].
  The decision field is structured dependent on the target plan.

[125] This reasoning lies beyond the model of normative decision analysis. Normative decision analysis relies on the assumption that the complexity of a decision situation can be conclusively reduced (Scherm and Frey 1999). In such a case, requirements on the descriptive power of models developed within a decision situation are not foreseen in the framework of normative decision analysis.

[126] There has been a broad methodological discussion on the valuation in the impact assessment step during the last years (see, *e.g.*, Fava and Denison 1991; Goedkoop 1995; Heijungs *et al.* 1992; Hofstetter 1998; Hofstetter *et al.* 1999; Hofstetter and Scheringer 1997; Hofstetter *et al.* 2000; Miettinen and Hämäläinen 1997).

Target Plan	
Preferences	**Occurrence definition**
Preferences are stated in the: - Choice of the temporal and financial scope of the study and the goal definition - Definition of the 'equivalency' of the functional units to be compared - Attitude towards risk arising from ignorance on future development within the technosphere - Choice of an environmental impact assessment method stating environmental preferences expressed in its structure - Overall target state of 'sustainable development' and its environmental dimension	The occurrence definition specifies the target state as: - Definition of 'environmental implications' expressed in the definition of 'environmental impact' categories and their eventual aggregation - Definition of environmental interventions to be reported as the linkage of the models of the life cycle of a product with the model of ecosphere In a broader sense, the occurrence definition also contains methodological requirements to be properly addressed in LCA, *e.g.*: - Mental model of the 'life cycle of a product' - Mental model of the structure and behaviour (including the organisational principles) of material and energy flows of technosphere as a basis for the modelling of the life cycle of a product - Mental model of the structure and behaviour of the ecosphere as a basis for the setting up of the valuation function, including the definition of the boundary between the technosphere and the ecosphere

Decision Field					
Scope for action	**Set of Environmental Conditions**	**Occurrence Function**	**Occurrence matrix**	**Valuation Function**	**Decision matrix**
Definition of similar functional units or design options to be considered as the alternative actions	Set of environmental interventions from the ecosphere to the technosphere (resources) or vice versa (emissions)	Rules for the attribution of material and energy flows to the product, *e.g.*: - Guidelines of ISO/EN 14041 - Mental model of the structure and dynamics of the socio-economic system - Attitude towards risk, related to the unpredictable behaviour of the socio-economic system - Biasing elements, *e.g.*, economic interests	Matrix of the life cycle inventories for each alternative	Rules for the attribution of environmental interventions to environmental impacts, commonly made by LC impact assessment methods: - Structure and behaviour of ecosphere	Results of the life cycle impact assessment as matrix or vector

Figure 3-5. Merging LCA into the model of normative decision making after Sieben and Schildbach (1994).

Note that the interpretation step of LCA usually serves to re-evaluate the assumptions made and models applied in view of the scope of the study, and to reflect the results of an LCA in the broader context of the decision situation. Therefore, it does not show up in this figure. As an extension of the model by Sieben and Schildbach (1994), the valuation function has been introduced. The valuations needed to determine the 'environmental implications' (or disutility) of the life cycle inventories are currently made by applying impact assessment methods.

The *scope for actions* is set by the different alternatives such as different design options or different functionally equivalent products, defined as 'functional units'. The alternatives are defined as viable options under the perceived constraints of the decision situation.

The set of *environmental conditions* of the decision situation in the sense of parameters that cannot be influenced by the decision-maker are the set of environmental interventions from the ecosphere to the technosphere (resources) or vice versa (emissions). They are the parameters that will be related to the action alternatives.

For each of the action alternatives and under consideration of the environmental conditions, an *occurrence function* is set up, which attributes occurrences (the LCIs as collections of emissions, resources, wastes) to the action alternatives. Attribution rules are stated in, *e.g.*:

- Guidelines of ISO/EN 14041.

As argued in Chapter 2.3.1 the enclosure of the life cycle of a product from the real socio-economic system – the first part of the attribution problem – is an artifice. It cannot be unambiguous due to the complex character of the socio-economic system.

For decisions such as setting system boundaries and allocation of multifunctional (joint) processes, no objectively correct decision rules can be stated (for details, see Chap. 5). These decisions can require the use of either:

- Mental models of the structure and dynamics of the socio-economic system;
- Values, manifested, *e.g.*, in the decision-maker's attitude towards risk in decisions related to the modelling of future processes[127].

It is important to notice the structure and behaviour of the socio-economic system and of the ecosphere are partial mental models of the real world and of its organisational principles. As such they are part of the occurrence definition and have to be properly reflected in the occurrence function.

In LCA, it is assumed that the structure and behaviour of the socio-economic system and of the ecosphere are not changed by other influences than the decision to be made: the *ceteris paribus*-assumption (see Chap. 2.2). However, as the product system is a partial model of the socio-economic system, uncertainties on the behaviour of the socio-economic system are inherent if statements on future conditions are to be made.

The occurrence function as sum of the attribution rules can also contain further (biasing) elements, such as, *e.g.*, economic interests if LCA is conducted only for marketing purposes.

[127] for the influence of further values on the life cycle inventory analysis, see Chap. 7

The combinations of each alternative with its respective environmental conditions constitute the *occurrence matrix* as a set of LCIs.

The valuation of several hundred emissions and resources in the occurrence matrix to determine their 'environmental implications' (or disutility) is very difficult. Therefore, a *valuation function* is set up in the form of environmental impact assessment methods in analogy to the occurrence function. This valuation function is the result of mental models of the structure and behaviour of the ecosphere and of the definition of 'environmental implications' and contains classification, characterisation and normalisation steps. The valuation function reflects environmental preferences.

The combination of the occurrence matrix with the valuation function produces the *decision matrix* with the weighted environmental interventions of the life cycle inventories as life cycle impact assessment. In the case of single-score impact assessment methods such as the Eco-indicator 95 and Eco-indicator 99[128], or the Ecoscarcity-points-method[129], the decision matrix produces an unambiguous ranking. In the case of multiple-score impact assessment methods such as the effect-oriented classification (CML-method)[130] results are not necessarily unambiguous and require further interpretation.

The following conclusions can be drawn:

- The explicit parts of LCA-methodology such as the definition of the functional unit, the definition of the product system and its summary, the life cycle inventory, as well as the impact assessment method and the results of the life cycle impact assessment can be located in the decision field of the model of normative decision analysis[131];

[128] Goedkoop 1995; Goedkoop and Spriensma 2000

[129] Brand *et al.* 1998

[130] Heijungs *et al.* 1992; Guinée *et al.* 2001

[131] A more formal way of fitting LCA-methodology into the model of normative decision analysis after Sieben and Schildbach 1994 can be made based on the matrix-calculation approach of Heijungs 1997. This more formal way is outlined below.
Based on the linearity assumption for economic and environmental processes Heijungs 1997 postulates a 'master equation' for the attribution of environmental problems to a cluster of economic processes, such as to a product system in LCA (Heijungs 1997:151):

$$g = \boldsymbol{g} \cdot \boldsymbol{R} \cdot \boldsymbol{Q} \cdot \tilde{\boldsymbol{B}} \cdot \tilde{\boldsymbol{A}}^{-1} \cdot \boldsymbol{a}$$

The scalar g as one-dimensional measure for the environmental problems is a product of:

- $\boldsymbol{a}$ vector of external demand of economic commodities, *e.g.*, induced by the product system under study
- $\tilde{\boldsymbol{A}}^{-1}$ inverted technology matrix containing economically valued technical coefficients; named after Koopmans 1951:37; also Leontief 1966; Chenery and Clark 1959

- Implicit parts of LCA-methodology such as mental models and values, *e.g.*, stated in preferences, can also be located within this framework;
- This interdependency of the target plan and the decision field cannot be negated by LCA as a decision support tool. Inconsistencies between the target plan, the decision situation ('reality') as perceived by the decision-maker and the models provided by or applied in decision support tools as descriptions of the decision field will most probably limit the usefulness of an LCA in the decision-making process. This is the case if:
  - The scope for actions as perceived by the decision-maker is not properly addressed in the models and results of LCA;
  - The assessment methods are applied which do not represent the decision-maker's values;
  - The setting up of the product system is made based on the mental model of the organisational principles and based on value choices, *e.g.*, because of differing temporal preferences that are not shared by the decision-maker (see Chap. 5.8.5);

$\tilde{\boldsymbol{B}}$ intervention matrix containing (mostly) economically not valued technical coefficients such as emissions or resources *in situ*, also called 'environmental commodities'; named after Heijungs 1997:61

$\boldsymbol{Q}$ the matrix that contains the combination of the fate coefficients of transient stressors (degradable emissions) and of the damage coefficients of intransient stressors (persistent emissions and resource consumption).

$\boldsymbol{R}$ the matrix that contains the impact factors corresponding to the characterisation factors in LCA

$\boldsymbol{g}$ the vector that contains the problem factors, also called weighing factors for the environmental impact categories, representing the environmental preferences

Within the model of Sieben and Schildbach 1994, the vector $\boldsymbol{a}$ represents one action alternative (a demand for economic commodities) that is multiplied with the model of the economic system $\tilde{\boldsymbol{A}}^{-1}$ and its related environmental interventions $\tilde{\boldsymbol{B}}$. The product $\tilde{\boldsymbol{A}}^{-1} \cdot \tilde{\boldsymbol{B}}$ corresponds to the occurrence function. The resulting vector $\boldsymbol{b} = \tilde{\boldsymbol{B}} \cdot \tilde{\boldsymbol{A}}^{-1} \cdot \boldsymbol{a}$ corresponds to the LCI and would fill one column (or row) of the occurrence matrix. The product of the matrices $\boldsymbol{Q} \cdot \boldsymbol{R}$ represents the model of ecosphere and its cause-effect nets resulting in environmental impacts $\boldsymbol{r} = \boldsymbol{R} \cdot \boldsymbol{Q} \cdot \boldsymbol{b}$. As time and space is usually compressed in LCA, $\boldsymbol{Q}$ is disregarded in LCA leading finally to the calculation of 'potential impacts'. The product of the matrices $\boldsymbol{g} \cdot \boldsymbol{R} \cdot \boldsymbol{Q}$ corresponds to the value function introduced to the model of Sieben and Schildbach 1994. The resulting scalar $g = \boldsymbol{g} \cdot \boldsymbol{R} \cdot \boldsymbol{Q} \cdot \boldsymbol{b}$ corresponds to one coefficient in the decision matrix (in fact a vector) of Sieben and Schildbach 1994 attributed to one action alternative.

The matrix calculation procedure makes the input-output calculation routine of LCA clearly visible with the inverted technology matrix $\tilde{\boldsymbol{A}}^{-1}$ and the intervention matrix $\tilde{\boldsymbol{B}}$ as model of the socio-economic system (see also Frischknecht 1998; Heijungs 1997:59ff:32ff).

- The attribution of material and energy flows to the product system is representing a 'causality' that is not shared by the decision-maker as a consequence of the above-stated factors.

In such cases, recommendations derived from the LCA-model lead to sub-optimal solutions if re-evaluated from a meta-perspective by the decision-maker (see Figure 1-1). Even worse, such inconsistencies might take the decision-maker to the point of completely ignoring the indications and results provided by an LCA (see also Chap. 3.2.3) or reject LCA as a decision support tool in general[132].

Chapter summary

The explicit and implicit elements of LCA-methodology can be integrated very well into the framework of normative decision analysis:

- The LCA model represents the decision field. The developed LCA model is a partial model of the socio-economic system as well as of the 'ecosphere', of the interconnections and of 'adverse environmental impacts';
- The setting up of the LCA model on the one hand is guided by preferences stated in choices made concerning temporal and financial scope of the study, concerning the 'equivalency' of functional units to be compared, the choice of the impact assessment method and its inherent value structure;
- The setting up of the LCA model on the other hand is guided by the occurrence definition. The occurrence definition in an LCA contains the parameters of the LCI and the definition of 'environmental implications';
- The attribution rules of material and energy flows to a product as its life cycle (the 'occurrence function') cannot be made in an unambiguous way. Besides the guidelines of, *e.g.*, ISO/EN 14041, the decision-maker needs to find depicted:
  - His/her mental model of the structure and dynamics of the socio-economic system;
  - His/her values manifested *e.g.,* in temporal preferences in modelling decisions related to temporal changes of the socio-economic system;
- Inconsistencies between the target plan, and the decision situation ('reality') as perceived by the decision-maker and the models provided by or applied in decision support tools as descriptions of the decision field limit the usefulness of LCA as a decision support tool.

[132] This is the main reason why Hofstetter 1998 proposes the explicit modelling of the valuesphere of the decision-maker and, based on it, consistently model the product system as part of the technosphere and the environmental cause-effect relation (ecosphere) (Hofstetter 1998:27ff).

3.2.4 Whose values and mental models are to be depicted in LCA?

LCA as a decision support tool is situated between, on the one hand, the requirement of being representative for the decision field and the decision-maker and of being 'objective' in the sense of allowing unbiased comparison of the environmental implications of products on the other hand. So the question arises:

a) Whose values, and
b) Upon whose mental models should the modelling in LCA be based?

Should it be based on the decision-maker's values and mental models to be *'descriptive' of the problem environment* (see Chap. 3.2.1) and thus to raise the chance of an LCA becoming relevant in the decision-making situation? Or should 'generally agreed upon' accounting and assessment rules be applied, risking the outcome of an LCA being ignored by the decision-maker for not being *'descriptive'*?

Ad a) LCA has been developed as a decision support tool within the philosophy of environmental management, which basically refers to the concept of sustainable development. Hence, sustainable development can be considered the overall target direction and underlying value system for environmental decision support tools. However, sustainable development as a normative concept will never be fully operational, nor will an interpretation of it ever be generally agreed on (see Chap. 3.3.1).

A second set of values influencing the conduct of an LCA is the decision-maker's personal preferences and in particular the person's temporal preferences. They are influencing how ignorance on future processes is dealt with and relate to the decision-maker's attitude towards risk (see Chap. 5.8.5). For this set of values, the same reasoning is valid as outlined below for mental models.

Ad b) the answer to the question of whose mental models should be depicted in LCA is less obvious. Mental models constitute a particular worldview and can be the result of particular interests. Conflicts can arise between the representation of the decision-maker's mental models and the claim of LCA to provide new insight and knowledge in a 'reliable' way.

Obviously, particular (economic) interests should not bias LCA[133]. On the other hand, LCA should provide recommendations on environmental improvement options that do not obviously contradict market and material characteristics. Furthermore, LCA-models should not be based on assumptions on the organisational principles of the socio-economic system that are

[133] see Rescher 1993:7f for the distinction of 'good reasons' as basis for efficient and effective decision-making and just motives derived from wishes and desires

not shared by the decision-maker. In fact, a common sense between the different stakeholders of a particular LCA should be obtained on any assumption (including mental models and values) made during the conduct of an LCA (see Chap. 23 on LCA as a group model-building process). Stakeholders of a particular LCA are, *e.g.*, the decision-maker(s), the modeller(s), the members of an eventual steering committee, the representatives of other material groups in comparative studies or eventually the internal or external peer reviewer(s), etc[134].

Some of the mental models applied during the conduct of an LCA can be backed up by the literature of resource economics, material sciences, environmental sciences, etc. However, some decisions during the setting up of the product system require the application of mental models, on which several lines of reasoning can be justified from scientific literature (see Chap. 6). In these cases, the choice can only be justified on an argumentative basis if the optimisation option derived from the LCA model is re-evaluated from a meta-perspective[135].

This methodological approach is chosen in the case studies where different attributions and allocation procedures are compared and evaluated.

3.3 Descriptive power of LCA and consequences for the life cycle inventory analysis

In the next chapters, the decision-maker's values are looked at more closely: the decisions that make him/her choose LCA as a decision support tool and that he/she should find included in an LCA-model. The descriptive power of current impact assessment methods is investigated in view of the normative basis of the environmental dimension of sustainable development.

This is done with the aim of viewing:

- To which extend current LCA methodology is able to address and reflect the values of the environmental dimension of sustainable development;
- How management rules of the environmental dimension of sustainable development as a consequence affect the attribution of material and energy flows to the life cycle of a product.

These questions are of crucial importance if improvement options derived from LCA models should not lead to sub-optimal solutions if re-evaluated from a meta-perspective.

[134] see also Lundie 1996:112ff

[135] because of the impossibility of verifying LCA-results or of validating the product system as model of the life cycle of a product

From these chapters, a third set of requirements of product systems and their respective LCI shall be established considering the 'descriptive' power of LCA in view of its normative basis: the environmental dimension of sustainable development.

3.3.1 Sustainable development as overall target direction

'Sustainability', 'environmental soundness', or other concepts and guiding principles[136] set up in the environmental discussion of the last decades are outlining a target state in very general terms. To apply these concepts to real life decision situations they need further refining. 'Sustainable development' as the most prominent principle of today will never be fully operational in that sense.

Minsch (2000) calls sustainable development a 'regulative idea' comparable to the Human Rights. Regulative ideas inspire and guide societal learning and forming processes; they will never be fully and conclusively operational. Like human rights, sustainable development will never be an accomplished achievement with regard to civilisation but an everlasting challenge[137]. This has already been recognised by Brundtland *et al.* (1987) who brought the term into the political debate:

> "...sustainable development is not a fixed state of harmony, but rather a process of change in which the exploitation of resources, the direction of investments, the orientation of technological development, and institutional change are made consistent with future as well as present needs...Thus, in the final analysis, sustainable development must rest on political will" (Brundtland *et al.* 1987:9).

Scholz and Tietje (2001) address these types of decision situations with unclear targets as ill-defined problems, which are typical for environmental problems. Their basic structure consists of the initial state that can be described, of a target state that is insufficiently known, and of a problem

[136] such as 'Spaceship Earth' (*e.g.,* Boulding 1966); 'Steady-State-Economy' (*e.g.,* Daly 1974), 'Eco-Cycle-Economy' (*e.g.,* Chadwick 1997; van Weenen 1995, S. 97f), 'Gaia' (*e.g.,* Lovelock 1995), 'Industrial Ecology' (*e.g.,* Frosch *et al.* 1997; Manahan 1999), and 'Industrial Metabolism' (*e.g.,* Ayres and Simonis 1994) or the recommended strategies for the environmentally sound redirection of anthropogenic material flows: effectiveness and efficiency in production and consumption, sufficiency in consumption, consistency or the anthropogenic flows with natural flows (*e.g.,* Braungart and Mcdonough 1999; Frei and Züst 1998:125; Hannsen 1999; Kimura and Suzuki 1996:84; Mertz 1999; van Weenen 1995; Vester 1980:85ff)

[137] after Minsch 2000:33

solving strategy by identifying and passing barriers by applying old and newly developed methods[138].

The assumed target state can even change over time, as scientific progress and changes in public values require re-evaluation of the current state, the assumed target state, and the methods applied for problem solving. As the target state is known not only insufficiently but can also change over time, it may even seem appropriate to talk about an 'insufficiently known, changing target state' of environmental problems, or just of a target direction. The assumed target state of 'sustainability' in its current understanding is a temporary, normative concept.

At the current state of the environmental debate, environmental tools can only provide hints on the way towards the environmental dimension of sustainability. The results achieved during problem solving, the assumed target state *per se* and the adequacy of the methods applied for problem solving have to be monitored and evaluated during the whole problem solving process. This is, of course, also true for LCA and implies a continuous evaluation and methodological development of LCA as a decision support tool.

3.3.2 Normative basis of sustainable development

In this chapter, the normative basis, on which sustainable development and the management rules to be derived in Chapter 3.3.3 are based on, is assessed.

LCA as a decision support tool is situated within the environmental dimension[139] of sustainable development. The social and economic dimensions are not taken into account in LCA. For the following, emphasis is put therefore on the environmental dimension of sustainable development.

Since the 1992 UNCED conference, held in Rio de Janeiro, the term 'sustainable development' has increasingly become the guiding principle of national and international policy making. Despite the popularity of the subject, the term 'sustainable development' remains vague within the political and scientific discourse[140]. It is commonly referred to as being a development:

> "...that meets the needs of the present without compromising the ability of future generations to meet their own needs" (Brundtland *et al.* 1987:8).

[138] after Scholz and Tietje 2001:20; see also Dörner 1976:10; Duncker 1974:1
[139] see, *e.g.*, Beck and Bosshart 1995; Hofstetter 1998:18ff
[140] for a variety of differing definitions of the term 'sustainable development', see Ninck 1997:51ff

Sustainable development is essentially based on the three pillars of economic efficiency, environmental responsibility and social solidarity that will allow socially, economically and environmentally sound long-term development[141].

One of its strongest moral fundaments is *intra- and inter-generative justice*. Economists will be reminded of the egalitarian theory of justice developed by Rawls (1971) in an individual, intra-generative context: Rawls claims that due to the fact that no individual can be sure about the personal consequences of any concept of justice applied (as they live under a 'veil of ignorance'), all individuals have an interest to improve the destiny of the subject being worst off, as it might be themselves[142]. This 'maximin-principle' has been transferred to the dynamic intergenerational context by Solow (1993) and forms the basis of current understanding of sustainability:

> "I will assume that a sustainable path for the national economy is one that allows every future generation the option of being as well off as its predecessors" (Solow 1993:168)[143].

The maintenance of a constant level of wealth and welfare implies not living from the stocks but from the flows. This point has been worked out by Hicks (1946) who links the (individual) income to the *demand of preserving the stocks* as the source of income:

> "The purpose of income calculations in practical affairs is to give people an indication of the amount which they can consume without impoverishing themselves. Following out this idea, it would seem that we ought to define a man's income as the maximum value that he can consume during a week, and still expect to be as well off at the end of the week as he was at the beginning. Thus, when a person saves, he plans to be better off in the future; when he lives beyond his income, he plans to be worse off" (Hicks 1946:172).

In the discussion on sustainability, the model of individual income by Hicks has been transferred to the intergenerational context: the claim to limit

[141] De Montmollin and Altwegg 1999:7; also Anonymous 1996f:9ff; Mauch *et al.* 1995:15f

[142] after Radke 1999:121

[143] The equitarian concept of justice is of course only one among others. Within economic theory, utilitarian concepts have also been proposed to determine the optimal path of resource consumption. Based on the assumption that a disadvantage of a person or generation, caused by not consuming, can be compensated by a higher utility derived from consumption by other persons or generations. Therefore utilitarian optimisation consists in maximising utility over all times (for a discussion, see, *e.g.*, Messner 1999:77ff).

the consumption to Hicks' income is considered equivalent to the claim to preserve the aggregated (natural and anthropogenic) capital of society[144].

There is, of course, no consensus on what the stocks are, from which income can be generated. Neo-classical economic theory considers the scarcity of all resources to be relative, as *substitution through technological progress* (*higher efficiency, other technologies*) will alleviate all scarcities. As long as the productive capital is maintained by reinvesting all parts of income attributable to natural production factors ('scarcity rents') into anthropogenic capital, a constant level of consumption is possible (a principle known as the Hartwick-rule[145]).

Environmental economy and environmental sciences have strongly questioned the neo-classical utilitarism (after Messner 1999:143ff; Radke 1999:122):

- *The non-consideration of natural utility sources*: consumption of man-made goods and services cannot be considered the only source of human utility. Natural resources have an intrinsic value, independent of its contribution to human consumption. Additionally, some components of natural capital have life support functions, to which no price can be attributed;
- *The non-consideration of social utility sources*: not only real income but also health, education or the rootedness in a culture are part of the societal target plan[146];
- *The non-consideration of complex interdependencies of nature*: the existence of critical minimal levels of certain natural resources (thresholds);
- A general *scepticism towards non-market valuation techniques* such as contingent valuation methods[147].

But one of the most central aspects criticised is the:

- Consideration of complete substitutability of different natural resources.

A fundamental uncertainty is closely linked to inter-generative justice, as a time span has to be specified within which the inter-generative justice paradigm has to be applied. It can easily be concluded that an unlimited existence of mankind on earth is not possible if substitutability is considered pessimistically. In contrast, an optimistic view of substitutability will lead to the justification of a never-ending time span.

144 see also Daly 1974 for his economics of the steady-state

145 see Hartwick 1977; also Dixit *et al.* 1980; Messner 1999:73ff; Radke 1999:120; Withagen 1996

146 see also Giarini 1980:25ff and its concept of 'utilisation value'. Utilisation value is defined as net contribution to 'our Dowry & Patrimony' as the general source of wealth and welfare, thus combining monetarised and non-monetarised elements of wealth.

147 see, *e.g.*, Patterson 1998:110

The point of view taken with regard to the time span of human existence and thus the perception of the risk of ignoring eventual non-substitutability of natural resources has essential implications for the definition of strategies towards sustainable development[148]. In a risk-taking position (as a consequence of an optimistic view of the substitutability of natural resources) technological progress will allow infinite human existence and a constant consumption level over time (if the Hartwick-rule is followed).

A risk adverse position would imply a pessimistic view of the future substitution options provided by technological progress. From the limited time span of human existence, the *sufficiency strategy* is considered adequate to keep the same living standard as long as possible.

If the substitutability is seen optimistic enough, considerations can be made how a just distribution of living quality would look. One of the possible alternatives would be the 'maximin' concept of justice by Rawls (1971) and Solow (1993) as outlined above.

Apart from differing estimations of the substitutability of natural resources, environmental economists and environmentalists have pointed out further prevailing conditions of a sustainable path:

- *Minimising maximal losses* suggests game theory (the 'minimax' concept), as the needs and technical possibilities of future generations cannot be known. This concept urges to choose a safe (safer) alternative if costs are not unacceptably high[149]. From this concept originally developed in the context of biodiversity, economists have recommended a cautious proceeding in decision situations that involve irreversibilities with potentially high future costs, and plead for leaving options open[150]. They argue that irreversible man-made destruction of natural resources as a basis of life makes future generations worse off;
- *Safety margins according to the minimum standards principle*: whereas knowledge about current machines tends to be complete, knowledge about environmental assets is seriously incomplete. Uncertainty consists, *e.g.*, in the unknown number of species or in the exact working of the carbon or the hydrological cycle[151]. Whilst uncertainty undoubtedly declines through time, the natural world will remain imperfectly understood. Assuming risk aversion, there is good reason not to reduce natural resources unless the benefits from doing so are known and are substan-

[148] An extensive discussion on substitutability of resources has been made by Dasgupta and Heal 1979 who distinguish between unnecessary vs. necessary and non-essential vs. essential resources for production (see, *e.g.*, Messner 1999:68ff; Radke 1999:82f).

[149] after Messner 1999:200ff

[150] Bishop 1978:16; see also Messner 1999:114ff/200ff

[151] Odum 1991:121

tial. This is akin to the safe minimum standards principle by Ciriacy-Wantrup (1952) and Bishop (1978)[152];

- *Breaking up path dependencies*: the evolutionary character of economic processes implies path dependencies of technological development. These path dependencies arise because of the endogenic sources of technological innovation and because of the inertia of social and economic systems, given by the time necessary from the invention of a technology to its market penetration[153];
- *Thermodynamic efficiency of the economic processes*: All economic processes are entropic[154] and thus irreversible. As energy of the universe is constant (1st law of thermodynamics) the entropy of the universe tends towards a maximum (2nd law of thermodynamics). The earth can be approximated as a subsystem connected to the cosmos by a steady state of energy flows[155]. In an economic process, low entropic energy is converted into high entropic energy and dissipated into the natural system, into which the economic system is embedded. Solar radiation is by far the most important source of energy on earth, and thus the only mean to lower the overall entropy of earth. As assimilation capacity of the autotrophic biosphere is limited (about 1% of the incoming radiation energy[156]), the thermodynamic efficiency of the economic processes is a basic claim for a long-term perspective of human existence[157,158].

[152] after Pearce and Atkinson 1995:169; see also Randall and Farmer 1995

[153] see, *e.g.*, Dosi *et al.* 1994; Schumpeter 1993/1939; also Messner 1999:171ff

[154] after Georgescu-Roegen 1971:281; see also Messner 1999:167ff; Rechberger 1999:35

[155] see Boulding 1966; Daly 1974; Constanza 1981; Georgescu-Roegen 1971; Vester 1980:29ff

[156] The technical assimilation of solar energy will also require space and will also be entropic. Their technical efficiency is currently at about 10-12% of the solar energy for photo-voltaic cells and at about 50% for solar heat production on the surface of earth. In these numbers, the energetic pay-back time is considered.

[157] Goergescu-Roegen postulated a further interpretation of the entropy law by claiming it to be valid for material also. He claimed that material in its daily use is dispersed by friction on a molecule by molecule basis and thus gets lost over time. Even for recyclable material the time needed makes it impossible to achieve complete recycling by recollecting the dispersed molecules and fitting them together again. These considerations imply that, despite the sentence of mass conservation, material gets unavailable for human uses as wear and tear of material takes place, which has to be compensated by the use of additional material. This means that a recycling rate of 100% of a material is not possible even under the use of unlimited energy (Georgescu-Roegen 1977:301; Georgescu-Roegen 1979:1029ff; see also Ayres 1999).

These considerations have been called the 4. law of thermodynamics, although thermodynamics has not provided any prove that would give this postulate the rang of a law (Messner 1999, footnote 90; Rechberger 1999:37).

The closing of material flows towards material cycles by recycling, the concentrated disposal of waste enabling future waste mining as well as trends toward dematerialisation and substitution of materials are starting points among others for the implementation of this principle[159];

- *Consistency of anthropogenic flows with natural flows*: the environmental resilience of biological systems differs. Environmental resilience is the capability of a biological system to bear and survive external disturbance without losing its internal organisational structure. Environmental resilience indicates how far a biological system can deviate from normal conditions without changing the internal structure of the system. Very resilient biological systems therefore can stand larger disturbances[160].

  Seen from an economic (anthropocentric) point of view, nature, with its biological systems, has several functions:
  - The function of resource provision;
  - The function of a sink for anthropogenic, highly entropic residual streams;
  - The function of providing natural habitat with apt conditions regarding temperature, climate, availability of food, etc.[161];
  - An 'aesthetic' or indirect utility function of nature[162].

  Environmental economists have identified the first three functions of nature (a-c) to set absolute limits to economic growth[163]. As the system behaviour of natural systems is largely unknown due to their complex structures, the consistency of anthropogenic flows with natural flows is another risk-reducing guiding principle[164].

---

[158] Vester 1981 follows another line of reasoning coming to the same conclusion. Based on the analogy of biological systems as reference point for the management of complex systems he claims multiple use to be a basic organisation principle of complex systems (Vester 1980:47f/54/83/348). As industrial processes have generally a low efficiency compared to nature, *e.g.,* in the conversion of energy (technical systems: about 30-60%; natural systems: 80-90% efficiency), multiple use and the closing of anthropogenic flows is a basic strategy for the management of the socio-economic system (after ibid., p. 68).

[159] see, *e.g.*, Manahan 1999:26ff

[160] see Messner 1999:151; also Odum 1991:71

[161] The function of providing lebensraum can further be specified as a production function mainly attributed to the autotrophic part of the biosphere and especially the settlement function of all material- and energy flows within nature. This function enables the complex and interdependent combination of the biosphere with its three natural systems atmosphere, hydrosphere and lithosphere (Messner 1999:187).

[162] Messner 1999:185ff

[163] Messner 1999:191f

[164] see, *e.g.*, Mertz 1999; 'compatibility' after Vester 1980:85f

The normative basis of sustainable development outlined above must be scaled down to make sustainable development more operational. It is only on the basis of management rules, investigated in the next chapter, that the 'descriptive' power of LCA in view of the environmental dimension of sustainable development can be assessed.

3.3.3 Management rules for sustainable development

During previous decades, many efforts have been made to develop the concept of sustainable development. One of the most extensive and theoretically best-founded scaling down of the environmental dimension of sustainable development is made in Messner (1999). Messner (1999) proposes seven management rules (MR) concerning environmental sustainability and material use based on the normative basis outlined above:

> "*First Management Rule*: For the long-term preservation of useable biomass and of the potential of biomass from *renewable natural resources*, it has to be guaranteed:
>
> - That fertile soils and watersheds apt for the production of biomass are kept available;
> - That natural production and life support functions remain intact and are not affected by mechanical interventions or emissions from the production of biomass;
> - That the consumptive use of renewable resources does not exceed the growth rate of usable biomass minus the resources needed for the maintenance of the renewable resource basis; and
> - That the genetic potential of renewable resources is protected and is assigned an adequate habitat.
>
> "*Second Management Rule*: The *material inputs from economy into nature* may not exceed the regionally and timely differentiated natural assimilation and absorption capacities reduced by a safety margin, which shall guarantee that the conditions of life are not made worse and that the prevailing average human lifetime is not reduced. In doubt and in view of high potential threat deriving from a material, its input has to be completely avoided.
>
> "*Third Management Rule*: *Non-renewable resources* are to be used in a way which maintains its energetic potential and guarantees the long-term availability of high-quality material resources for a spectrum of material applications as broad as possible, especially for all fields of applications to satisfy basic needs, without exceeding the maximum tolerable level of nature for emissions or mechanical interventions. Therefore:

- The rate of material dispersion and the rate of energetic dissipation of non-renewable resources on a superordinate area of functional applications, should not be higher than:
  - The rate of substitution with (additional) renewable resources; or
  - The rate of increase in resource productivity because of technical progress and resource saving; or
  - The rate of substitution with still abundant non-renewable resources;
- The exploitation and use of non-renewable resources shall not lead to emissions or mechanical interventions on the natural life support functions above maximum tolerable levels;
- Concerning the principle of resource conservation, all activities aimed to ensure the availability of resources should not lead to a higher level of environmental interventions;
- A complete consumption or non-accessibility of single chemical base elements have to be avoided to ensure a long-term innovation potential in the material use;
- The thermodynamic potentials of subsequent uses of non-renewable material shall be used and corresponding future options shall be kept open bearing in mind the more restrictive economic limitation of material resources;
- Non-renewable material shall be kept in circular flows and its dispersion shall be avoided as much as possible. With the aim to reach high thermodynamic efficiency in the circular flow economy, material flows are to be aspired to where the change of entropy of the material is kept as low as possible over subsequent uses. In this sense, the principle of low entropic differences in material cycles is valid;
- The exploitation of non-renewable resources from irreplaceable 'critical' nature protection areas has to be avoided to ensure, besides life support functions, the conservation of as high as possible levels of biological diversity, which has to be considered a potential for future resource substitutes;
- The destruction of interconnected biotopes for the sake of the exploitation of non-renewable resources has to be compensated by the definition of ecologically equivalent compensation areas of the same size.

"*Fourth management rule*: Human interventions on nature related with *high threats or unacceptable risks* are to be avoided. If insecurity and uncertainty dominate, cautious proceeding is advisable and a sufficient safety margin is to be introduced.

"Fifth management rule: In a *combined use of renewable and non-renewable resources* shocks in resource availability, quality losses in the resource use and forms of inefficient resource use are to be avoided in the long run. Renewable resources will be the only resource basis for future generations, thus the following guidelines have to be considered for the realisation of an inter-temporal continuous transformation path in resource use:

- The use of high-quality resources shall be reduced to a minimum level as soon as possible to ensure a transformation without major difficulty towards an economy based on renewable resources. The principle of resource conservation is valid;
- The economic processes of an economy are to be organised and co-ordinated in a way that the lowest possible material and energy losses, the lowest possible affection of natural resources and natural areas and the lowest possible dispersion of material to reach the production targets occur. In this sense, the principle of thermodynamic efficiency in the production is valid;
- The demand side has to be integrated into the transformation process. Therefore, thermodynamically efficient forms of use of products and services as well as a generally resource conserving lifestyle has to be aspired to. In this sense, the principle of sustainable consumption is valid;
- Material-related and technological innovations are essential components for the transformation process in the resource use. Bearing in mind the environmental risks associated with the inter-temporal trends for higher entropy differences of innovations, a concept for material recycling should be in place for each product innovation at market entry, according to the principle of low entropic differences.

"Sixth management rule: Facing the continuous *growth of the population* and having in mind the limited space and resource availability bearing the risk of frequent famines and the death of huge masses of human beings, a stabilisation of the size of population has to be aspired to.

"Seventh management rule: The pressure on nature caused by *mechanical interventions* may not lead to a degradation of subordinate regional or global life support functions. To secure and preserve the natural functions and the biodiversity, unmodified and regionally interconnected nature protection areas shall be assigned. The claim for nature protection areas might be possible in single cases if compensation is made in form of an ecologically equivalent compensation area. Due to ecological uncertainties in determining the size and structure of nature protection areas,

safety margins have to be considered" (Messner 1999:400ff; translated from German by WF).

3.3.4 Descriptive power of LCA in view of the management rules of sustainable development

LCA by itself does not generate values. It is based on underlying concepts of 'environmental soundness' or 'sustainability', commonly considered the target state in the actual public discussion. LCA as a methodology has been developed within these normative concepts and should depict them as tool used for environmentally conscious decision-making. In this chapter, the 'descriptive' power of LCA in view of the management rules of sustainable development is assessed. Reference is made to current impact assessment methods such as the CML-method[165]; the ecoscarcity-points method[166]; the critical volumina-method[167]; the EPS-method[168] or the Eco-indicator 95 and Eco-indicator 99[169].

Considering the diversity of aspects of the environmental dimension of sustainable development covered by the seven management rules, it is obvious that only a few aspects are covered by current LCA practice. Among these are an improved environmental performance through modifications of the system related to a higher effectiveness and efficiency, which:

- "Reduce the reference mass flows within the product system, *e.g.*, by making the product more efficient or by reducing consumption of some raw material;
- "Reduce emission factors or resource consumption factors, or change to other types of emissions or resources, *e.g.*, by reducing emissions and energy consumption per kilogram of product produced;
- "Change types of emissions or types of resources used in the system, *e.g.*, by technology changes or by using other types of raw materials" (Hannsen 1995:182).

LCA with its input-output model very well addresses environmental improvement options related to more effective and efficient use of resources in process optimisation[170] (under the assumption that 'causality' is properly reflected during attribution of material and energy flows!). LCA can be a

165 Heijungs *et al.* 1992; Guinée *et al.* 2001
166 Ahbe *et al.* 1990; Brand *et al.* 1998
167 Habersatter 1991
168 Steens and Ryding 1992
169 Goedkoop 1995; Goedkoop and Spriensma 2000
170 incorporating, *e.g.*, management rules (MR) MR 3, point 3 and MR 5, point 3

very effective tool for relative comparisons when a combination of energy related, material application related and process related aspects play a role[171]. This is especially true for global impact categories like potential greenhouse gas emissions, or potential ozone depletion.

LCA is less 'descriptive' in situations in which toxic and/or hazardous substances are involved because of the compression of time and space. This compression does not allow the integration of regional background contamination or interventions, environmental thresholds, temporally and spatially resolving use patterns of resources, risk, etc. into impact assessment methods[172].

Some further aspects of the above-stated management rules are not covered at all or covered in a rudimentary way by current impact assessment methods, *e.g.*:

a) Demand-related issues, *e.g.,* the aspect of sufficiency[173];
b) Resource consumption;
c) Mechanical impacts such as soil compression[174], risk-related considerations[175] or effects on genetic biodiversity[176];
d) Management rules for material and energy flows within the technosphere, *e.g.,* the entropic aspects of material use.

Ad a) None of the impact assessment methods stated in the introduction to this sub-chapter addresses demand-related issues such as the aspect of sufficiency.

Ad b) For all of the impact assessment methods discussed on an international level, the valuation of 'resource consumption' has to be considered unsatisfying[177]. Several attempts have been made to integrate the valuation of 'resource consumption' into LCA-methodology, *e.g.,* by:

- Heijungs *et al.* (1992; background, p.69), building up on the ratio of actually consumed to recoverable ('mineable') reserves of resource stocks for

[171] after Brezet *et al.* 1999:36

[172] see, *e.g.,* Schaltegger 1997a; affecting MR 1, points 1-4; MR 2; MR 3, points 2, 4, 7, 8; MR 4; MR 5, points 1 and 5; MR 7

[173] affecting MR 5, point 4; partially MR 6

[174] MR 3; MR 7
Land use change is considered in the Eco-indicator 99-method operationalised by the dependency of species encountered on a certain land use type (Goedkoop and Spriensma 2000:67ff). However, land use change does not only mean mechanical interventions but physical modifications of different scale and extension and also includes different immissions of (toxic) substances caused by different uses. Species encountered are thus not a precise indicator for mechanical or physical impacts.

[175] MR 4

[176] MR 1, point 1

[177] see, *e.g.*, Brezet *et al.* 1999:36; Goedkoop and Spriensma 2000:79ff

a-biotic resources and on the ratio of actually consumed to annual production for renewable resources;

- Brand *et al.* (1998:87ff), giving the consumption of non-renewable energy carriers (including hydropower) resource-scarcity points based on Swiss political reduction targets for CO_2 and electricity;
- Goedkoop and Spriensma (2000:79ff) build the valuation of resource consumption around the effort for extracting resources assuming that all resources are equally important to mankind;
- Steens and Ryding (1992) weight the consumption of a non-nrenewable resource according to the economic impacts on other safeguard subjects. This impact is determined with the willingness-to-pay approach[178], assuming the substitution of the resource on a sustainable path. The renewable resources – as assumed sustainable processes – are weighted by their market price. Weighting factors are available for some metals, some fossil fuels, fresh water in areas of deficiency and some renewable resources[179].

The main problem for all the methods quoted (except Steens and Ryding 1992; see below) is that LCA is primarily based on an a-temporal inventory of flows ('in-out') and not on a balance sheet principle considering sources and sinks. As a consequence, 'resource scarcity', which as a valuation parameter implicitly or explicitly underlies all resource consumption assessment methods, cannot be properly addressed. This is a severe limitation, not only in the context of the assessment of resource consumption[180].

Apart from this structural deficit, the definition of 'scarcity' is ambiguous, as various possible definitions of 'resource scarcity' are available[181]. They all face the following problems:

- Defining 'recoverable' (available) stocks;
- Price dependency of the 'mineable' stocks;
- Ignorance on future substitutability and thus demand of a resource;
- Ignorance on future technological improvement of extraction and processing;

[178] see, *e.g.*, Keeney and Raiffa 1976:125ff

[179] after Lindfors *et al.* 1995a:90/167

[180] In analogy, the same argument holds for the assessment of emissions and their 'environmental implications'. The inventoried emission flows lack specific information on their sinks and sources. Valuation of their relevance can thus only be made on (abstract) potential impacts.

[181] Cleveland and Stern 1993; Fritsch 1980; Hertwich *et al.* 1996; Lee 1998; Siebert 1983; Smith 1978; Stern 1999; Streissler 1980; Guinée and Heijungs 1995; see also Wellmer 1997; Wellmer 1998

- Ignorance on type and time of the introduction of a backstop technology[182], etc.

The EPS-method after Steens and Ryding (1992) provokes a particular problem, not only in the context of resource consumption. The method is based on the willingness-to-pay for the restoration of the normal status of pre-defined safeguard objects. The validity of willingness-to-pay surveys and the anthropocentric approach underlying the EPS-method have often been questioned by (environmental) economists (see also Chap. 3.3.1). Furthermore, the determination of the substituting sustainable processes is highly speculative.

It must be concluded that none of the above methodological propositions for the inclusion of 'resource consumption' into current impact assessment methods is fully satisfying.

Ad c) None of the above-mentioned impact assessment methods considers mechanical impacts, risk-related considerations, or effects on genetic biodiversity.

Ad d) Several management rules address the management of material and energy flows within the technosphere[183]. These rules are not *a priori* covered by an LCA-model as LCA relies on simple input-output accounting, *e.g.*:

- The thermodynamic efficiency of circular flows for non-renewable resources (MR 3, points 5 and 6);
- Greatest possible limitation of entropic losses related to material use throughout the life cycle of a product (MR 3, point 6);
- Current and future substitution rates of materials, especially the rate of (the intended) substitution of non-renewable materials with renewable ones (MR 5, point 1).

Such aspects must be respected (if possible) during the implementation of the 'causality' principle while attributing material and energy flows to the product under study during the life cycle inventory analysis.

3.3.5 Consequences for the attribution of material and energy flows to a product

In view of the management rules by Messner (1999) covering the environmental dimension of sustainable development, the 'descriptive' power of LCA has to be considered rather weak, independently of the impact assessment method chosen. Current impact assessment methods address the

[182] Backstop technologies substitute an existing technology, cutting the demand for a resource in a specific area of application (almost) completely (see, *e.g.*, Solow 1974:4).

[183] MR 3, points 5 and 6; MR 5

environmental dimension of sustainable development in a very selective way:

- LCA accounts as flow inventories. Therefore, issues related to effective and efficient use of materials and energy are addressed suitably (under the precondition that attributed material and energy flows represent 'causal relationships');
- Time or space dependent aspects such as toxicity or risk issues use patterns of resources, entropic aspects of material use, etc. are not *a priori* properly addressed by LCA models;
- Demand-side related issues (*e.g.*, sufficiency of consumption or growth of human population), genetic biodiversity or rates of technological change etc. are not covered by impact assessment methods at all.

In addition to the management of environmental interventions from the technosphere on the ecosphere, the management rules quoted also cover the management of the material and energy flows *within* the technosphere. From the viewpoint of resource management within the technosphere it has to be stated that:

- Management rules for a sustainable use of renewable and non-renewable resources differ, though they are interconnected as it is assumed that renewable resources will increasingly substitute non-renewables[184];
- Management rules for sustainable material flows have to be specified for each resource and the materials gained from it in its specific context of current (and future) use, as the context of extraction, production, use/consumption, waste treatment and recycling can vary for each resource.

Management rules concerning material and energy flows within the technosphere, *e.g.*, the entropic aspect of material use, are not *a priori* reflected in LCA models. These aspects must be integrated into LCA models with case-specific (material-specific) setting up of the product system if inconsistencies between the LCA model and the decision-maker's value system – and thus sub-optimal recommendations – shall be avoided. Only in this way, the 'causality' underlying the first attribution is properly met (see Figure 2-5).

A further requirement on the product system within LCA can be derived from the above:

A product system that supports the process of efficient and effective decision-making best provides improvement options that are in line with the sustainable management rules referring to the:

a) Material and energy flows within the technosphere;

[184] based on MR 1; MR 3; MR 5

b) Consequences of the environmental interventions in the ecosphere.

The following chapter summarises and complements the requirements of a product system established in the life cycle inventory analysis outlined in Chapter 3.2.4.

3.4 Summary: Requirements of product systems and their LCIs

The requirements of a product systems and of its respective life cycle inventory (LCI) can be summarised as follows, based on Chapters 3.1 to 3.3, taking into account the modelling characteristics of LCA outlined in Chapter 2:

A product system with its respective LCI that supports the process of efficient and effective decision-making best – and thus the most 'descriptive' and most adequate model of the life cycle of a product in LCA is:

- Complete, operational, decomposable, non-redundant, minimal and comparable;
- As simple, manageable, transparent, cheap and quick but still as 'adequate' as possible under the given economic constraints depending on the goal and scope of the study and allows instant re-calculation (for direct applications);
- Actor-based, *i.e.* reflects the action space and the decision-maker's mental model of his/her range of responsibility in view of the management rules of sustainable development;
- As site- and case-specific as possible, *i.e.* uses as much site-specific information as possible.

It respects:

- Material and market characteristics of the materials involved in the life cycle of a product;
- The decision-maker's mental model of the organisational principle of the socio-economic system;
- The decision-maker's attitude towards risk.

It provides:

- Improvement options that are in line with the sustainable management rules referring to:
  - The material and energy flows within the technosphere;
  - The consequences of the environmental interventions in the ecosphere.

Decisions in the life cycle inventory analysis related to the attribution of environmental interventions to the product under study must be made in accordance to the above-stated characteristics.

In addition, the marginal costs of collecting and computing more and better information about environmental impacts must not exceed the marginal benefits of information for the natural environment; the ratio of environmental benefits to the economic costs of the tool must stand in an efficient relationship compared to other investment options.

In the next part of this book, a closer look is given to the modelling procedure for the setting up of the product system. The modelling process is analysed for decision situations using the series of standards ISO 14040 as reference. Furthermore, the role of the decision-maker's mental models and values during the setting of system boundaries and allocation is investigated.

III.

MENTAL MODELS AND VALUE CHOICES IN THE LIFE CYCLE INVENTORY ANALYSIS

Chapter 4

INTRODUCTION

"Causality is not the question but the answer" (Heijungs 1994:8)

The setting up of the product system as life cycle of a product is looked at more closely in this part of the book. This modelling is commonly thought to be part of the life cycle inventory step according to ISO/EN 14041.

The setting up of the product system has been called the first part of the attribution problem of LCA (Chap. 2.3.1): the causal attribution of material and energy flows to the product under study. The resulting model represents the environmental interventions from production, consumption and end-of-life for which a product is held responsible.

In Chapter 5, the influence of the decision-maker's mental models and value choices on decisions related to the first part of the attribution problem is investigated. The process of establishing a product system with its respective LCI is analysed, following the guidelines of the series of standards ISO 14040. Inconsistencies of the modelling procedure in the standards are worked out, considering the different lines of reasoning found in literature.

In Chapter 6, literature about current allocation procedures for open-loop recycling is reviewed. The focus lies on different approaches to handling reuse & recycling in LCA as a consequence of different mental models and values. Recommendations are given for the case-specific choice of an allocation procedure for reuse & recycling.

Finally, the role of the decision-maker's mental models and values for the attribution in the life cycle inventory analysis are summarised based on the previous chapters of this section (Chap. 7).

The guidelines of ISO 14040 for the definition of a product system and its LCI are analysed. Different decision situations are distinguished and the necessity of applying mental models and values during the life cycle inventory analysis is worked out.

Chapter 5

ANALYSIS OF ISO 14041 FOR MENTAL MODELS AND VALUES

5.1 Reference ISO 14040: current status

Various publications[185] have been dealing with different aspects of defining the product system, including the setting of system boundaries and allocation in multi-input/-output processes or recycling. In these publications, procedures for the conduct of LCA are proposed.

LCA has been subject to standardisation by the International Standard Organisation (ISO), leading to the series of standards ISO 14040. These International Standards describe principles of conduction and reporting of LCA studies with certain minimal requirements. Here, the series of standards ISO 14040 is taken as consensus of the international discussion and will therefore be used as a reference.

Table 5-1. The series of standards ISO 14040

Number	Title
ISO/EN 14040:1997	Life Cycle Assessment – Principles and framework
ISO/EN 14041:1998	*Life Cycle Assessment – Goal and scope definition and inventory analysis*
ISO/EN 14042:1999	Life Cycle Assessment – Life cycle impact assessment
ISO/EN 14043:1999	Life Cycle Assessment – Life cycle interpretation
ISO/TR 14047	Life Cycle Assessment – Examples of application of ISO 14042
ISO/TR 14048	Life Cycle Assessment – Data documentation format
ISO/TR 14049:2000 (E)	*Life Cycle Assessment – Examples of application of ISO 14041 to goal and scope definition and inventory analysis*

[185] such as Fava and Denison 1991; Fava *et al.* 1991; Lindfors *et al.* 1995a

The series of standards ISO 14040 treating life cycle assessment currently consists of four standards and three technical reports or its drafts. Table 5-1 shows the current status of standardisation for the series of standards ISO 14040.

Modelling the life cycle of a product is dealt with in the standard ISO/EN 14041:1998 'Life Cycle Assessment – Goal and scope definition and inventory analysis', as well as in the Technical Report ISO/TR 14049:2000 'Life Cycle Assessment – Examples of application of ISO 14041 to goal an scope definition and inventory analysis'. The Technical Report provides supplemental information to the International Standard, ISO/EN 14041.

5.2 Life cycle inventory analysis

The definition of the product system as model of the life cycle of a product and of its summary, the life cycle inventory (LCI), is the second step within LCA-methodology. It follows the 'goal and scope definition', where the functional unit, the system boundaries, allocation procedures and data requirements, etc. are provisionally fixed.

According to ISO/EN 14040, the life cycle inventory analysis

> "...involves data collection and calculation procedures to quantify relevant inputs and outputs of a product system. These inputs and outputs may include the use of resources and releases to air, water and land associated with the system" (ISO/EN 14040, chap. 5.2.1).

The process of setting up the product system is iterative[186].

> "As data are collected and more is learned about the system, new data requirements or limitations may be identified that require a change in the data collection procedures so that the goals of the study will still be met. Sometimes, issues may be identified that require revisions to the goal or scope of the study" (ISO/EN 14040, chap. 5.2.1).

From a theoretical standpoint, the setting up of the product system requires various attributive steps:

[186] The positioning of the modelling of the life cycle of a product as product system is not very clear within ISO 14040. Although the description of the products and its corresponding product systems as well as the first system boundary setting are situated in the first step 'scope definition' (ISO/EN 14040, chap. 5.1.2), the final system definition and system boundary setting are considered being part of the second step 'life cycle inventory', similar to allocation. This is the consequence of the iterative definition of a finally consistent LCA-model (see also Guinée *et al.* 2001, part 3, pp. 49ff).

1. Differentiation of the technosphere (the socio-economic system) and the ecosphere (nature);
2. The modelling step per se, the enclosure of the system attributed to the functional unit from the technosphere;
3. Allocation of unit processes to be attributed to more than one product, which can be considered the third level of attribution[187].

Ad 1) The differentiation of the technosphere and the ecosphere is a subjective distinction.

This point is not further explored in this book although this issue is highly depending on mental models[188]. This book focuses on the question of how a product system and its LCI should be best set up with the impact assessment methods currently available. As partial models of the ecosphere, these impact assessment methods predetermine the boundary between the life cycle of the product as part of the technosphere and the ecosphere.

This line of reasoning is further supported by the assumption that modelling characteristics of LCA (see Chap. 2.2) will more or less remain the same over the next years[189]. Fundamental modifications of impact assessment methodology, *e.g.*, the dynamisation and site-specific consideration and valuation of material and energy flows require huge amounts of data. On top of gathering it such data would also have to be kept up-to-date over the years. The costs would not stand in an economically and possibly environmentally justifiable relationship to the results deducible[190].

Ad 2) et 3) The modelling step per se, including the setting of system boundaries and allocation of multifunctional processes is the heart of the life cycle inventory analysis and is further analysed in the following chapters. The modelling – the enclosure of the material and energy flows attributed to the functional unit – is the first part of the attribution problem described in Chapter 2.3.1.

Two attribution principles can be distinguished, namely descriptive and marginal LCA.

In descriptive LCA, it is asked: *Which environmental interventions are 'caused' on average by the product under study?*

In marginal LCA, it is asked: *Which environmental interventions are 'caused' by an additional (or avoided) product under study?*

In the next chapter, different methodological decision situations within the life cycle inventory analysis according to ISO/EN 14041 are distin-

[187] after Huppes 1994; Tillman *et al.* 1994:21ff.
[188] for literature, see Guinée *et al.* 2001, part 3, pp. 87ff
[189] see also Chap. 2.3.3
[190] see also Schaltegger 1996

guished and analysed for subjective elements such as the use of mental models or value choices. Emphasis is put on the allocation of multifunctional processes, especially related to reuse & recycling.

The decision situations in the life cycle inventory analysis which are looked at in more detail are:

- The definition of modules and unit processes while setting up the product system as a process tree (Chap. 5.3);
- The primary attribution principles for material and energy flows to the product under study, including the setting of system boundaries (Chap. 5.5);
- The allocation principles for multifunctional processes (Chapters 5.6 and 5.7);
- The choice of the functions (or products) that get allocated material and energy flows (Chap. 5.6.3);
- The setting of a system boundary related to reuse & recycling (Chap. 5.8.2);
- The identification of processes related to reuse & recycling that are shared by more than one product (Chap. 5.8.3);
- The determination of 'equivalency' of in- and output material when modelling 'closed loop' material flows (Chap. 5.8.4);
- The handling of time-dependent aspects of reuse & recycling (Chap. 5.8.5):
    - Recycling processes that will be used for the treatment of the post-consumer material;
    - Technology or process that will be substituted in the future by the material to be recycled;
    - Percentage of the material collected for recycling and the yield of the recycling processes;
    - Service life of the product assumed;
    - Stability of the chosen allocation factor over time.

Several decisions related to the life cycle inventory analysis are not looked at any further, such as:

- Selection of environmental interventions to be reported (ISO/EN 14041, chap. 5.3.5) and their structuring in categories (ISO/EN 14041, chap. 4.4 and chap. 5.3.4);
- Data quality issues (coverage, precision, completeness, uncertainty and representativeness) (ISO/EN 14041, chap. 5.3.6);
- Dealing with data shortfalls and differing data quality (ISO/EN 14041, chap. 6.4.2, chap. 7);
- Selection of data collection procedures (ISO/EN 14041, chap. 6.2);
- Selection of data sources (ISO/EN 14041, chap. 6.3).

All of these five points are not directly related to the attribution problem in the life cycle inventory analysis.

5.3 Aims of the life cycle inventory analysis: model representing the life cycle of a product

The aim of the life cycle inventory analysis is the definition of a product system as a model of the life cycle of a product (or other 'functional unit'). This product system is the result of the first part of the attribution problem: the attribution of material and energy flows to a product under study.

> "A product system is a collection of unit processes connected by flows of intermediate products which perform one or more defined functions (the service(s) of the functional unit under study; WF). A product system description includes unit processes, elementary flows and product flows across the system boundaries ...and intermediate product flows within the system" (ISO/EN 14041, chap. 4.2).
>
> "The system should be modelled in such a manner that inputs and outputs at its boundaries are elementary flows" (ISO/EN 14040, chap. 5.1.2.2).

Product systems are modelled in a modular way[191]. They commonly consist of modules representing different life cycle stages of the product. These modules are subdivided into sub-modules and finally 'unit processes'. The boundary to the ecosphere is crossed by 'elementary flows', also called 'environmental interventions'.

> "Product systems are subdivided into a set of unit processes... Unit processes are linked to one another by flows of intermediate products and/or waste for treatment, to other product systems by product flows, and to the environment by elementary flows... The boundary of a unit process is determined by the level of modelling detail that is required to satisfy the goal of the study" (ISO/EN 14041, chap. 4.3).

Bearing in mind the modelling characteristics of LCI and the resulting model, a first – almost trivial – preference indicating a corresponding value can be stated: less environmental impact is better.

It should be noted that the life cycle of a product is not tangible but a theoretical construct. Therefore, it has to be looked at primarily before further attribution principles can be developed and applied.

[191] see, *e.g.*, Frühwald and Wegener 1996:91

5.4 Theoretical construct 'life cycle of a product'

The 'life cycle of a product' is a theoretical construct. As a first approach to this construct, ISO/EN 14041 states the goal and scope dependency of what is to be considered the life cycle of a product (see Chap. 5.4.1). Furthermore, it proposes several processes that can be part of such a product life cycle (see Chap. 5.4.2).

5.4.1 Goal and scope dependency

According to ISO 14040, the attribution of material and energy flows to a functional unit is goal and scope dependent:

> "LCA studies are conducted by developing models that describe the key elements of physical systems. It is often not practical to study all the relationships between all of the unit processes in a product system, or all of the relationships between a product system and the system environment. The choice of the elements of the physical system to be modelled is dependent on the definition of the goal and scope of the study" (ISO/EN 14041, chap. 4.5).

This can be interpreted that the shape of a product system and the attribution rules to be applied can vary in dependency of the goal and scope of the study.

5.4.2 The product system as a list of possible processes to select

Apart from goal and scope dependency, ISO/EN 14041 is rather unspecific about the shape of a product system. Processes and life cycle stages to be possibly included into the definition of a product system are listed:

> "Several life stages, unit processes and flows should be taken into consideration, *e.g.*:
> - Inputs and outputs in the main manufacturing/processing sequence
> - Distribution/transportation
> - Production and use of fuels, electricity and heat
> - Use and maintenance of products
> - Disposal of process water and products
> - Recovery of used products (including reuse, recycling and energy recovery)
> - Manufacture of ancillary materials
> - Manufacture, maintenance and decommissioning of capital equipment
> - Additional operations, such as lighting and heating

- Other considerations related to impact assessment (if any)" (ISO/EN 14041, chap. 5.3.3)

Plus:

"Any decisions to omit life cycle stages, processes or inputs/outputs shall be clearly stated and justified. The criteria used in setting the system boundaries dictate the degree of confidence in ensuring that the results of the study have not been compromised and that the goal of a given study will be met..." (*ibid.*).

By defining and further refining the product system out of the above-mentioned processes, the life cycle of a product is constructed as a mental model itself.

5.5 Setting of system boundaries based on relative effects

The system boundaries determine which unit processes are included in an LCA. In practice, there are life cycle stages and processes that can a priori be considered part of the product system to be defined for the goal and scope of the study. Based on this knowledge, a first attribution principle can be applied that is related to the setting of system boundaries: the relative effects of processes to the total effects.

ISO/EN 14041 proposes three criteria for the initial inclusion of in- and outputs and for its revision during the conduct of the inventory analysis. These principles are based on relative contribution:

"a) *mass*: an appropriate decision, when using mass as a criterion, would require the inclusion into the study of all input that cumulatively contribute more than a defined percentage to the mass input of the product system being modelled;
b) *energy*: similarly, an appropriate decision, when using energy as a criterion, would require the inclusion in the study those inputs that cumulatively contribute more than a defined percentage of the product system's energy inputs;
c) *environmental relevance*: decisions on environmental implications criteria should be made to include inputs that contribute more than an additional defined percentage to the estimated quantity of each individual data category of the product system. For example if sulphur oxides were selected as data category, a criterion could be established to include any inputs that contribute more than a predefined percentage to the total sulphur oxide emissions for the product system.

> These criteria can also be used to identify which outputs should be traced to the environment, *i.e.* by including final waste treatment processes" (ISO/EN 14041, chap. 5.3.5).

These criteria, also called 'cut-off criteria', serve as attribution principles based on *relative effects*. With the choice of a level of relevance for cutting off 'irrelevant' material and energy flows, a decision-maker's preference and corresponding value is manifested.

5.6 Multifunctional processes and allocation based on 'causal relationship'

So far the guiding principles for attribution in ISO/EN 14041 are:

- The scope and goal dependent choice of life cycle stages and processes;
- The relevance of a particular flow in relation to the total flows or to total environmental impact.

They do not go far enough when dealing with multifunctional processes. Processes can be relevant according to the above-mentioned criteria but provide several functions. An additional attribution principle – in a multifunctional context called allocation procedure – has to be applied if multifunctional processes provide functions that are not covered by the scope of the study.

The question to be answered for the attribution of multifunctional processes might be: What determines the share of the products or functions of a shared process? Or in even more general terms: What determines the material flows within the socio-economic system?

Multifunctional processes require the application of allocation procedures based on a *causal relationship*.

The question of 'causal relationship' related to the allocation of multifunctional processes brings us back to the first part of the attribution problem of LCA described in Chapter 2.3.1 as well as to the question of the organisational principle of the socio-economic system. Different mental models of causal relationships of material and energy flows can be distinguished in the discussion on LCA-methodology (Chap. 5.7).

5.6.1 Definition of the allocation problem

In the standard ISO/EN 14041, the allocation problem in LCA related to the partial attribution of environmental interventions from a multifunctional process to the product under study is defined as follows:

> "Life cycle inventory analysis relies on being able to link unit processes within a product system by simple material or energy flows. In practice,

few industrial processes yield a single output or are based on a linearity of raw materials input and outputs. In fact, most industrial processes yield more than one product, and they recycle intermediate or discarded products as raw materials. Therefore, the materials and energy flows as well as associated environmental releases shall be allocated to the different products according to clearly stated procedures" (ISO/EN 14041, chap. 6.5.1).

Multifunctional processes can be found in, *e.g.*:

- Shared coproduction: processes simultaneously providing more than one product in an unfixed ratio;
- Joint coproduction: processes necessarily and simultaneously providing more than one product in a fixed ratio;
- Combined waste treatment such as, *e.g.*, waste incinerators;
- Processes related to reuse & recycling.

Coproduction has also been called a multi-output process; combined waste treatment processes have also been addressed as multi-input processes[192]. Recycling processes have been addressed as single input-single output processes that are shared by two product systems. However, recycling requires additional considerations, which will be discussed in Chapter 5.8.3.

5.6.2 Allocation principles

The standard ISO/EN 14041 specifies several principles of how, as previously mentioned, situations requiring allocation should be approached:

- *Causal relationship*: the consideration of the attribution of input and output data have to reflect fundamental input-output characteristics or approximate them as much as possible (ISO/EN 14041, chap. 6.5.2, para. 1, and chap. 6.5.3, point b);
- *100%-rule*: all input- and output data have to be completely considered, double-counting has to be avoided (ISO/EN 14041, chap. 6.5.2, point 2);
- *Consistency of the rules applied*: the same allocation rules have to be applied for similar input and output data (ISO/EN 14041, 6.5.3, para. 3);
- *Sensitivity analysis*: the conduct of a sensitivity analysis if several allocation procedures seem applicable (ISO/EN 14041, chap. 6.5.2, point 3);
- *Documentation*: the requirement for the documentation of the unit processes affected and of the allocation rules applied (ISO/EN 14041, chap. 6.5.3, para. 2).

[192] see, *e.g.*, Tillman *et al.* 1994

A further set of requirements for the resulting LCI was established in Chapter 3.4, which should guide the definition of the product system and as such also the allocation of multifunctional processes.

5.6.3 Identification of multifunctional processes

As a basic decision situation during the modelling of the product system, multifunctional processes have to be identified. The following quotation refers to the identification of products or functions in coproduction processes, whereby environmental interventions should be allocated:

> "Some outputs may be partly coproducts and partly waste. In such cases, it is necessary to identify the ratio between coproducts and waste since the inputs and outputs shall be allocated to the coproducts part only" (ISO/EN 14041, chap. 6.5.3).

Coproducts are defined as:

> "any of two or more products from the same unit process" (ISO/EN 14041, chap. 3.2).

This means that coproducts are the intended outcome of a process, whereas waste is "any output from the product system which is disposed of" (ISO/EN 14040, chap. 3.20).

Other authors distinguish coproducts, by-products and waste by positive, slightly positive/insignificant and negative market prices[193]. Nonetheless, all authors propose a finally market-priced based classification of the outputs of a process for the identification of outputs to carry environmental interventions. The preferences on which the above classification is made have to be considered part of the decision-maker's mental model of the organisational principle of the socio-economic system.

This point seems noteworthy, as various exponents of the LCA community strictly oppose the use of monetary data for the setting up of the product system, *e.g.*, for the allocation of joint coproduction processes in descriptive LCA. Their opposition is curious: the choice of products or functions that get environmental interventions allocated and the question of how much they should get allocated are obviously closely linked.

The definition of waste and coproducts is time-dependent and thus has to be made depending on the scope of the study:

[193] see, *e.g.*, Lindfors *et al.* 1995a:59; Heintz and Baisnée 1991:38ff; Tillman *et al.* 1994:24; Frischknecht 1994; Förster and Gilgen 1997; Heijungs *et al.* 1992, background, pp. 23f

> "When one scope of an LCA is a (non-incremental and therefore nonlinear) change towards sustainability, monetarized values of outputs both of processes and life-cycles may lead to wrong conclusions concerning the classification of outputs (*e.g.*, co- or by-product, waste) and to wrong allocation of environmental burdens because the allocation would then be based on today's economy and their implicit valuation criteria. As in today's economy natural capital (still) is a free good, another valuation scheme should be chosen. Otherwise LCA might lose its purpose of showing a way towards sustainability" (Frischknecht 1994:128).

The consideration of time-dependent changes of the socio-economic system and its dependency on the decision-maker's attitude towards risk is further discussed in Chapter 5.8.5.

Concerning recycling, the identification of functions or products to be allocated is closely linked to the identification of multifunctional processes itself. It depends strongly on the decision-maker's mental model of his/her range of responsibility in view of the management rules for sustainable management as well as his/her mental model of the organisational principle of the socio-economic system. This issue is discussed in detail in Chapter 5.8.3.

In the case of waste treatment as multi-input process, an allocation problem in analogy to multi-output processes occurs if one output has to be allocated to various inputs, *e.g.*, dioxin emissions to different waste fractions burnt in a municipal waste incinerator.

5.6.4 Step 1: Can allocation be avoided?

For the allocation of multifunctional processes ISO/EN 14041 suggests a three-step procedure. The avoidance of allocation is suggested as a first step to define a product system as objectively as possible:

> "Wherever possible, allocation should be avoided by:
> - Dividing the unit process to be allocated into two or more sub-processes and collecting the input and output data related to these sub-processes
> - Expanding the product system to include the additional functions related to the coproducts taking into account the requirements of chap. 5.3.2" ("Function, functional unit and reference flow"; WF) (ISO/EN 14041, chap. 6.5.3).

Dividing unit processes is only possible in situations when the same output is used in two separate following processes – in short, in non-multifunctional processes. Subdividing unit processes can help reduce the allocation problem by further refining the real multifunctional process. By this, the relevance of the allocation problem can be reduced and the choice of an allocation procedure loses in importance. Subdividing unit processes avoids allocation by setting of system boundaries.

System expansion leads to two different models, each for descriptive and marginal LCA, although not explicitly mentioned in ISO/EN 14041[194].

For descriptive LCA, expanding the product system avoids allocation by fully accounting the (multifunctional) process and redefining the functional unit while adding additional functions.

Another procedure is also subsumed under system expansion in ISO/EN 14041, the 'avoided-burden' procedure: the (multifunctional) process is also fully accounted; however surplus-functions are eliminated by subtracting equivalent (average) mono-functional processes to obtain a mono-functional functional unit[195].

System expansion and the avoided-burden procedure are theoretically feasible in marginal LCA. In the case of system expansion, multifunctional processes lead to the inclusion of further products into the functional unit. The provisionally defined product system during scope and goal definition is gradually expanded to a whole-system model finally covering various functions. Such whole-system models allow one to determine effects of marginal changes in one product system throughout the whole-system (*e.g.*, based on linear programming). Processes could partially be attributed to the additional (or avoided) product assessed. In view of the complexity of the material and energy flows within the socio-economic system, marginal LCA-models built on system expansion are hardly operational (for further details on preconditions and requirements for using LP, see Chap. 5.6.5).

The 'avoided-burden' procedure also fully accounts for the multifunctional process, but subtracts surplus functions, assuming their substitution by marginal equi-functional mono-functional processes[196]. The resulting product system becomes a mono-functional system.

For marginal LCA, the avoided-burden procedure is the more realistic procedure. No additional calculations such as linear programming are needed and substitutions are handled on a process-for-process basis. The resulting model remains reasonably small in most cases[197].

The avoided-burden procedure for marginal LCA requires:

[194] Note that ISO/EN 14041 does generally not distinguish between descriptive (average) and change-oriented (marginal) approaches to LCA. However, comments are made in this book on the consequences of the procedure for both approaches.

[195] ISO/EN 14041 subsumes the 'avoided-burden' procedure under system expansion; see ISO/EN 14041, annex B2

[196] see, *e.g.*, Weidema 1999; Weidema *et al.* 1999

[197] see footnote 196

- Explicit modelling of substitution effects and as such the modelling of the behaviour of the socio-economic system based on the decision-maker's mental model of the organisational principle of the socio-economic system (see Chap. 5.7);
- Reasonable determinability of substituting processes.

5.6.5 Step 2: Physical relationship for unavoidable allocation

There are cases where subdivisions of processes, system expansion or the avoided-burden procedure are not feasible for the scope and goal of the study. In such cases, allocation is unavoidable.

Unavoidable allocation of multifunctional processes has to be based on 'causal relationship'.

> "Allocation procedures should ...approximate as much as possible such fundamental input-output relationships and characteristics" (ISO/EN 14041, chap. 5.2.1).

Hence, ISO/EN 14041 is suggesting a second step:

> "Where allocation cannot be avoided, the system inputs and outputs should be partitioned between its different products or functions in a way that reflects the underlying physical relationships between them; *i.e.* they shall reflect the way in which the inputs and outputs are changed by quantitative changes in the products or functions delivered by the system. The resulting allocation will not necessarily be in proportion to any simple measurement such as the mass or molar flows of coproducts" (ISO/EN 14041, chap. 6.5.3).

This second step according to ISO/EN 14041 has also been called 'direct partition' to distinguish allocation based on physical (or chemical or biological) causal relationship from 'general partition' based on arbitrarily chosen parameters like mass, mole, volume, profits, revenues, etc. (see below)[198].

> "A partition is called direct when processes, energy- and material-flows are allocated by breaking up the system in subsystems which allows to consider functional, physical, chemical or biological causalities" (Frischknecht 1994:129).

The above-mentioned section of the standard brings modelling techniques such as linear programming (LP) to mind, which is commonly applied in engineering sciences. LP and the determination of effects induced

[198] see also Chap. 5.7.1

by marginal changes are, of course, a valid option for marginal LCA when determining causal relationship[199].

A crucial precondition for the use of LP is that a system (or a process) can be completely modelled. This means that all material flows have to be tracked back to elementary flows or to parameters not requiring any further allocation (such as quantities of waste if they are considered in an impact assessment method). In other words, the entire life of all materials involved in a product system has to be tracked back to their extraction and forward to their final disposal. Thus, the entire net of product systems, connected with material and energy flows to the product system under study, has to be completely defined when using LP. In such a case, allocation is made by expanding the system to a whole-system model and by determining the effects of marginal changes within the system (refer to the part on system expansion for marginal LCA in the previous Chap. 5.6.4).

Unfortunately, such whole-system models are hardly buildable in the case of recycling materials such as metals, where a multitude of different, often unknown product systems can be involved along the material cascade. Additionally, modelling a cascade of product systems can be against the goal and scope of LCA-studies, *e.g.*, in the case of environmental product design. In such cases, the 'avoided-burden' procedure discussed as a first step procedure in Chapter 5.6.4 is a possible way out.

For marginal LCA, the mentioned principle is applicable in two cases:

a) For whole-system models based on system expansion as outlined above if physical relationship or price elasticities are assumed to determine the changes within the whole-system; and
b) For shared multifunctional processes. In this case, environmental interventions to be allocated can be determined directly by changing the ratio of products in dependency of the functional unit (the change) to be assessed. Additional considerations are necessary to integrate the consequences of the changed additional outputs properly (!). This problem is illustrated in the following examples for marginal LCA, where 1) a shared bi-functional process on process-level with variable product ratio, and 2) a bi-functional joint coproduction process with fixed product ratio are allocated (Figure 5-1).

The marginal change of the environmental interventions (ΔI) of a multi-functional process caused by the additional production and consumption of a product x_1 can be determined. However, the difference in the production and consumption of product x_2 has to be dealt with consistently, possibly with a

199 for linear programming (LP) and LCA, see, *e.g.*, Azapagic and Clift 1994; Azapagic and Clift 1999b; Ekvall and Finnveden 1999:13f; see also and Chap. 6.4

substitution approach adding additional impacts from the production of the marginal substitute (for (1)), or subtracting avoided production processes from the marginal product substituted (for (2)).

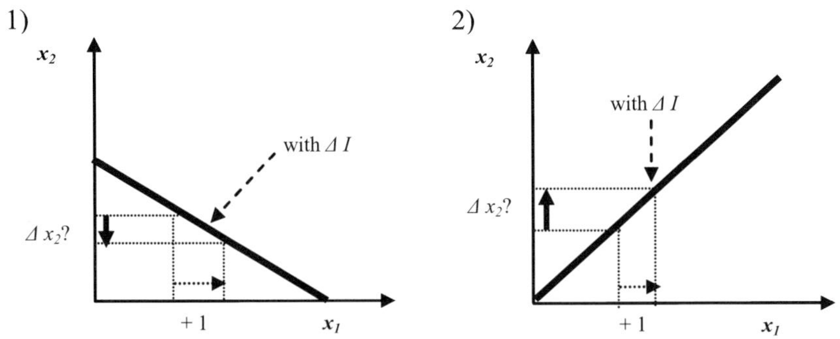

Figure 5-1. Examples of implementing the second step of the ISO allocation procedure for 1) a shared bi-functional process, and for 2) a bi-functional joint coproduction process for marginal LCA
x_1: quantity of product 1; x_2 quantity of product 2; ΔI: changed environmental interventions induced by the change in the product ratio

The applicability of the above-mentioned second step of the ISO-allocation procedure is highly limited in the case of descriptive LCA. This will be illustrated in the following example.

Figure 5-2 illustrates a possible procedure for the determination of the share of environmental interventions to be attributed to a product resulting from a bi-functional shared process for descriptive LCA[200].

The determination of the share of environmental interventions to be attributed to a product resulting from a bi-functional shared process for descriptive LCA is based on various assumptions:

- The linear relationship of the outputs of the products x_1 and x_2;
- The linear change of environmental interventions if the product ratio is changed;
- The determinability[201] of the environmental interventions if only one of the products is produced.

The mathematical determination of the share of environmental interventions to be attributed to one product resulting from a bi-functional shared

[200] In contrast to shared coproduction processes, the product ratio cannot be changed in joint coproduction processes.
[201] measured or interpolated

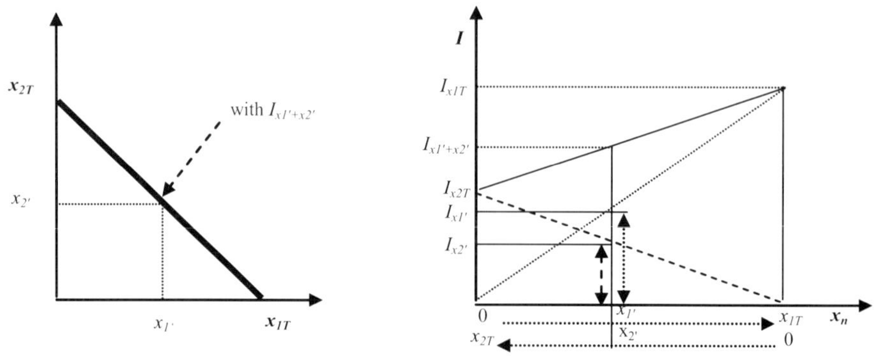

Figure 5-2. Determining the share of environmental interventions to be attributed to a product resulting from a bi-functional shared process for descriptive LCA considering quantitative changes according to ISO/EN 14041, chap. 6.5.3, point 2
I: environmental interventions related to the process, I_{x1T}: environmental interventions if only product x_1 is produced; I_{x2T}: environmental interventions if only product x_2 is produced; $I_{x1'}$: environmental interventions attributed to quantity of products $x_{1'}$; $I_{x2'}$: environmental interventions attributed to quantity of products $x_{2'}$; x_{1T}: quantity of products x_1 if no product x_2 is produced; x_{2T}: quantity of products x_2 if no product x_1 is produced; $x_{1'}$: quantity of products x_1 considered for attribution; $x_{2'}$: quantity of products x_2 considered for attribution

process based quantitative changes gets very complicated if any of the mentioned points is not given. In most real cases, the production functions will be more complex than the above example. Thus in most cases, the allocation procedures applicable to shared multifunctional processes in descriptive LCA are either third-step procedures (Chap. 5.6.6) or the avoided-burden procedure discussed previously (Chap. 5.6.4; see also Chap. 24).

5.6.6 Step 3: 'Causal relationship' in other cases

As the third and last step ISO/EN 14041 suggests:

"Where physical relationship alone cannot be established or used as the basis for allocation the inputs should be allocated between the products and functions in a way which reflects other relationships between them. For example, input and output data might be allocated between coproducts in proportion to the economic value of the products" (ISO/EN 14041, chap. 6.5.3).

This third step has also been called 'general partition'.

"A partition is called general when processes, energy- and material-flows are allocated by a common criteria to be determined (*e.g.*, mass, energy or exergy content, concentration, prices) partly without considering functional, physical, chemical or biological causalities (or realities) for differentiation" (Frischknecht 1994:129)[202].

This allocation principle remains as the last resort option if either of the other two previous steps are not possible.

In the case of descriptive LCA, several allocation factors are applicable for the allocation of coproduction processes, such as revenues, mass, volume, profits, etc. The choice of an allocation factor largely depends on the decision-maker's mental model of the organisational principle of the socio-economic system (see Chap. 5.7).

Figure 5-3 illustrates a simple example for the implementation of the third step according to the ISO-standard for: 1) a shared bi-functional process with a variable product ratio; 2) a joint coproduction process with a fixed product ratio for descriptive LCA:

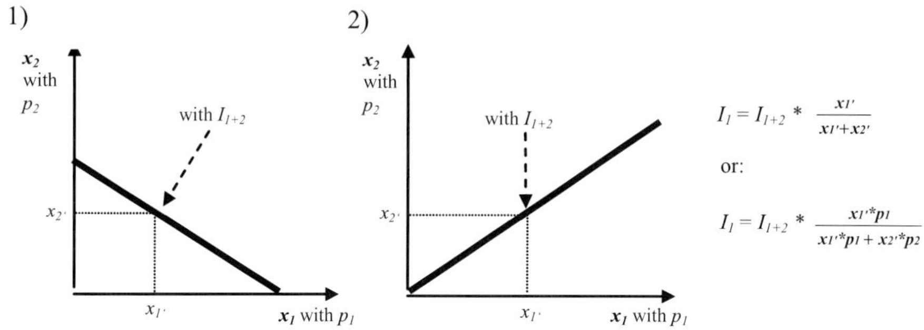

Figure 5-3. Example for in implementation of the third step for 1) a bi-functional shared process and 2) a bi-functional joint coproduction process for descriptive LCA; based on mass or based on revenues

x_1: quantity of product 1; x_2 quantity of product 2; p_1: price of product 1; p_2: price of prod. 2; I_{1+2}: total environmental interventions of the coproduction process; I_1: environmental interventions attributed to quantity x_1

202 The above interpretation of ISO/EN 14041 is also shared by Ekvall *et al.* (1994): "If the last-option recommendation in ISO 14041 is generously interpreted, it allows also for allocation based on a relationship which is not causal. This includes allocation in proportion to an arbitrary physical property of the products, such as mass, volume or energy content. Allocation based on physical properties of the products is the predominant allocation method in recent LCI practice because data on these properties are generally readily available and easily interpreted. In some cases, this allocation may coincide with allocation based on physical, causal relationship. However, when the allocation is not based on an accurate model of causal relationships, it will not provide reliable information on the consequences of our actions" (Ekvall and Finnveden 1999:14f).

For marginal LCA, the third step is NOT applicable in a strict sense as it contradicts the basic accounting principle of marginal LCA. The allocation of environmental interventions based on, *e.g.*, the mass ratio of several products or on the relative share of revenues corresponds to the accounting principle of descriptive LCA.

5.6.7 Remarks on the three-step procedure of ISO/EN 14041

Apart from the three-step procedure outlined above, Lindfors *et al.* (1995b:60ff) mention the option of choosing an arbitrary ratio as an allocation factor. For example, the attribution of the material and energy flows can be made half and half to both of the coproducts (50/50 allocation) or 100% to one product and 0% to the other one (100/0 allocation).

Note that the allocation procedure according to ISO/EN 14041 does not distinguish between descriptive (average) and change-oriented (marginal) LCA[203]. The decision rules of ISO/EN 14041 thus do not properly address decision situations encountered during the conduct of a marginal or descriptive LCA. As outlined above, the applicability of the allocation procedures stated in ISO/EN 14041, chap. 6.5.3 differs considerably for marginal and descriptive LCAs. They are not suited to any of the two approaches. This is also true for allocation related to recycling (see Chap. 5.8.1).

A proposal for attribution principles differing for descriptive and marginal LCA is made in Chapter 22, referring as much as possible to above-mentioned allocation principles. It is highly recommended to integrate the distinction of marginal and descriptive LCA in a future revision of the series of standards ISO 14040.

Chapter summary

- According to ISO/EN 14041 the attribution of material and energy flows to the product under study has to be made according to both of the following criteria:
  - The scope and goal dependent choice of life cycle stages and processes;
  - The significance of a particular flow to the total flows or to total environmental implications (allocation based on effects);
- Having in mind the modelling characteristics of LCI and the resulting model, a first and almost trivial preference indicating a corresponding value can be stated: less environmental impact is better;

[203] nor different applications of LCA, *e.g.*, the assessment of products or the evaluation of waste management systems!

- With the choice of a level of relevance for cutting off 'irrelevant' material and energy flows, a decision-maker's preference and corresponding value is manifested;
- In the case of multifunctional processes an allocation of the related material and energy flows needs to be cause-related. Allocation based on effects is not possible, as not the inclusion or exclusion of a process but its subdivision and partial attribution of its environmental interventions is needed. Hence, the decision-maker's mental model of the organisational principle of the socio-economic system has to be considered;
- The determinations of coproducts that get environmental interventions attributed as well as the choice of the allocation factor are closely linked. In this perspective, the choice of economic allocation factors is a valid option if coproducts are determined out of the outputs of a process based on their market prices (for descriptive LCA);
- The allocation procedures according to ISO/EN 14041 do not distinguish between the descriptive (average) and change-oriented (marginal) LCA. The applicability of the allocation procedures in ISO/EN 14041 differs considerably for both descriptive and marginal LCAs. The ISO-guidelines cannot be considered suitable to any of the two approaches.

5.7 Mental models of the causal relationship of material and energy flows

The causal relationship of material and energy flows is a key element of the decision-maker's mental model of the socio-economic system as it refers to its organisational principle.

The decision-maker's mental model of the organisational principle of the socio-economic system is decisive if changes of the socio-economic system caused by a decision are to be modelled. The modelling of changes is followed in marginal LCA.

But also in average (descriptive) LCA, the mental model of the organisational principle of the socio-economic system influences the attribution. This becomes obvious if joint coproduction processes have to be allocated. In joint coproduction processes, the share of outputs is fixed. Hence, no physical, chemical, biological or technical causal relationship can be established for the allocation of such processes. Only the 'arbitrary' choice of an allocation factor can solve this allocation problem, finally reflecting the decision-maker's mental model of the organisation principle of the socio-economic system[204].

Summarising, the question of the 'causal relationship' of material and energy flows is raised in LCA:

- During the identification of products of multifunctional processes that will have environmental interventions attributed;
- During coproduct allocation in descriptive LCA; or
- In marginal LCA as a basic attribution principle if marginal effects are determined throughout the socio-economic system.

Within LCA-literature and literature related to environmental management, several opinions have been exposed or aspects stressed on the organisational principle of the socio-economic system. Some literature emphasises the physical/chemical or technical determination of material and energy flows; amongst them are the series of standards ISO 14040 (see Chap. 5.7.1).

Other mental models of causal relationship start from the reasoning that ultimately human decisions determine the material and energy flows within the socio-economic system. These mental models of causal relationship differ in the assumptions on the preferences and values that guide human decision-making (see Chapters 5.7.2 to 5.7.5).

Depending on the decision-maker's mental model of the organisational principle of the socio-economic system, differing data and assumptions will be used for the modelling, resulting in differing models of the product system.

In the following chapters, the most common mental models of the organisational principle of material and energy flows of the socio-economic system are discussed. They start with the implicitly stated mental model of physical/chemical causation upon which the series of standards ISO 14040 is built.

5.7.1 Physical/chemical causation

ISO/EN 14040 favours physical/chemical causation. It describes the product system as a collection of materially or energetically connected unit processes, which performs one or more defined functions (ISO/EN 14040, chap. 3.15).

On causal relationship the standard states:

> "…the underlying physical relationships ...shall reflect the way in which the inputs and outputs are changed by quantitative changes in the pro-

[204] In such cases; ISO/EN 14041 prescribes the conduct of a sensitivity analysis. For the usefulness of sensitivity analysis in such situations (see Chap. 1.2).

ducts or functions by the system. The resulting allocation will not necessarily be in proportion to any simple measurement such as the mass or molar flows of coproducts" (ISO/EN 14041, chap. 6.5.3).

For descriptive LCA, allocation based on a physical, chemical, biological or technical causal relationship reaches its limits if the ratio of products or functions is fixed by the nature of the process, such as in joint coproduction. Such allocation problems can be handled by:

- Expanding the system while adding additional function to the functional unit if this is in accordance with the goal and scope of the study;
- The avoided-burden procedure, while the substituting processes have to be determined based on (most probably) economic arguments;
- The selection of an arbitrarily chosen allocation factor, *e.g.*, mass, volume, revenues, etc.

If partial attribution of joint coproduction processes is necessary for the goal and scope of the LCA, allocation principles based on mental models of causal relationship other than physical/chemical have to be found.

Considering marginal LCA, it is highly doubtful whether a change induced to the socio-economic system by a decision can be reasonably modelled excluding economic data, *e.g.*, for the determination of products and processes to be substituted or additionally caused.

Emphasising a physical/chemical causation of material flow is based on the view that processes of a product system are linked in a technical sense, as in the perspective of engineering sciences. If interpreted strictly, this perspective disregards economic reality. However, it is shared by a fair proportion of the LCA community. This is manifested in their general scepticism towards economic data being integrated into the definition of the product system[205].

5.7.2 Economic causation

The following lines of reasoning start from the point that – at least to a certain degree – material flows within the technosphere are caused by human decisions. The system is changed as a consequence of decisions, resulting in changed material and energy flows and thus in changed environmental interventions.

The economic causal relationship as a basis for allocation in LCA had already been proposed early on in the methodological discussion by Huppes (1992):

[205] although economic data is currently used for the identification of functions to be allocated to (see Chap. 5.6.3)

> "In a social sense all impacts of a process are caused by the value the process creates, based on all the functions the process helps to fulfil. Allocation should reflect this state of affairs. In multiple processes physical properties seldom indicate shares in the total value created. Thus, in principle, the values created by the function of a process should be the basis for allocation" (Huppes 1992:65).

Following economic reasoning it could be argued that human actors following economic rationality determine material and energy flows within the socio-economic system. (Micro-) Economic theory distinguishes two types of actors representing supply and demand: producers and consumers. Each of these actors directs his/her action towards a single target.

The primary goal of a producer is the maximisation of profits while searching the minimal factor cost combination of their production factors, *e.g.*, labour, land and capital[206].

The driving force and thus the 'raison d'être' for every production process is the expected profit for the producers running the operation. Huppes (1994) states about this topic:

> "In a social sense, the value created causes the process. Its costs, *i.e.* the remaining economic inputs required and the unwanted outputs that cannot be disposed of freely, are caused socially by the value created, as are all its environmental inputs and outputs" (Huppes 1994:78).

On the other hand, it can be argued that processes and services are made available because consumers ask for them as they provide satisfaction of need. Informed, rational consumers would strive to gain as much utility as possible under the given constraints of their budget[207].

Thus, enterprises and consumers are regarded as making decisions with a single target, which is:

- Maximising profits and utility respectively by minimising costs.

In this approach, purely economic considerations determine the material and energy flows within the socio-economic system; the physical world is disregarded[208].

To summarise, several market situations can be distinguished, depending on the respective market power of vendors and buyers. Participants of a perfect market would strive for the marginal revenues being the marginal price,

[206] see, *e.g.*, Dinkelbach and Rosenberg 1996:77ff; Varian 1995:305ff/323ff

[207] see, *e.g.*, Varian 1995:90ff/253ff

[208] see, *e.g.*, Georgescu-Roegen 1971 who has brought entropic considerations into economic reasoning; see also Ayres 1999

whereas a monopolist would restrict his/her output to maximise his/her marginal revenues[209].

Similar considerations can be made for the demand side.

In practice, the reasoning behind the use of economic data differs:

a) for descriptive and marginal LCA;
b) for allocation related to multi-output processes and recycling in descriptive LCA.

Ad a) In marginal LCA, price elasticities, especially cross-elasticities, can be used to model changes in supply and demand as a consequence of the change to be assessed in environmental terms[210]. The underlying assumption is that all actors base their actions purely on prices. In descriptive LCA, economic data is used for the attribution of environmental impacts to several products or functions (see *ad b*)).

Ad b) In descriptive LCA, economic data is used for allocations related to multiple output processes and related to recycling.

The underlying assumption in the case of allocation of multiple output processes is that profit guides the actions of a company. From this perspective, the environmental impacts related to a production process can be attributed to the different coproducts based on the relative share to total profits of the different coproducts. This procedure is applicable if data on profits is available. However, in most cases, data on profits is confidential and thus rarely obtainable, especially if LCA-results are intended for the public. Instead of profits, the gross sales value can be used. The underlying assumption is that all coproducts of a process contribute to total profits relative to their revenues. Thus, the relative share of revenues is taken as approximation of the relative contribution of the different coproducts to total profits[211,212].

For allocation related to recycling, several procedures are possible, using economic data in different ways (see Chap. 6.2 to Chap. 6.6, also Chap. 10 and Chap. 18.7).

Several aspects should be taken into account if economic data is used for allocation purposes in LCA, *e.g.*, taxes, subsidies and dividends, different

[209] see, *e.g.*, Varian 1995:387ff; for LCA, see, *e.g.*, Frischknecht 1998:104ff; Guinée *et al.* 2004

[210] see, *e.g.*, Ekvall 2000

[211] for details parallelities to cost accounting, see Huppes 1992/1994

[212] The analogy assumption of relative contribution to profits and relative gross sales values is a simplification. Price forming is much more complex, *e.g.*, depending on total production cost, market power, expectations of readiness-to-pay (see, *e.g.*, Wöhe 1996:458ff). Correspondingly, not all products of a company (or process) necessarily contribute to total profits in relation to their particular gross sales values (see, *e.g.*, the portfolio-analysis of a company developed by the Boston Consulting Group (Kleindorfer *et al.* 1993:309ff)).

currencies, unstable prices, inflation or regulatory based market failures. These aspects have to be dealt with consistently depending on how economic data is used for allocation[213].

Decision-makers bound to economic theory would probably heavily rely on economic data such as profits or revenues (gross sales values)[214] and elasticities for the modelling of changes of the product system[215], instead of choosing 'arbitrary' physical or chemical allocation factors.

Apart from the above considerations, economic data is also used in LCA even if modelling 'purely' on the physical, chemical, biological, or technical causal relationship:

- For the definition of 'coproducts' and waste, which is required for the identification of functions (or products) that will have environmental interventions attributed (see Chap. 5.6.3);
- For the choice of substituting processes and products, particularly for marginal but also for descriptive LCA.

5.7.3 Physical/chemical and economic causation

A further line of reasoning starts from the point that material and energy flows are closely related to decisions made during the design of products.

One of the main questions in product design is material selection, as material, processing, material properties/performance and total costs are interlinked[216].

> "Because of the interplay between material choice, product performance and manufacturing economics, an effective designer must carefully balance the properties and cost that result from a specific material choice against the willingness of the market to pay the price that such a choice will require" (Clark *et al.* 1997:255).

Decisions about materials are made throughout the design process. A model of the material selection process is presented in Figure 5-4.

In the first step, a subset of candidate materials is selected from all materials available by applying non-negotiable primary constraints – about

[213] see, *e.g.*, Lindeijer and Huppes 1999:19ff; Frischknecht 1998:84ff

[214] proposed for the allocation of infrastructure in analogy to the allocation of fixed costs based on the partial contribution to total gross sales value, and as a first approximation in cases where data on profits of single products is not available (see Huppes 1992; Huppes 1994)

[215] see, *e.g.*, Ekvall 2000

[216] see, *e.g.*, Askeland 1996

100'000 engineering materials are currently available[217]. These constraints are imposed by technical requirements derived from the function and application of the product.

Out of this subset of candidate materials, a short-list of materials is selected by applying partially negotiable technical requirements and by maximising performance selecting the best combination of properties.

Final selection of a material for a specific part is achieved by applying further constraints such as processability, appearance, cost, etc. A rational choice is made if maximal technical performance is achieved at minimal total cost[218].

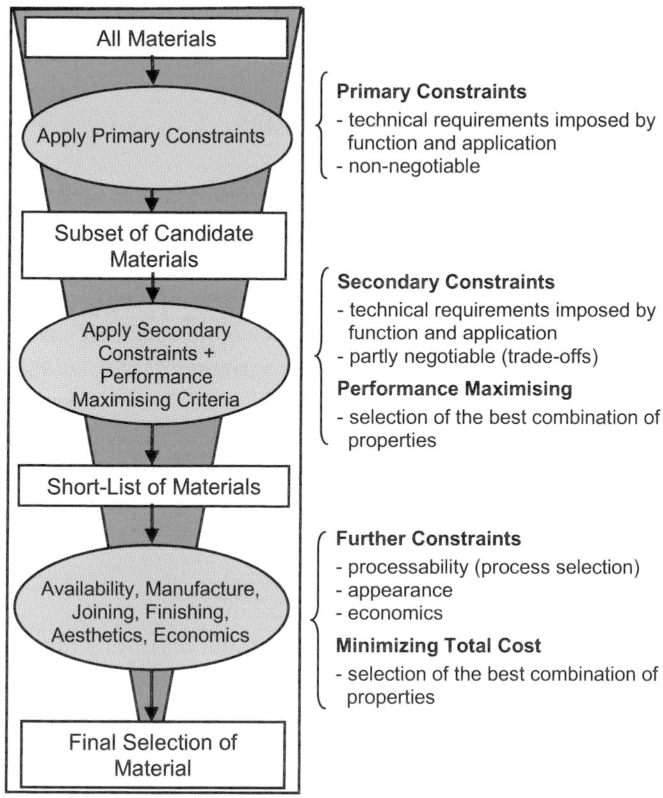

Figure 5-4. Process of rationale material selection (from Ashby 1992:57; modified according to Dieter 1997a:244; reprinted with permission from Elsevier)

[217] Dieter 1997a:243
[218] see Ashby 1992; Datsko 1997:8; Dieter 1997b; Ovens 1997

The same criteria, the same decision path and the same weighting factors as illustrated above have also been implemented in computer-based decision-support tools for rational material selection[219].

Dieter (1997) points out that one could also start with the selection of the manufacturing processes instead of selecting a material prior to the manufacturing processes[220]. Nonetheless,

> "...both approaches end up at the same decision point" (Dieter 1997a:244).

This line of reasoning combines elements of the mental models of purely physical/chemical causation (Chap. 5.7.1) and of purely economic causation (Chap. 5.7.2). Thus, the integration of economic data into product systems is acceptable within this line of reasoning.

5.7.4 Economic and environmental costs

Frischknecht (1998:41ff) argues that environmentally conscious decision-making not only follows purely economic logic but also includes environmental aspects. He therefore introduces a disutility function claiming that rational decision-making should be based on economic as well as on environmental information. The disutility of a decision option is calculated by adding up private costs (including internal environmental costs) and external environmental costs weighted by an 'environmental exchange rate'. This single score is used for allocation and system modelling, as disutility is considered the determining parameter for the material flows of a product system.

In analogy to the purely economic consideration in Chapter 5.7.2, the decision would be to:

- Maximise profit or utility while minimising disutility.

The disutility approach allows one to integrate environmental aspects in models of decision-making. However, the explicit integration of external

[219] see, *e.g.*, Chen *et al.* 1995:69; Anonymous 1996a:37; Liao 1996:6f; Mohamed and Celik 1998:337.
In deviation of the three-step procedure proposed by the above authors, Giachetti 1998 allows one to give different weighting factors in his multi-attributive decision model for performance parameters and material cost. According to him, the importance of material performance and cost varies depending on the product to be produced (*e.g.*, aeroplane vs. consumer good).
See also Weiss 1997 for general considerations on computer-aided material selection.

[220] see, *e.g.*, Schey 1997 or Dieter 1997b for the selection of manufacturing processes

costs into the decision-making process seems to be far from reality[221]. External environmental costs as such are rarely taken into account for decision-making, although environmental considerations and their consequences on stakeholders are often integrated into the decision-making process.

A major problem of external environmental costs is their quantification. As external environmental costs cannot be completely known, they will always indicate just the lower limit of the costs for repairing or avoiding environmental damages[222].

5.7.5 Socio-economic causation

In management theory, the economic single-target model of maximising profits (or utility respectively) outlined in Chapter 5.7.2 has been replaced by the much broader stakeholder model (*e.g.*, Thompson 1967; Freeman 1984, Pfeffer and Salancik 1978; Staehle and Nork 1992; Dyllick *et al.* 1997; Züst 1998; Schaltegger and Sturm 1992:8; Schaltegger and Figge 2000; Mac 2002).

It starts from the point that enterprises are embedded in a socio-cultural, political/legal, technological and economic environment. Thus, decisions of an enterprise are multi-objective: the aim of an enterprise lies not only in the production and distribution of a product or service or in gaining profit, but also in the satisfaction of expectations and demands of interested parties. Thus, (after Schaltegger and Sturm 1992:12ff referring to Hill 1991) rational decisions are made by an enterprise to reach:

- Maximum efficiency in the economic environment;
- Maximum effectiveness in the technological environment;
- Maximum social legitimacy;
- Maximum autonomy in the political/legal environment.

The stakeholder model also forms the ideal basis for the series of standards for environmental management systems ISO 14001ff where stakeholders are addressed as 'interested parties' (ISO/EN 14001, chap. 3.11).

The stakeholder model is suited to describe rational behaviour of an enterprise in general terms. This approach has not been used in life cycle inventory analysis as an organisational (and allocation) principle.

[221] although some international research projects are treating this topic, such as ExternE investigating externalities of energy, financed by the European Commission DGXII

[222] see Freeman 1993 or Keeney and Raiffa 1976:125 for methods to determine external environmental costs and their theoretical and practical limitations

5.7.6 Consequences for the definition of the product system

The mental model of 'causal relationship' within processes is strongly related to the decision-maker's mental model of the socio-economic system and its organisational principle. Within LCA-literature and literature related to environmental management, propositions have been made concerning the organisational principle of the socio-economic system. Some literature stresses physical/chemical or technical determination of material and energy flows. Among this literature is the series of standards ISO 14040. If (huge) whole-system models covering many products are to be avoided, then the attribution problem in the inventory analysis cannot be solved relying solely on the physical/chemical determination of material and energy flows.

Another line of reasoning states that human decision finally determines the material and energy flows within the socio-economic system. This approach differs in the assumptions of the preferences and values that guide human decision-making. Depending on the decision-maker's mental model of the organisational principle of the socio-economic system, differing data and assumptions will be used for the modelling and differing product systems will result.

5.8 Allocation for reuse & recycling

A further decision situation encountered during the setting up of a product system is the multifunctionality issue related to recycling. For recycling, the attribution problem cannot be solved by solely relying on the relative effect of a process or by applying a cause-related allocation procedure as outlined above for multifunctional processes. The particularity of reuse & recycling ies in a twofold attribution problem caused by:

- The bi-functionality of the recycling processes as waste treatment and secondary material production;
- The multifunctionality of primary material production processes and waste disposal, as these processes may[223] be seen as joint multifunctional processes with subsequent functions[224]. These subsequent functions are the multiple use of a material.

These attribution problems related to recycling cannot be solved based either on effect-related criteria or by applying a mental model of causal relationship (as functions are inherently joined). A third guiding principle has to be applied for dealing with recycling in LCA: a mental model of the deci-

[223] according to ISO 14041, chap. 6.5.4, point a)
[224] in differentiation to common coproduction with (almost) simultaneous outputs

sion-maker's '*range of responsibility*' defining the life cycle steps and processes the decision-maker feels (or should feel) responsible.

This book claims that the range of responsibility should be chosen in a way that LCA supports best decision-making according to the general and material specific management rules for sustainable development (see thesis II; Chap. 1.5).

5.8.1 Some allocation principles for reuse & recycling

The allocation principles and procedures mentioned above in Chapter 5.6

> "also apply to reuse and recycling situations" (ISO/EN 14041, chap. 6.5.4).

According to the standard ISO/EN 14041, reuse & recycling situations require additional elaboration for the following reasons:

- "Reuse and recycling ...may imply that the inputs and outputs associated with unit processes for extraction and processing of raw materials and final disposal of products are to be shared by more than one product system
- Reuse and recycling may change the inherent properties of materials in subsequent use
- Specific care is needed for system boundaries definition regarding recovery processes" (ISO/EN 14041, chap. 6.5.4).

All three of the above mentioned points have to be carefully considered in a consistent allocation procedure. They basically refer to the third attribution principle: the decision-maker's mental model of his/her range of responsibility in view of the management rules for sustainable development.

Considering reuse & recycling, allocation procedures according to ISO/EN 14041 do not distinguish descriptive (average) or change-oriented (marginal) approaches.

5.8.2 Setting of system boundaries for reuse & recycling

The setting of the system boundaries is a key issue for modelling recycling in LCA. Recycling processes can be seen as a secondary material production as well as waste treatment. Therefore, recycling processes are attributable to at least two product systems in sequence. The setting of system boundaries becomes more complicated if primary material production of secondary material and waste disposal of downstream product systems are identified to be attributable to the product under study (see Chap. 5.8.3).

ISO/EN 14041 states on the issue of system boundaries:

> "The system boundaries define the unit processes that will be included in the system to be modelled. Ideally, the product system should be modelled in such a manner that inputs and outputs at its boundary are elementary flows. In many cases there will not be sufficient time, data or resources to conduct such a comprehensive study" (ISO/EN14041, chap. 5.3.3).

This is particularly relevant for recycling, where material can be used subsequently for many products.

Should for any reason system expansion not seem practical, material flows entering and leaving a product system need to be cut off. The standard suggests that in- and outputs shall be treated in the same way:

> "Allocation procedures shall be uniformly applied to similar inputs and outputs of the system under consideration. For example if allocations made to useable products (*e.g.*, intermediate of discarded products) leaving the system, then the allocation procedure shall be similar to the allocation procedure used for such products entering the system" (ISO/EN 14041, chap. 6.5.3. para. 3).

In the case of recycling, three possible locations for setting the system boundaries can be distinguished[225]:

a) After the use phase/after demolition and before any recycling activity;
b) During the recycling processes (collection, sorting, refining); the setting of system boundaries depends on the specific situation and the allocation procedure chosen;
c) After reprocessing the material before the production of intermediate products (*e.g.*, after remelting/refining scrap as ingot production for aluminium).

ISO/EN 14041 stresses:

> "…specific care is needed for system boundary definition regarding recovery processes" (ISO/EN 14041, chap. 6.5.4).

An interpretation may be that the setting of system boundaries can only be done in combination with the choice of an allocation procedure and with the identification of the multifunctional processes related to reuse & recycling. This identification relies on the decision-maker's mental model of his/her range of responsibility.

[225] see also Doka 2000:13f

5.8.3 Identification of multifunctional processes related to reuse & recycling

The next step in the modelling process is to identify the multifunctional processes related to the reuse & recycling of materials. ISO/EN 14041 states:

> "Reuse and recycling ...may imply that the inputs and outputs associated with unit processes for extraction and processing of raw materials and final disposal of products are to be shared by more than one product system" (ISO/EN 14041, chap. 6.5.4).

The following processes can potentially be identified as shared, multifunctional processes related to the issue of reuse & recycling, (for descriptive LCA)[226]:

- The primary material production processes providing virgin material to the system;
- The primary material production processes providing – perhaps many products earlier – the input secondary material;
- The recycling processes providing the input secondary material;
- The recycling processes processing the post-consumer material (output secondary material);
- Waste material disposal in the product system;
- Disposal of waste material occurring during the life cycle of the product;
- The final disposal of the material after the last application at the end of the material cascade.

These material flows are relevant in relation to the total mass flows or the total environmental implications of a product system. Furthermore, the availability of secondary material is inherently linked to primary material production processes and final disposal. Thus, the two attribution criteria stated so far (the relative effect to the total material and energy flows or environmental impacts as the 'cut-off criterion' (Chap. 5.5) and the cause-related mental model of the organisational principle of the socio-economic system (Chap. 5.6)) are not sufficient for decisions which, of the above-mentioned processes, should (partially) be included in a product system. A third criterion has to be applied to solve the attribution problem related to reuse & recycling: the range of the decision-maker's responsibility for the material and energy flows related to one product.

The range of the decision-maker's responsibility is not only an ethical issue. Moreover, it expresses the decision-maker's conception of the role of information for leading our society towards environmentally sustainable

[226] see also Ekvall and Tillman 1997:156

conditions. The crucial point in this respect is the question actors need to consider in asking what kind of information is necessary to direct their actions toward an environmentally effective and efficient socio-economic system[227].

Two approaches can be distinguished in literature for the identification of multifunctional processes related to reuse & recycling: the 'sunk-effects approach' and the 'asset approach'.

Sunk-effects approach

Inspired by the economic approach of sunk-cost, Lipsey and Steiner (1972) on the sunk-cost approach state:

> "'bygones are bygones', and they should have no influence on deciding what is currently the most profitable thing to do. The 'bygones-are-bygones' principle extends well beyond economics, and is often ignored in poker, war, and perhaps in love" (Lipsey and Steiner 1972:177, quoted after Frischknecht 1998:59).

The economic argument behind this approach is that accounting for past costs leads to path-dependencies causing suboptimal resource allocation.

The sunk-cost approach can be transposed to environmental considerations as the sunk-effects approach. The sunk-effects approach claims that only environmental interventions directly related to a decision should be attributed to the alternatives. For instance, recycling metal should not carry any environmental burdens from its former primary material production as they already occurred. The environmental burdens related to recycling and reprocessing are allocated to the further users of the material if 'waste' or post-consumer material is further used by others. In this sense, only environmental interventions caused by the recycling process itself should be accountable. It is argued that the decision alternatives are the use of primary material vs. the use of secondary material. Therefore, the secondary material substitutes virgin material[228].

The sunk-effects approach is commonly applied for the setting up of public databases[229]. The sunk-effects approach is, *e.g.*, the basis for the 'cut-off allocation procedure'[230].

[227] see von Hayek 1945 for the role of economic data in society

[228] note the parallelism to the reasoning on the closed-loop procedure of ISO/EN 14041, chap. 6.5.4

[229] see, *e.g.*, Doka 2000:18; Förster and Gilgen 1997

[230] see Boguski *et al.* 1994:151ff; Ekvall and Tillman 1997:160; Fava *et al.* 1991:79ff; Klöpffer 1996:27ff; Rydberg 1995:236; Vogtländer *et al.* 2001:345

Some scientists with strong ties to economics[231] favour the sunk-effects approach. It is also applied in the environmental accounting schemes of companies[232]. The sunk-effect approach intends to depict environmental interventions occurring in the future with the aim of minimising them.

In the context of descriptive product LCA this approach can lead to suboptimal recommendations. A simple example illustrates this. In Figure 5-5 the added primary energy consumption of two products is compared over several life cycles.

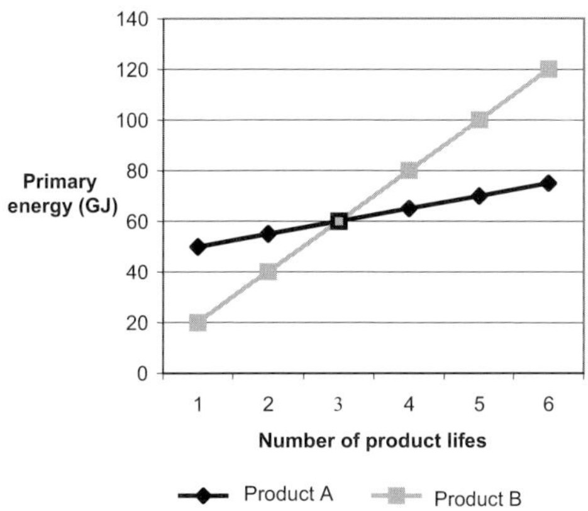

Figure 5-5. Added primary energy consumption of two products compared over several life cycles (fictitious data)

Product A is made from a material with high-energy consumption during the production of the primary material. The product is recycled with a recycling yield of 90%; primary material substitutes the lost material. Recycling processes require around 5% of the energy consumption of primary material production (fictitious data).

Product B is made from a material, which is not recycled but burnt, requiring primary material for each life cycle.

Results show that product A would never have been produced/bought as its first production requires more energy than product B. Applying the sunk-effect approach the subsequent production/use of product B would be favoured although after the third product life, the added primary energy con-

[231] see, *e.g.*, Schaltegger and Sturm 1992:157/187ff; Huppes 1992; Huppes 1994 Braunschweig and Müller-Wenk 1993:73; see also Chap. 6.2

[232] Schaltegger and Sturm 1992:157/187ff; Braunschweig and Müller-Wenk 1993:73

sumption of product A is lower compared to product B. The sunk-effect approach would thus lead to an environmentally suboptimal result.

Product A would only be produced if the environmental interventions of primary energy consumption were shared with subsequent product systems; therefore, subsequent product systems have to carry environmental interventions from the energy consumption of primary material production.

A further example is presented in the first case study on aluminium windows (Chap. 8ff). In this example, an allocation procedure that is based on the sunk-cost approach leads to suboptimal results.

The above reasoning is only valid for descriptive LCA; it does not hold for marginal LCA. Marginal LCA depict the future environmental consequences of a decision; the sunk-effect principle has to be applied to provide reasonable results. Thus:

- Ignoring past environmental interventions in the a-temporal model of descriptive LCA leads to systematic errors;
- Taking into account past environmental interventions leads to systematic errors in change-oriented, marginal LCA.

Asset approach

In the asset approach, the material is considered an asset, consisting of the environmental interventions related to its primary material production and its final waste disposal. These environmental interventions are allocated to each of the product systems in the cascade, based, *e.g.*, on the number of subsequent uses or on devaluation and material loss over the product system under study.

The recycling processes directly preceding and following the system under study are allocated depending on the chosen allocation procedure and on the way in which system boundaries are set.

This approach extends the range of responsibility – at least as far as primary material production and final disposal is concerned – over the whole material cascade. It stresses the entropic character of economic processes and the need of thermodynamic efficiency of a (sustainable) circular flow economy (see also Chap. 3.3.3, 3rd management rule, points 5 and 6).

All allocation procedures referring to the material cascade model by Sirkin and ten Houten (1994), (see Chap. 9.5.2) or to the model of material pools (if different grades are distinguished; Chap. 9.5.3) are committed to the asset approach[233]. The same is true for the 50/50-allocation and its adaptations[234].

As a consequence of applying the asset approach and the cascade model, environmental interventions can be accounted for that which has already occurred. From a decision perspective, doing so leads to path-dependencies that impede optimal resource allocation and prevention of environmental impacts respectively.

Nonetheless, applying the asset approach can lead to more appropriate recommendations than the sunk-effects approach if re-evaluated according to requirements of product systems and their LCIs (Chap. 3.4) and based on material-specific management rules for sustainable development. This is at least true for descriptive LCA[235,236]. This has to be considered the consequence of the modelling characteristics of LCA-resulting in an a-temporal, spatially undifferentiated input-output model.

This book concludes that the choice of the best-suitable approach concerning the range of the decision-maker's responsibility should be made dependent on the material and market characteristics for each of the materials involved, taking into account the corresponding management rules for sustainable development.

Chapter summary

- For the identification of shared processes related to reuse & recycling, the two criteria stated so far – a) the relative effect to the total material and energy flows or environmental impacts as 'cut-off criterion', and b) the cause-related mental model of the organisational principle of the socio-economic system – are necessary but not sufficient. A third criterion has to be applied to solve the first part of the attribution problem related to reuse & recycling: the range of responsibility for the material and energy flows related to one product;
- For the modelling of the decision-maker's range of responsibility, two approaches are applicable: the sunk-effects approach ('bygones are bygones') and the asset approach containing considerations on thermodynamic efficiency. This book claims that the choice of the best-suitable

[233] such as Lindeijer 1994; Wenzel *et al.* 1996; Werner and Richter 2000b after Buhé *et al.* 1997; Ekvall and Tillman 1997:159; Karlsson 1994; Kim *et al.* 1997; Newell and Field 1998; for aluminium, see Rombach 1998:424

[234] see Ekvall 1994; Fava *et al.* 1991:80; Lindfors *et al.* 1995a:63ff/217ff

[235] It is of course not rational to base decisions on past environmental interventions (or cost). Basically, these models are based on (hypothetical) present and future substitution processes considering material losses and devaluation.

[236] see also the case study on aluminium window frames, Chap. 8ff

range of the decision-maker's responsibility should be made dependent on the material and market characteristics for each of the materials involved in a product, taking into account its management rules for sustainable development (in descriptive LCA);

- Ignoring past environmental interventions in the a-temporal model of descriptive LCA leads to systematic errors. Taking into account past environmental interventions leads to systematic errors in change-oriented, marginal LCA;
- Taking into account past environmental interventions in descriptive LCA can be right from a model perspective. If used for decision support, descriptive LCAs can lead to path dependencies causing environmentally non-efficient resource allocation.

5.8.4 Open-loop recycling and closed-loop recycling

The standard ISO/EN 14041 distinguishes between allocation for closed material cycles (closed-loop) and allocation for open material cycles (open-loop):

> "A closed-loop allocation procedure applies to closed-loop product systems. It also applies to open-loop product systems, where no changes occur in the inherent properties of the recycled material. In such cases, the need for allocation is avoided since the use of secondary material displaces the use of virgin (primary) material. However, the first use of virgin materials in applicable open-loop product systems may follow an open-loop allocation procedure outlined below.
> An open-loop allocation procedure applies to open-loop product systems, where the material is recycled into other product systems and the material undergoes a change to its inherent properties. The allocation procedures for the shared unit processes mentioned in chap. 6.4.2 should use, as the basis for allocation:
> - Physical properties
> - Economic value (*e.g.*, scrap value in relation to primary value) or
> - The number of subsequent uses of the recycled material..." (ISO/EN 140041, chap. 6.5.4).

From a decision-maker's perspective, the above definitions lead to a rather unsatisfying situation: for the easiness of the calculation routine only, an open-loop material flow can be considered closed-loop, although the material is actually leaving the system under study. This can be applied "if the import and export of secondary raw material ... are equivalent" (ISO/TR 14049; chap. 8.3.2).

The distinction between a technical description of a product system and the allocation procedures for recycling is illustrated in ISO/EN 14041 (Figure 5-6):

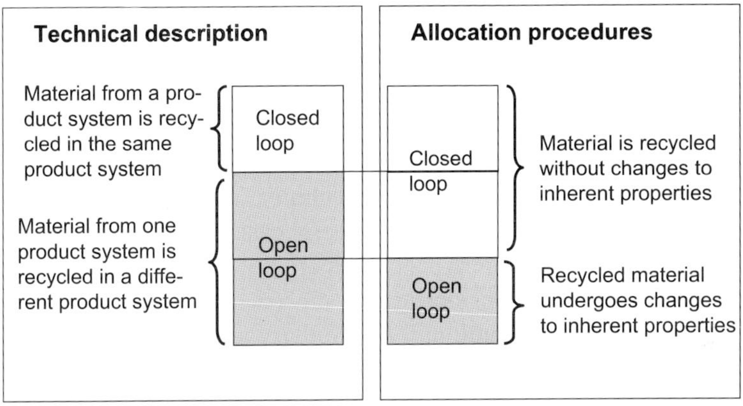

Figure 5-6. Distinction between technical description of a product system and allocation procedures for recycling made in ISO/EN 14041, chap. 6.5.4; reprinted with permission from the European Organisation for Standardisation, Brussels

The distinction of 'closed-loop', 'open-loop' and 'open-loop considered closed-loop' situations bears some inherent problems. True closed-loop situations seldom occur in reality, apart from reuse in a strict sense[237] and in catalytic reactions. The difficulty remains in distinguishing between 'open-loop' and 'open-loop considered closed-loop' situations. This distinction requires a decision about which conditions the loop can be considered 'closed'. It should be specified:

- Which parameters – physical, chemical, biological, technical or economic – describe the equivalency of inputs and outputs of a process or product?
- For outputs and inputs still considered equivalent, which deviation of the parameters is tolerable?

The first question is trickier to answer than it may seem. The parameter selected has to be valid for the whole cascade of product systems through which a material runs. This is necessary if environmental interventions shall be allocated to all product systems in the cascade on the same basis to follow the 100%-rule (Chap. 5.6.2).

The choice of a parameter for the assessment of equivalency of inputs and outputs is another subjective choice.

[237] see Chap. 5.8.6

The distinction of 'open-loop' and 'open-loop considered closed-loop' situations has an important implication from a methodological point of view: it can lead to an unsteady allocation function ('jumps'). Numeric results can vary depending on whether inputs and outputs are considered equivalent or not. Even small differences can lead to substantial differences in LCA-results if two similar products are compared (see also Chap. 9.6). This is the starting point for the 'value-corrected substitution' developed for aluminium in building applications, described in Chapter 10.

From a methodological point of view:

- The closed-loop situation in both the technical description and the allocation procedure can be considered a special case of an open-loop situation where the material's inherent properties are not changed; or
- It can be argued that considering the closed-loop situation as a special case of the open-loop situation, the secondary material leaving the system is replacing primary material in subsequent product systems. In a closed-loop situation (as assumed in the allocation procedure), the replacement is complete. In the open-loop situation the material undergoes changes in its inherent properties during the product system. Therefore, the replacement of primary material can only be partial. Environmental burdens associated with this 'incomplete' substitution because of the changed material properties (or material losses) should be allocated to the product system causing them. If an acceptable parameter for the devaluation of a material over a product system can be found, the value-corrected substitution provides a steady allocation function (for details, see Chap. 10).

5.8.5 Temporal changes and the decision-maker's attitude towards risk

Before going into more detail about different allocation procedures for reuse & recycling, some comments are necessary about the integration of the temporal aspect inherently related to recycling.

Recycling and reuse are imperatively related to the course of time.

> "This necessary sequence of activities or functions (of open-loop recycling; WF) fulfilled within material or energy cascades is due to the entropy law and the irrevocable tendency of increasing disorder, and may be observed in nearly any product system" (Frischknecht 1998:97).

Coproduction and reuse & recycling are also distinguished in the allocation rules according to ISO/EN 14041. The step-wise allocation procedure established for coproduction also applies to recycling (see Chap. 5.6.4ff) although reuse & recycling situations require additional considerations related to temporal aspects (see Chap. 5.8.1).

Temporal aspects are relevant for descriptive (attributive) LCA as well as for planning (marginal) LCA. In planning LCA, the necessity to consider future changes in the socio-economic system is obvious. But also in descriptive LCA, assumptions related to the temporal course of the socio-economic system have to be made (see below).

Frischknecht (1998) distinguishes four different types of LCA, depending on the time span of the planning to properly address temporal changes of the socio-economic system (Table 5-2).

This distinction is more on a theoretical than a practical level. The definition of 'short run', 'long run' and 'very long run' depends strongly on the rate of change of the industrial sectors involved in LCA. However, being systematic, Table 4-2 gives a good starting point for detecting time-dependent aspects that need to be included in a specific LCA.

Table 5-2. Four different types of product systems and its characterisation in terms of the temporal structure of change in demand and system model properties (Frischknecht 1998:XX)

			System model properties	
Name of the system model	Goal of the study	Temporal structure of a change in demand	Variability of the factors of production	Technologies available/ technical performance
Status quo	Environmental reports Statements to authorities	No change	All fixed	No choice/ fixed
Short run	Short-term system optimisation, *e.g.*,: - Changes in demand, - Negotiations with supplies	One time only	Capital equipment fixed	No choice/ fixed
Long run	Hot spot identification and elimination: - Product system optimisation - Product development - Product system comparison	Long-term trend	All variable	Free choice/ fixed
Very long run	Very long-term (strategic) planning: - Technology development - Technology optimisation - Technology comparison	Very long-term trend	All variable	Free choice/ variable

reprinted with permission from the author

Additionally, the handling of changes over time of the socio-economic system depends on the decision-maker's attitude towards risk. This attitude is an expression of his/her value system and indicates his/her temporal preferences[238].

[238] Keeney 1992:7

The attitude towards risk influences also the handling of reuse & recycling in LCA. In the case of reuse & recycling, temporal aspects affect the modelling of:

a) Recycling processes which will be used for the treatment of the post-consumer material;
b) Technology or processes which will be substituted in the future with the material to be recycled;
c) Percentage of the material collected for recycling and the yield of the recycling processes;
d) Service life of the product assumed;
e) Stability over time of the allocation factor chosen.

Three different types of attitude towards risk are distinguished for the discussion of the influence of the decision-maker's attitude[239]:

- Risk-averse, as consequence of a pessimistic world-view;
- Risk-calculating, as a consequence of a realistic, appraising world-view;
- Risk-taking, as consequence of an optimistic world-view.

Not all temporal aspects related to reuse & recycling as stated above need to be assessed for their risk in the same way[240]. Due to this, the following points intend to show the range of possible decisions related to recycling in LCA where temporal aspects are involved. It is not intended here to provide a generally applicable, conclusive framework for dealing with temporal aspects related to recycling in LCA[241]. In view of the broad range of pro-

[239] after Rescher 1993:64; see, *e.g.*, Wildavsky 1993 for factors determining the attitude towards different types of risk

[240] see, *e.g.*, Wildavsky 1993:196 for different attitudes towards risk, depending on the specific situation

[241] The distinction of the three attitudes towards risk would allow to integrate the above reasoning into Eco-indicator 99-methodology (see Goedkoop and Spriensma 2000; Hofstetter 1998), which distinguishes different cultural perspectives (after the Cultural Theory by Thompson *et al.* 1990). The three different attitudes towards risk are attributed to three of the four archetypes considered actively involved in environmental decisions and depicted in Eco-indicator 99: 'hierarchist' = 'risk accepting/calculating', 'egalitarian' = 'risk-averse', 'individualist' = 'risk-taking' (based on Hofstetter *et al.* 2000; van Asselt *et al.* 1995:17:164).
Hofstetter 1998 has lined out some relation of attributing environmental interventions related to recycling and cultural perspectives: "Individualists may argue that as long as there is a market demand or recycled plastic the environmental impacts from the initial production can be allocated to some extend to the whole product cascade. Hierarchists may agree with this procedure only for products with short life times because they consider the development of the recycling market as too uncertain in the long run. Egalitarians would argue that those that use virgin material should carry the whole responsibility for all the environmental impacts caused by its production because the hypothetical re-use is too uncertain" (Hofstetter 1998:314f).

ducts, technologies, branches, particularities and the complex human attitude towards risks[242], the elaboration of a generally applicable and conclusive framework for the integration of temporal aspects related to recycling in LCA is considered fairly arbitrary; it would reflect, more than anything else, the author's own perception.

As a more pragmatic approach, the range of possible choices for each of the temporal aspects related to reuse & recycling is highlighted by discussing them in relation to each of the three types of attitude towards risk. As such, a framework is provided to explicitly address and discuss the temporal aspects in LCA with the decision-maker and commissioner of an LCA[243]. Only in this way can the decision-maker's attitude towards risk be specifically addressed in the life cycle inventory analysis. In this sense, the following tables are intended as a basis for discussion.

Ad a) Recycling processes that will be used for the treatment of the post-consumer material

A first question arises in how future-recycling processes should be integrated into an LCA. Risk-averse attitudes would probably model a system

Table 5-3. Attitude towards risk and the assumption of future recycling processes

	Future recycling processes assumed	
Attitude towards risk	Average technology-mix considerations	Marginal technology considerations
Risk-averse	No future processes considered	No future processes considered
Risk-calculating	Actual average technology mix as worst-case Actual average technology mix as worst-case assumption; Or partial consideration of technological improvement; determination of a future technology mix based on marginal technology considerations;	Actual marginal technologies as worst-case assumption; Or best guess marginal technology based on statistical data or market insight
Risk-taking	Consideration of technological improvement: best available technology	Best available technology

However, for reasons given in the main text, it is abstained from elaborating this option.

242 see Rayner 1993; Wildavsky 1993; Jungermann and Slovic 1993; Keeney and Raiffa 1976:148ff

243 remember the requirements on decision support tools by von Winterfeld and Edwards 1986 and Keeney and Raiffa 1976 stated in Chap. 3.2.1

and exclude any future processes, setting the system boundary accordingly Risk-calculating attitudes would possibly assume actual technologies or include technological improvement based on marginal technology considerations, market insight or statistical data. A risk-taking position would probably use data on best-available technology for the integration of future recycling processes (see Table 5-3).

Ad b) Technology or process that will be replaced in the future by the material to be recycled

Concerning recycling, one might consider technologies or processes that will be replaced by the material to be recycled. Again, risk-averse behaviour would presumably not consider any future processes or technologies and model the product system accordingly. Risk-calculating behaviour would presumably use actual technologies as worst-case assumptions or take technological improvement options into account. In a risk-taking position, worst available technologies might be considered for substitution (see Table 5-4).

Table 5-4. Attitude towards risk and the assumption of processes that will be substituted

	Assumptions on processes that will be substituted	
Attitude towards risk	Average technology-mix considerations	Marginal technology considerations
Risk-averse	No future processes considered	No future processes considered
Risk-calculating	Actual average technology mix as worst-case assumption; Or partial consideration of technological improvement; determination of a future technology mix based on marginal technology considerations	Actual worst marginal technologies as worst-case assumption; Or best guess marginal technology based on statistical data or market insight
Risk-taking	Consideration of technological improvement: worst available technology will be substituted	Actual worst available technology will be substituted

The decision over whether substitution takes places within the proper material cascade, or if substitution effects are taken into account throughout the socio-economic system (*e.g.*, recycled PVC substitutes for primary aluminium) is probably less dependent on the decision-maker's attitude towards risk but on the goal and scope of the study.

Ad c) Percentage of the material collected for recycling and the yield of the recycling processes

Another temporal aspect of reuse & recycling is the percentage of the material to be collected for recycling and the yield of the recycling processes. Risk-averse attitude would presumably not consider any future processes and model the product system accordingly. Risk-calculating attitude would use actual recycling data or statistical interpolations for short-living products, whereas in a risk-taking attitude statistical interpolations would be also applied to long-living products (see Table 5-5).

Table 5-5. Attitude towards risk and the assumption of future recycling percentages assumed

Attitude towards risk	Future recycling percentages assumed (no differentiation between average technology-mix and marginal considerations)
Risk-averse	No future data considered
Risk-calculating	Actual average recycling percentage; Or statistical interpolation of actual recycling coefficients for short-living products
Risk-taking	Statistical interpolation of actual recycling coefficients

Ad d) Service life assumed

For comparison with other products, usually assumptions on the service life of products under study have to be made. The uncertainty about the future destiny of the product can be integrated differently, depending on the basic values of the modeller or commissioner. Risk-taking persons would assume the technical service life, whereas risk-averse persons would base calculations on the economic payback-time (Lalive d'Epinay *et al.* 2000:28). Risk-calculating persons would possibly take statistical average data[244] (see Table 5-6).

Table 5-6. Attitude towards risk and the assumption of service life

Attitude towards risk	Assumption of service life (no differentiation between average technology-mix and marginal considerations)
Risk-averse	Around economic pay-back time
Risk-calculating	Around statistical average data
Risk-taking	Up to technical service life

Ad e) Stability over time of the allocation factor chosen

As the socio-economic system changes over time, time dependent allocation factors such as prices or price ratios will also change. Attitude towards risk determines to which extent ignorance on the future allocation factors are acceptable in a specific LCA or not. Whereas risk-averse behaviour would probably lead to a priori excluding time dependent allocation factors, risk-calculating and risk-taking behaviour would possibly allow time-dependent allocation factors for short-living products or also for long-living products (see Table 5-7).

[244] see, *e.g.*, Anonymous 1994a and Anonymous 1995b for building materials and products

Table 5-7. Attitude towards risk and the stability over time of the allocation factor chosen

Attitude towards risk	Stability over time of the allocation factor chosen (no differentiation between average technology-mix and marginal considerations)
Risk-averse	No time-dependent allocation factor chosen
Risk-calculating	Time-dependent allocation factor chosen for short-living products
Risk-taking	Time-dependent allocation factor chosen also for long-living products

In the following, on an exemplary basis, the choice of an allocation procedure and its corresponding allocation factors in dependency of the attitude towards risk are illustrated. The allocation procedures mentioned are further described in Chapter 10 and Chapter 11.

Table 5-8. An example: attitude towards risk and selection criteria for allocation procedures and allocation factors

Attitude towards risk	Selection criteria for allocation procedures and allocation factors (no differentiation between average technology and marginal considerations)
Risk-averse[245]	No future data used: - no future recycling yield - no future price-ratio (or other time-dependent allocation factors) - no recycling credits shall be granted (on processes or materials) => Cut-off procedure
Risk-calculating	For short term considerations and based on statistical extrapolation, future data used: - future recycling yield - future price-ratio (or other time-dependent allocation factors) - recycling credits shall be granted (on processes or materials) => Value-corrected substitution complemented with the closed-loop procedure based on material pools For long term considerations, no future data used as statistical extrapolation becomes uncertain: - no future recycling yield - no future price-ratio (or other time-dependent allocation factors) - no recycling credits shall be granted (on processes or materials) => Cut-off procedure
risk-taking	Future data used: - future recycling yield - future price-ratio (or other time-dependent allocation factors) - recycling credits shall be granted (on processes or materials) => Value-corrected substitution complemented with the closed-loop procedure based on material pools

[245] Note that risk-averse behaviour does not imply pessimistic assumptions and worse environmental profiles, as things do not necessarily improve. There is also a risk of not wanting to take any risk...

In a risk-averse position, no future data on recycling yield or any time dependent allocation factor would be considered; recycling credits would not be granted. Hence, the cut-off procedure (see Chap. 11.1) only using data of the current situation would probably be chosen to deal with the issue of open-loop recycling.

In a risk-calculating position, the cut-off procedure would probably only be applied to long-term considerations if statistical extrapolation or any assumption becomes more uncertain. For short-term considerations, future data on recycling yield or time dependent allocation factors might be used; also recycling credits could be granted. The allocation procedure applied would possibly be the value-corrected substitution (see Chap. 10) complemented by the closed-loop procedure based on material pools (Chap. 11.2).

A risk-taking attitude would consider estimations on future recycling yields and on time dependent allocation factors; recycling credits for the substitution of future processes would be granted. The value-corrected substitution complemented by the closed-loop procedure based on material pools would also be applied to long-term applications (see Table 5-8).

The influence of the decision-maker's attitude towards risk, *e.g.*, when selecting an allocation procedure for reuse & recycling, implicitly brings temporal preferences into modelling. This attitude contrasts with the a-temporal, spatially undifferentiated character of LCA-models (see Chap. 2.3.2)!

Finally, a more general risk-related aspect should be mentioned which is not tied to the temporal aspects of recycling: the distinction between systematic and unsystematic risks. In the modelling context of LCA, errors during data input are unsystematic risks, as each datum can be the 'victim' of human error. These types of risks can be limited by systematic internal control procedures.

In contrast to unsystematic risk, *e.g.*, the use of average data from public databases constitutes a systematic risk. Such public data suffers from a lack of representation, relevance and precision for various reasons (after Schaltegger and Burritt 2000:249; added). Such average data is systematically biased as it:

- Often stems from well-organised firms;
- Collected by interest-guided branch organisations;
- Hides the whole spectrum of 'good' and 'bad' producers within a branch;
- Remains unspecific about the product, hiding the whole spectrum of products behind one data set;
- Remains sluggish compared to the small changes in production and environmental impacts of a company (or even branch);
- Abstracted from their site-specific (geographic) and temporal context.

This kind of risk cannot be avoided, except by the renunciation of the use of average data stemming from public databases. Given the high resource requirements of data collection in terms of manpower and financing, this is rather illusory from a practical perspective. For decision-making purposes, the use of data from public databases has to be restricted to processes that lie beyond the action space of the decision-maker and which are assumed to be not affected by the decision. The more an LCA is actor-based and site-specific (and the respective data used!), the more suitable the resulting model becomes (see also Chap. 3.1.2).

In the next chapter, a review of literature is provided concerning allocation procedures for open-loop recycling. The chapter concentrates on different approaches for the handling of reuse and open-loop recycling in LCA and its relationship to the decision-maker's mental models and values influencing the life cycle inventory analysis and LCA in general.

Prior to the review, an excursus is given on the examples of allocation related to reuse & recycling in the Technical Report ISO/TR 14049:2000, to treat the issue from the perspective of the series of standards ISO 14040.

Chapter summary

- The modeller's (and the decision-maker's) attitude towards risk determines to a certain extent how time-related issues concerning allocation related to reuse & recycling are handled in the life cycle inventory analysis. This affects the handling of, *e.g.*:
  - The future recycling processes;
  - The technology or process that will be substituted with the recycled material;
  - The percentage of the material recollected and the yield of recycling;
  - The service life of the product assumed;
  - The stability of the allocation factor chosen over time;
- The influence of the decision-maker's attitude towards risk, *e.g.*, when selecting an allocation procedure for reuse & recycling, implicitly brings temporal preferences into modelling. This attitude contrasts with the a-temporal, spatially undifferentiated character of LCA-models.

5.8.6 Excursus: allocation for reuse & recycling in ISO/TR 14049:2000

The Technical Report ISO/TR14049:2000 must also be consulted to completely treat reuse & recycling according to the series of standards ISO 14040. The technical report provides three examples of allocation procedures for reuse & recycling (ISO/TR 14049, chap. 8.3.1).

All examples represent descriptive LCAs:

Closed-loop recycling of ethylene in the production of HFC-134a

The ethylene is re-circulated without losses within the same production process. The ethylene of the output displaces an equivalent amount of the input ethylene needed for each batch: the net consumption of ethylene is almost zero. The closed-loop procedure according to ISO/EN 14041, chap. 6.5.4 is applied.

Open-loop recycling with closed-loop procedure applied to aluminium packaging

A material pool model is applied. It is assumed that the product system delivers secondary raw material into that pool and is supplied with secondary material by the pool (see also Figure 9-2). As the amount of secondary raw material delivered to the pool is larger than the amount taken out of the pool, the product system has a net output of scrap. It is argued that the net output of scrap increases the amount of available secondary aluminium that displaces virgin aluminium metals in other systems. Thus, the recycling process is attributed to the system under study and a closed-loop procedure is applied for the net aluminium output, assuming that the output substitutes part of the virgin material input of the system. With this procedure, the difference between producing aluminium from secondary raw material and producing the same product from virgin material is calculated and allocated to the system under study.

Note that the aluminium pool is considered to be of homogenous composition equivalent to virgin aluminium. This procedure is further explained in Chapter 11.2.

Open-loop recycling applied to fibres for kraft bleached paperboard

The environmental interventions caused by the primary material production of fibres are allocated based on the use of subsequent uses of the fibre. A flow model based on mass flows and yield of the recycling processes determines the number of subsequent uses. The environmental interventions from the primary material production are allocated to the system under study based on the percentage of fibres going to recycling weighted by the inverse of the number of subsequent uses. The environmental interventions of the part not recycled are fully allocated to the system under study.

This example can be seen as an illustration of a system expansion. Note that the system has to be entirely known or approximated to allow one to use this approach. Furthermore, the applications should be similar to make 'the number of uses' an adequate basis for allocation. This example relies on the asset approach illustrated in Chapter 4.8.3, as the environmental interventions related to primary material production of fibres are distributed over several subsequent product systems.

Chapter 6

ALLOCATION PROCEDURES FOR OPEN-LOOP RECYCLING

Allocation related to open-loop recycling is looked at in more depth as an extension of the preceding chapters and as a preparation for the two case studies.

A variety of methodological propositions have been made for allocation related to reuse & recycling. The following chapters describe three approaches to allocation related to reuse & recycling. Furthermore, a systematic survey is presented on allocation procedures for reuse & recycling that have been described in scientific literature. Finally, recommendations for the choice of an allocation procedure for reuse & recycling in product design-related decision situations are given.

The existence of differing approaches to the attribution problems related to reuse & recycling must be interpreted to be the consequence of differing mental models and values accompanying the attribution problem related to reuse & recycling.

6.1 Introduction

The adequate handling of quantitatively relevant material and energy flows imported or exported from the product system under study is a common problem in LCA.

Open-loop recycling is encountered where:

- Recycled material or energy in cascade use is introduced to the product system from other applications;
- Material or energy is exported from the product system to other applications.

Both situations can commonly be found in one and the same product.

In 'real world' systems, traditionally primary and secondary materials are used as inputs (multi-input processes). At the output side several recyclable or reusable outputs with differing material's inherent properties (multi-output processes) can occur. As such, the product system can be part of a whole cascade or net of product systems.

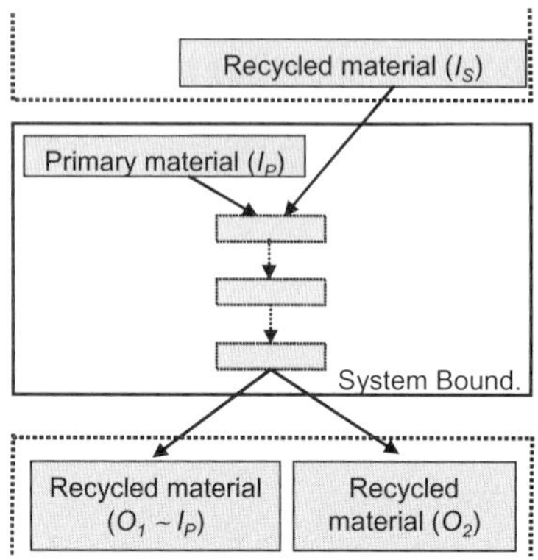

Figure 6-1. A product system with recycled material: a combination of multi-input/output processes in a cascade of product systems with changes of the material's inherent properties over the system under study

For modelling purposes, the generation of recycling materials can also affect the use of other materials within the same material cascade. For instance, recycling material can substitute for primary material production processes and avoid waste treatment. The generation of recycling materials can even affect the use of other materials throughout the socio-economic system.

Figure 6-1 illustrates such a system, indicating also the material's inherent properties (the 'material quality' in brackets).

Allocation procedures for open-loop recycling have to deal consistently with the following occurrences:

- Multi-input processes, *e.g.,* if primary and (different) grades of secondary materials are used for the production of intermediate products;
- Multi-output processes, *e.g.*:
  - When recycling processes provide different grades of recycled material; or
  - When primary material production processes are identified to be shared by several subsequent product systems (see Chap. 5.8.3);
- The bi-functional character of the recycling processes (waste treatment for the previous product system and secondary material extraction for the following product system);

- Changes to the material's inherent properties (of different material grades) used over the life cycle of a product;
- Cascade use of material over different subsequent product systems;
- Substitution effects related to the use of recycling material.

As already stated in Chapter 5.8.3, the following processes can be possibly shared by several product systems and are thus subject to allocation procedures (for descriptive LCA):

- Primary material production processes providing virgin material to the system;
- Primary material production processes providing – perhaps many product systems earlier – the input secondary material;
- Recycling processes including material losses providing the input secondary material;
- Recycling processes including material losses processing the post-consumer material (output secondary material);
- Disposal of waste material occurring during the life cycle of the product;
- Final disposal of the (downcycled) material after the last application at the end of the material cascade.

Marginal LCA does not rely on the 'descriptive' attribution of these processes but on accounting changes within the socio-economic system. All the processes mentioned above are subject to substitution processes.

Substitution effects can also affect material flows of other materials not physically related to the product system under study. Recycling of PVC can, *e.g.*, lead to substitution effects reducing the use of aluminium or wood.

Three approaches can be distinguished for the discussion of allocation procedures for open-loop recycling (Figure 6-2).

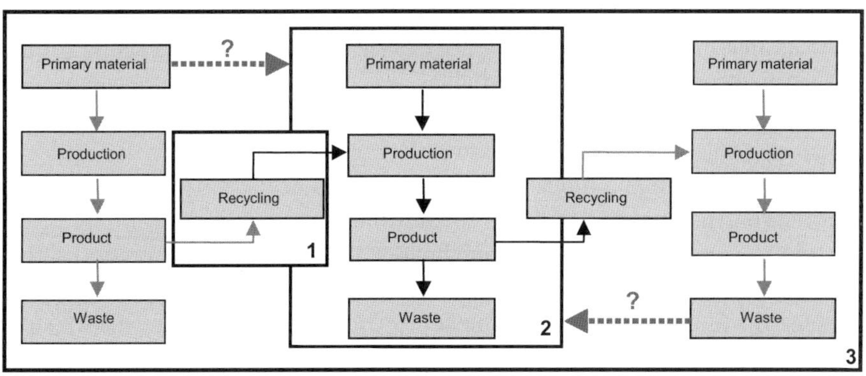

Figure 6-2. Three different approaches to tackle the attribution problem related to recycling: 1) on process level, 2) on product system level, 3) considering the whole cascade of product systems
Note that further material flows related to physically independent product systems can be considered for marginal LCA as a consequence of substitution effects (not depicted)

The three approaches differ in the level of aggregation on which the attribution problem is tackled:

- *Process level*: allocation is made strictly on the process level as a bottom-up approach (No 1 in Figure 6-2). The attribution problem concerning recycling is resolved by considering recycling a bi-functional process, serving for waste treatment as well as for secondary material production. All other processes are allocated according to procedures applicable to coproduction. This level can only be used for descriptive LCA (see Chap. 6.2);
- *Product system level*: the product system under study, a multifunctional system that provides the function related to the product system (No 2 in Figure 6-2). The product system serves for waste processing for the previous life cycle, but it profits from the occurring primary material production. On the other hand, it delivers secondary material to subsequent product systems as a source for secondary material but profits from the waste processing function of subsequent systems. Allocation procedures on the product system level are based on substitution and are applicable for descriptive LCA, allowing partial attribution of processes not directly associated to the product system under study. They can also be applied in marginal LCA allowing the substitution of whole processes (Chap. 6.3);
- *Whole-system level*: the whole cascade of product systems (the 'material life cycle'), a multifunctional system, unifying all functions of the cascade of product systems (No 3 in Figure 6-2). This approach solves the attribution problem related to recycling based on the top-down approach. Allocation procedures on whole-system models can be used in descriptive LCA by avoiding allocation with system expansion[246] and in marginal LCA, allowing partial attribution of processes (Chap. 6.4).

The choice of an approach certainly depends on the goal and scope of the study, respective to the object being assessed. A product assessment will possibly require a different approach than assessing a regional waste management system. Quick-scan evaluation during product design will require a different approach than a product assessment in an academic context.

Moreover, the decision-maker's mental model of the organisational principle of the socio-economic system, his/her mental model of range of the decision-maker's responsibility and his/her attitude toward risk related to the ignorance on future behaviour of the socio-economic system are manifested in the chosen approach. The choice can be interpreted as the decision-

[246] by adding additional functions to the system under study; this has to be distinguished from the 'avoided-burden' approach that is also subsumed under 'system expansion' by ISO/EN 14'041.

maker's conception of the role of environmental information for sustainable development for him and for all other actors within the socio-economic system.

This book advocates a further criterion for the choice of the approach for the consistent modelling of recycling in LCA: the consideration of material and market characteristics of the materials involved in a product system and the corresponding management rules for sustainable management (see Chap. 3.4).

The three approaches are described more in detail and discussed in the following chapters.

6.2 Approaching recycling on the process level

Examining recycling on process by process level follows a 'bottom-up' approach. This approach is based on the assumption that if the attribution problem is solved specifically for each process, the attribution problem is solved for the whole cascade of product systems and for the product system under study in particular[247].

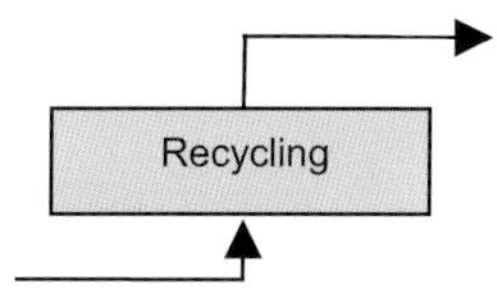

Figure 6-3. Solving the attribution problem related to recycling on process level

Looking at the attribution problem related to reuse & recycling on the process level reduces the attribution problem to the allocation of the recycling process (Figure 6-3). Other processes related to reuse & recycling – such as the primary material production as a primary source of the recycling material – are attributed to the system first using the primary material; environmental interventions of final disposal at the end of the cascade are attributed to the product system where the material is disposed.

If the attribution problem is handled on process level ('strict coproduct allocation'), the allocation of the environmental interventions of the recycling process can be made using allocation procedures developed for multifunctional (multi-output) processes. In this case, the two functions to be

[247] see, *e.g.*, Huppes 1994:75

weighted for the determination of the allocation factor are the function 'waste disposal' and 'provision of secondary material' . As these two functions are in fact inseparably joined, only economic allocation factors are applicable[248,249].

Huppes (1994), Lindeijer and Huppes (2001) and Guinée *et al.* (2004) propose to consider the recycling process as a waste treatment process until the negative market price of the material in process crosses the zero-point and becomes positive; all the processing with positive market price of the material is considered secondary material extraction that is attributed to the following product system (third option in Figure 6-4)[250].

Several situations for the setting of system boundaries and allocation can be distinguished, depending on the market price of the post-consumer material and the market price of the material after recycling/upgrading (Fig. 6-4):

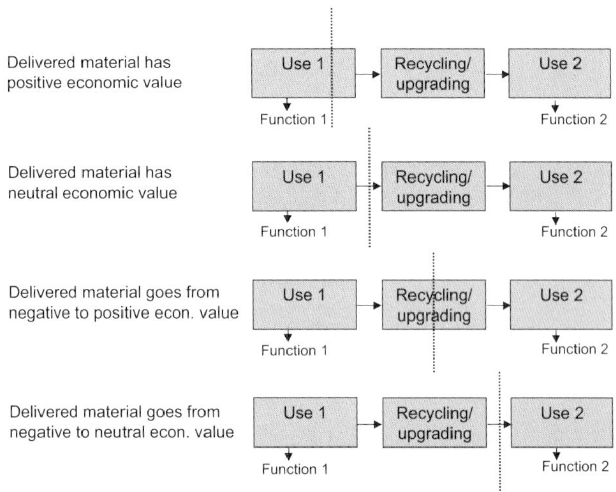

Figure 6-4. System boundaries and allocation according to the strict coproduct allocation (after Lindeijer and Huppes 2001:689, modified; reprinted with permission from the authors)

This approach becomes problematic if the material to be recycled has a positive market price (the first option in Figure 6-4), whereby no zero-point can be determined. In these cases and strictly applying the above favoured allocation rule for coproduction, the use phase has to be considered a joint

248 see Huppes 1992:64

249 Mass-based allocation would result in a 50/50-allocation, corrected by the yield of the recycling processes. It is doubtful if mass is an adequate common parameter for the description of the functionality of waste disposal and secondary material extraction.

250 Huppes 1994:84f and Lindeijer and Huppes 2001:309ff

process requiring allocation between the fulfilling of the function of the product and the provision of secondary material for the next life cycle[251]. This assumption can interfere with the decision-maker's mental model of the socio-economic system and its causalities.

Although a positive market price of material (*in situ*!) to be recycled is not very common nowadays, it is assumed it shall become more and more common, as 'waste' flows are more reintegrated into the material and energy flows of the socio-economic system as valuable sources of resources.

Important determinants of this approach are governmental waste management policies. Governmental interventions on waste flows can strongly influence waste and recycling material flows, thus having a large impact on the cost/price structure of recycling processes.

The above considerations are only valid for descriptive LCA, as the two functions of the recycling process of treating waste and providing secondary material are inherently joined. Thus, substitution processes will occur elsewhere in the socio-economic system; the recycling process itself is a precondition of these substitutions. Attribution rules for marginal LCA are subsumed under the following two levels.

Note that the second option in Figure 6-4 provides the same result as the cut-off procedure. The cut-off procedure solves the allocation problem by system boundary setting, excluding the recycling processes from the system.

The data requirements for this approach are limited to technical data on the recycling process as well as to data required for the determination of the allocation factors (*e.g.*, cost/price structure of the recycling process).

6.3 Approaching recycling on product system level

The examination of a single product system lies between the bottom-up approach of considering recycling on a process-by-process basis (Chap. 6.2) and the top-down approach considering the whole cascade (Chap. 6.4).

For descriptive LCA, the examination of recycling on the product system level allows one to apply the asset approach outlined in Chapter 5.8.3 without having to define a whole (possibly hypothetical) cascade of product systems. Environmental interventions of processes not directly linked to the product system can still be attributed (Figure 6-5). Such processes are, *e.g.*, primary material production of the secondary material used for production or final waste disposal at the end of the cascade. Substitution processes are assumed to take place only within the (unspecified) cascade of product systems, with which the product system under study is associated.

[251] reconfirmed by Huppes 2000b, personal communication

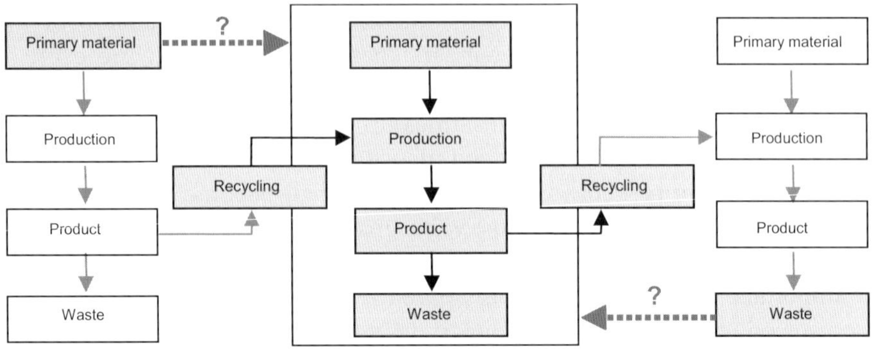

Figure 6-5. Solving the attribution problem related to recycling on product system level assuming substitution within the (unspecified) cascade of product systems

For descriptive LCA, substitution on a process-by-process level or partial substitution of various processes can be assumed. In descriptive LCA, substitution on a process-by-process level is, *i.e.* the basis for the 'closed-loop' procedure where recycling material is assumed to substitute the same amount of primary material (see below).

Substitution of processes can also be assumed to be partial. This means that the substitution effect is not calculated on a mass basis but corrected by a factor reflecting the change of the material's inherent properties of secondary material. The changes in a material's inherent properties usually come along with a change in its functionality and applicability, changed market price and thus with changed utility[252]. The changes in a material's inherent properties are one of the main reasons for the limited service life of a material. They determine the moment when a material is supplied for final disposal (*e.g.*, waste paper fibres too short for recycling).

In such cases, addressing the changes of the material's inherent properties is of crucial importance. These changes serve for the determination of an allocation factor, based on which the environmental interventions of primary material production and/or recycling and/or final disposal can be attributed to the system under study.

If recycling is treated at the product system level in descriptive LCA, substitution as an allocation principle is commonly applied. Several situations can be distinguished, differing in the scale of changes in the material's inherent properties:

[252] Heintz and Baisnée 1991:46 stress the importance of defining 'equivalency factors' for secondary material compared to primary material, referring implicitly to the functionality of secondary material compared to primary material.

- The closed-loop situation, where the material's inherent properties are not changed over the considered product system, and where the material is (or is assumed to be reused) in the same application. In this case, the closed-loop allocation procedure according to ISO/EN 14041 outlined in Chapter 5.8.4 is applicable. It is assumed that the substitution of primary material is complete. Thus, no environmental interventions from primary material production or final disposal are allocated to the system;
- The open-loop approach, assuming a closed-loop situation where the material's inherent properties are changed and where the material is (or is assumed to be reused) in the same application. In this case, the changed material properties are disregarded or considered irrelevant. Recycling is dealt with like a closed-loop situation by introducing an unspecified, homogenous material pool from which secondary material is taken from and to which material to be recycled (or recycled material) is re-exported[253];
- The open-loop situation, where the material's inherent properties are changed, and where the material is (or is assumed to be) reused in other applications. In these cases, the substitution of primary material is assumed to be only partial. Thus, environmental interventions related to primary material production and final disposal have to be partially allocated to the system under study based on a parameter describing the devaluation of the material caused by the changes in the material's inherent properties[254].

For practical reasons[255], the following parameters can be used for the description of the changes of a material's inherent properties:

1. Physical, chemical, biological or technical characterisation, *e.g.,* fibre length, length of molecules, content of impurities, elasticity, corrosion resistance, etc.;
2. Economic parameters such as price.

Both types of parameters cause methodological problems.

Ad 1) Technical functionality of a material is multi-criterial. On the one hand, more than one technical parameter can determine the material's inher-

[253] see, *e.g.*, Tillman *et al.* 1994:25; ISO/TR 14'049, chap. 8.3.2; see also Chap. 5.8.6, Chap. 9.5.3 and Chap. 11.2

[254] This is the starting point for the development of the value-corrected substitution described in Chap. 10 (see also Huppes 2000a; Werner and Richter 2000a; Werner and Richter 2000b)

[255] It is conceivable to define a parameter for 'functionality' or 'potential utility' of its own and to use it for the description of the inherent material properties. For the operationalisation of this multi-criterial parameter, reference has to be made at least partially to physical-chemical or economic parameters.

ent properties and thus its functionality, *e.g.,* in the case of alloys. On the other hand, the relevant physical-chemical parameters can change over the whole cascade of product systems. This can lead to arbitrary allocation factors not representing the decision-maker's criteria for material selection.

Ad 2) On the other hand, economic parameters contain more than technical aspects; they also reflect the supply-and-demand situation within the socio-economic system in which they are formed and are thus time- and site dependent. The use of supposedly more 'objective' physical-chemical parameters for allocation cannot hide that often the choice of a physical-chemical parameter is subjective and is made in relation to the functionality within the socio-economic system. This gets obvious if not SI-units are used for the physical-chemical characterisation but function-related units like elasticity, fat content, etc. Huppes (1992) concludes:

> "Function and value are quite intricately related. The value, quantified in financial terms as price in a market, is always based on the function a good or service has for the one acquiring it. The function will always have a material aspect to be expressed in SI-terms, at least in production and waste management. So there is not a choice between either the physical, the functions or the value approach" (Huppes 1992:61).

Substitution on a process-by-process level is also feasible for marginal LCA if it is assumed marginal processes are to be substituted. On the one hand, substitution can be assumed to affect only processes associated with the (undefined) cascade of product systems, *e.g.,* marginal primary material production processes or marginal waste treatment or recycling technologies. Following this 'avoided-burden' approach, all allocation procedures based on substitution are, in principle, also applicable in marginal LCA.

On the other hand, substitution effects can also be assumed to take place in product systems not physically linked to the material and energy flows of the product system under study. Assuming substitution of marginal processes throughout the socio-economic system on a process-by-process basis is favoured, *e.g.*, by Weidema (1999) (see Chap. 6.5.4).

For marginal LCA, partial substitution of processes can only be made assuming 'arbitrary' allocation factors, as is made with the 50/50-procedure where equal price elasticities of primary material and secondary material are assumed as a basis for allocation (see Chap. 6.5.4). Determining the effective allocation factors for partial substitution of processes in marginal LCA requires whole-system models (see below).

6.4 Approaching recycling including all product systems in a cascade

Recycling can also be examined by modelling the whole cascade of product systems. If the whole cascade of product systems is considered, all environmental interventions related to material extraction, recycling, and disposal can theoretically be attributed unambiguously and completely to the single functions (or product systems) provided by the cascade (top-down approach).

Figure 6-6 shows the life cycle of a material over several product systems with all it's functions provided.

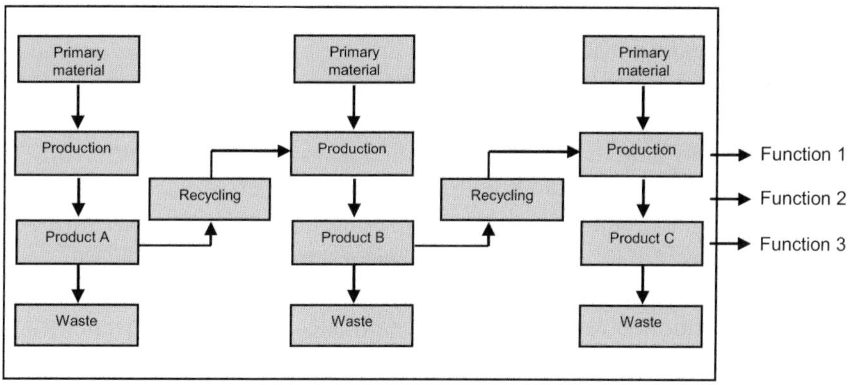

Figure 6-6. Cascade of product systems as a multifunctional whole-system

If the attribution problem related to recycling is tackled by considering the entire cascade of product systems as a multifunctional whole-system, then some of the procedures can be applied that have been developed for the allocation of coproduction or multi-output processes respectively.

For descriptive LCA, the attribution problem can be solved by:

- Avoiding allocation by system expansion[256] and thus inventorying and comparing multifunctional systems;
- Allocating the whole-system partially to all the functions provided by it, *e.g.*, by applying revenues of the functions as an allocation factor.

For marginal LCA, attribution can be made considering the effects induced to the whole-system model by marginal changes. Propositions have

[256] The avoided burden approach, also subsumed under 'system expansion', does not lead to a whole-system; subdivision of processes as a second option to avoid allocation according to ISO/EN 14'041, chap. 5.2.1 is in contradiction to examining the whole cascade.

been made to model whole-systems approaches based on physical-chemical determinations of material and energy flows as well as using economic data (elasticities):

- Allocation based on mathematical models, depicting changes of the whole-system induced by marginal changes of single parameters. This approach is based on linear programming. It has been proposed by Azapagic and Clift (1994/1999b)
- Allocation based on whole-system models, using price elasticities for the determination of the effects of marginal changes in a pre-defined whole-system model, proposed by Ekvall (2000)

So far, no attempt has been made in the context of LCA to include product systems into whole-system models which are not physically linked with the product system under study.

As a precondition of the whole-system approach, the whole cascade of product systems or the whole net of substitutions has to be definable; corresponding data has to be available.

The whole-system approach is favoured in cases where results with 'scientific rigour' should be achieved by avoiding subjective decisions as much as possible. In fact, considering the whole cascade allows for the attribution based on effects, while a mental model of the organisational principle of the socio-economic system is applied during the setting up of the whole-system. The above-mentioned procedures assume physical-chemical causation for an LP-model (Chap. 5.7.1) and economic causation for a model based on elasticities (Chap. 5.7.2).

6.5 Description of allocation procedures for open-loop recycling

The allocation procedures described in this chapter are based on differently structured product systems, which considerably impedes their comparability. For instance, some of the allocation procedures are based on a cascade of two product systems, whereas others are designed for three or more product systems in a cascade. Furthermore, they will differ if primary material is also introduced into subsequent product systems.

The following description of allocation procedures encountered in literature is structured according to the structure on the systems on which they are applied. Note the way the allocation procedures are described is applicable for descriptive LCA.

Allocation procedures relying on assumptions on the substitution of processes are marked with an **S** in the first row. These allocation procedures can also be applied for marginal LCA, using marginal data instead of average data. Some additional allocation procedures, which are exclusively applicable for marginal LCA, are described in Chapter 6.5.4.

The following colours and symbols are used for the characterisation of the allocation procedures described (Table 6-1):

Table 6-1. Level of treating recycling; symbols and their description used for the classification of allocation procedures used in Chap. 6.5.1 to Chap. 6.5.4

Process level	Product system (and associated processes)	Whole-system
No processes other than those directly related to the product system L_i are subject to allocation: V_i W_i R_{i-1} and/or R_i	Primary material production processes are shared by the product system using it, including all downstream systems. Waste processes are shared by all product systems upstream. For product system L_i the following processes are subject to allocation: all $V_{n \geq i}$ all $W_{n \leq i}$ R_{i-1} and/or R_i with n = total number of product systems in cascade	All the processes related to recycling are distributed to all product systems (n = total number of product systems) in the cascade. For product system L_i the following processes are subject to allocation: all V_n all W_n all R_n all m_n, f_n, q_n with n = total number of product systems in cascade m_i = mass of material i f_i = function of material i q_i = "quality of material i

Further symbols used:

S Procedure based on substitution

● Non-additive allocation procedure, not following the 100%-rule

Note that the procedures described refer to the cascading of *materials*. Open-loop recycling of energy or material cascades with final energetic use of a material cannot *a priori* be handled with these procedures, *e.g.*, if masses are used in an allocation procedure. In these cases, additional reasoning is necessary to adjust allocation procedures to the specific open-loop situation.

For the inclusion of energy in allocation procedures developed for materials, cut-off procedures, the avoided-burden approach or allocation procedures using economic allocation factors can solve the problems. Pure energy cascades can be addressed by considering, *e.g.*, the amount of energy, exergy or energy cost for descriptive LCAs[257] or with LP-modelling, or by considering price-elasticities (see Chap. 6.5.4).

[257] see, *e.g.*, Lindfors *et al.* 1995b, Technical Report No 7:16f

6.5.1 Two product systems without the import of primary material into the second product system

Early propositions for the allocation of open-loop recycling can be found in several SETAC-documents. They refer to a cascade of a two product system where recycled material completely substitutes primary material production of the second product system (Fava *et al.* 1991:79ff).

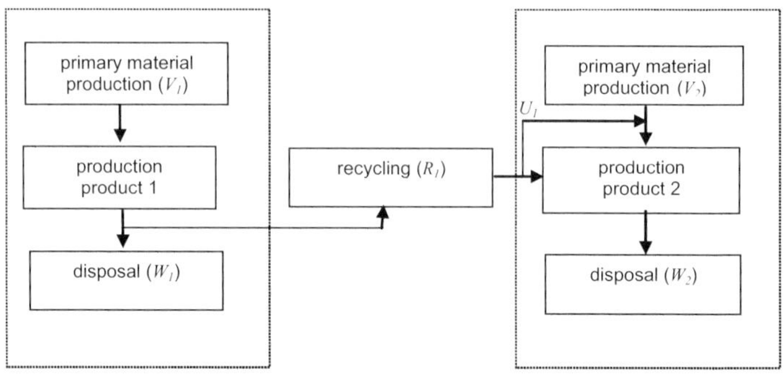

Figure 6-7. Simplified illustration of processes and material flows of a cascade of two product system (capital letters stand for the effective environmental interventions of a process)
U_1: upgrading; all expenditures necessary for allowing the use of secondary material instead of primary material

Table 6-2. Description of allocation procedures for a cascade of two product systems without input of primary material other than in the first product system

	Normative assumptions	Equations	Remarks
	Both product systems are equally necessary to enable recycling. Environmental interventions caused or omitted by recycling are equally distributed	$L_1 = L_1 - ½ (W_1 + V_2 - R_1 - U_1)$ $L_2 = L_2 - ½ (W_1 + V_2 - R_1 - U_1)$	50/50-procedure for two product systems; after Fava *et al.* 1991:79ff; see also Chap. 6.5.4
S	Each product system is attributed the environmental interventions *directly* caused by that system. The material to be recycled is considered raw material for the following product system. Recycling substitutes primary material production and avoids waste treatment.	$L_1 = L_1 - W_1$ $L_2 = L_2 - V_2 + R_1 + U_1$	Cut-off procedure for two product systems; after Fava *et al.* 1991:79ff; Ekvall and Tillman 1997:160. Operationally, the cut-off procedure solves the allocation problem related to recycling at the process level by the setting of system boundaries

continued

Table 6-2 continued

Normative assumptions	Equations	Remarks
Environmental interventions that are avoided and additionally caused due to recycling (except for the process of upgrading) are shared, based on the relative mass of the products produced in the cascade.	$L_1 = L_1 - (R_1 - V_2 + W_1) * \left(\frac{m_1}{m_1 + m_2}\right)$ $L_2 = L_2 + U_1 - (R_1 - V_2 + W_1) * \left(\frac{m_2}{m_1 + m_2}\right)$ with m_i = mass of product i	Franklin-methodology for two product systems, proposed by Boguski *et al.* 1994:152f

U_1: upgrading; all expenditures necessary for allowing the use of secondary material instead of primary material
Note: capital letters stand for the effective environmental interventions of a process

Such a situation is illustrated in Figure 6-7 and Table 6-2. Life cycle stages that are not relevant for the modelling of recycling are omitted. Note that a collection and recycling yield of 100% is assumed in the model.

6.5.2 Three product system with a single import of primary material into the first product system and final waste disposal after the last

Figure 6-8 and Table 6-3 describe allocation procedures applicable to a cascade of three systems. Primary material is only imported to the first product system, whereas final waste treatment occurs only in the third product system. Life cycle stages that are not relevant for the modelling of recycling are omitted.

Note that a collection and recycling yield of 100% is assumed in the models.

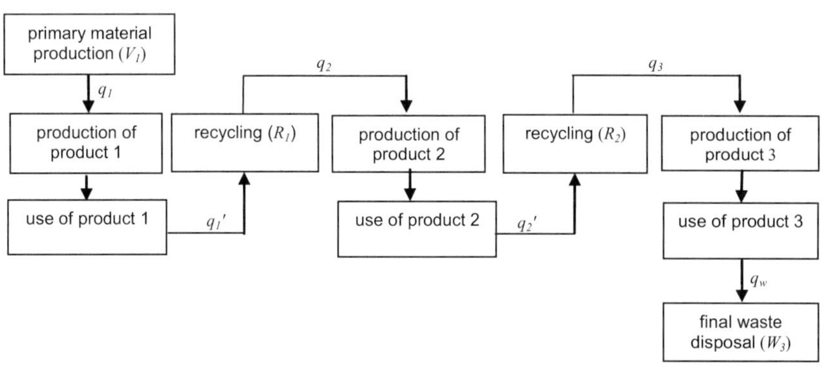

Figure 6-8. Simplified illustration of processes and material flows of an ideal cascade of three product systems with a single import of primary material into the first product system
Note: capital letters stand for the effective environmental interventions of a process

Table 6-3. Description of allocation procedures for a cascade of three product systems with a single import of primary material into the first product system (various references)

	Normative assumptions	Equations	Remarks
	Final waste disposal is considered an unavoidable consequence of primary material extraction.	$L_1 = V_1 + W_3$ $L_2 = R_1$ $L_3 = R_2$	Proposed by Östermark and Rydberg 1995, quoted after Ekvall and Tillman 1997:159
S	Both product systems linked by a recycling process are equally necessary to enable recycling. Waste treatment and primary material production are avoided with recycling,	$L_1 = V_1 + W_1 - \left(\frac{V_2 + W_1 - R_1}{2}\right) = \left(\frac{V_1 + W_3 + R_1}{2}\right)$ $L_2 = V_2 + W_2 - \left(\frac{V_2 + W_1 - R_1}{2}\right) - \left(\frac{V_3 + W_2 - R_2}{2}\right) = \frac{R_1 + R_2}{2}$ $L_3 = V_3 + W_3 - \left(\frac{V_3 + W_2 - R_2}{2}\right) = \frac{V_1 + W_3 + R_2}{2}$ Note: V_2 and V_3 as well as W_1 and W_2 do not show up in Figure 6-8. They are equivalent to V_1 and W_3 respectively	The 50/50-procedure for three product systems proposed by Fava *et al.* 1991:79ff; after Schneider 1996:267; recommended by Lindfors *et al.* 1995a:63 for 'key issue identification'
S	Each product system is attributed the environmental interventions *directly* caused by that system. The material to be recycled is considered raw material for the following product system. Recycling substitutes primary material production and avoids waste treatment.	$L_1 = V_1$ $L_2 = R_1$ $L_3 = R_2 + W_3$	Cut-off procedure for three product systems; after Fava *et al.* 1991:79ff; Ekvall and Tillman 1997:160; also Frischknecht 1994:127
S	Secondary material is a valuable resource. Primary material is needed to make this resource exploitable. Environmental interventions of primary material production are shared depending on quality losses. Recycling processes substitute waste processes.	$L_1 = \frac{q_1 - q_2}{q_1} * V_1 + R_1$ $L_2 = \frac{q_2 - q_3}{q_1} * V_1 + R_2$ $L_3 = \frac{q_3}{q_1} * V_1 + W_3$	Proposed by Karlsson 1994:24 Note that the quality of fractions *after* recycling is considered in this approach.

continued

Table 6-3 continued

	Normative assumptions	Equations	Remarks
	Each product system is attributed the environmental interventions *directly* caused by that system. The environmental interventions of the post consumer recycling processes are allocated to the system so long as the recycling material has a negative market price. Every additional process (market price of the recycling material > 0) is allocated to the system using secondary material as secondary material production processes.	$L_1 = V_1 + R_{1\,(p<0)}$ $L_2 = R_{1(p>0)} + R_{2\,(p<0)}$ $L_3 = R_{2\,(p>0)} + W_3$	A strict coproduct allocation procedure by Huppes 1994:74; see also Schneider 1996:256 Note that scrap and other waste prices can be very volatile and distorted by governmental waste-policy interventions (see Schneider 1994:42, also Sirkin and ten Houten 1994:222).
●	All product systems depend on primary material production. The environmental interventions of primary material production are allocated to the product systems in the cascade based on the quality reduction over the product system. Recycling processes are secondary material production processes; waste processes are attributed to the product system directly involved.	$L_1 = \frac{q_1 - q_{1'}}{q_1} * V_1$ $L_2 = \frac{q_2 - q_{2'}}{q_1} * V_1 + R_1$ $L_3 = \frac{q_3}{q_1} * V_1 + R_2 + W_3$	Proposed by Brouwer and Lindeijer 1993, quoted after Schneider 1996:247 Note that the quality of fractions *before* recycling is considered in this approach. This term is used for describing 'inappropriate use' in Lindeijer 1994 (see below). This procedure is not additive and thus not compliant with the 100%-rule
S	Secondary material is a valuable resource. Primary material and final disposal are needed to make this resource exploitable. Environmental interventions of primary material production are shared depending on quality losses. Recycling processes substitute waste processes.	$L_1 = \frac{q_1 - q_2}{q_1} * (V_1 + W_3) + R_1$ $L_2 = \frac{q_2 - q_3}{q_1} * (V_1 + W_3) + R_2$ $L_3 = \frac{q_3}{q_1} * (V_1 + W_3)$	Proposed by Wenzel *et al.* 1996; quoted after Ekvall and Tillman 1997:159

continued

Table 6-3 continued

	Normative assumptions	Equations	Remarks
S	Recycling substitutes primary material production and avoids waste treatment. The difference is allocated to the product system supplying material to be recycled.	$L_1 = V_1 - V_2 + R_1 = R_1$ $L_2 = V_2 - V_3 + R_2 = R_2$ $L_3 = V_3 + W_3 = V_1 + W_3$ Note: V_2 and V_3 do not show up in Figure 6-8. They are equivalent to V_1	Proposed by Karlsson 1994; Fleischer 1994:61ff; Klöpffer 1996; this procedure corresponds to a proposition by Östermark and Rydberg 1995
S ●	Material is considered to run an eternal material closed-loop, *i.e.* open-loop recycling is considered closed-loop.	$L_1 = R_1$ $L_2 = R_1$ *or* $L_2 = R_2$ *if* $R_2 > R_1$ $L_3 = R_2 + W_3$	Proposed by Lübkert 1991; quoted after Schneider 1996:245; this procedure is non-additive and thus not in line with the 100%-rule.
	Environmental interventions avoided and additionally caused with recycling are shared with the products produced in the cascade on relative mass basis.	$L_1 = (V_1 + R_1 + R_2 + W_3) * \left(\frac{m_1}{m_1 + m_2 + m_3}\right)$ $L_2 = (V_1 + R_1 + R_2 + W_3) * \left(\frac{m_2}{m_1 + m_2 + m_3}\right)$ $L_3 = (V_1 + R_1 + R_2 + W_3) * \left(\frac{m_3}{m_1 + m_2 + m_3}\right)$ with m_i = mass of product *i*	Franklin-methodology for three product systems, proposed by Boguski *et al.* 1994:153ff
	Environmental interventions avoided and additionally caused with recycling are shared with the products directly associated to the recycling processes on a mass basis.	$L_1 = V_1 + W_1 - \left(\frac{m_1}{m_1 + m_2}\right) * (V_2 + W_1 - R_1)$ $L_2 = V_2 + W_2 - \left(\frac{m_2}{m_1 + m_2}\right) * (V_2 + W_1 - R_1) - \left(\frac{m_2}{m_2 + m_3}\right) * (V_3 + R_2 + W_2)$ $L_3 = V_3 + W_3 - \left(\frac{m_3}{m_2 + m_3}\right) * (V_3 + R_2 + W_2)$ with m_i = mass of product *i*	Proposed by Vigon *et al.* 1993; after Schneider 1996:269
	Primary material production, final disposal and the recycling processes are equally necessary to enable the different functions of the material over the whole cascade of product systems.	$L_1 = \left(\frac{f_1}{f_1 + f_2 + f_3}\right) * (V_1 + R_1 + R_2 + W_3)$ $L_2 = \left(\frac{f_2}{f_1 + f_2 + f_3}\right) * (V_1 + R_1 + R_2 + W_3)$ $L_3 = \left(\frac{f_3}{f_1 + f_2 + f_3}\right) * (V_1 + R_1 + R_2 + W_3)$ with f_i = function of material *i* or product *i*	Described in Ekvall and Tillman 1997:159

continued

Table 6-3 continued

Normative assumptions	Equations	Remarks
Primary material production is distributed to all product systems depending on the number of subsequent product systems.	$L_1 = \frac{V_1}{n}$ $L_2 = \frac{V_1}{n} + R_1$ $L_3 = \frac{V_1}{n} + R_2 + W_3$ with: n = number of subsequent uses of the material	Based on ISO/TR 14049 chap. 8.3.3, where a more complex example is presented.
If in a product system a higher material grade ('material quality') than necessary is used, this grade is no longer available for applications requiring this grade.	$L_1 = \left(\frac{q_1}{q_1+q_2+q_3}\right) * V_1$ $L_2 = \left(\frac{q_2}{q_1+q_2+q_3}\right) * V_1 + R_1$ $L_3 = \left(\frac{q_3}{q_1+q_2+q_3}\right) * V_1 + R_2 + W_3$	Proposed by Lindeijer 1994:29, see also Schneider 1996:251 This procedure relies on the mental model of 'appropriate fit' proposed by Sirkin and ten Houten 1994.
The environmental relevance of 'inappropriate fit' and 'inappropriate use' are to be weighted equally.	$L_1 = \left(\frac{1}{2}\right) * \left(\frac{q_1 - q_2}{q_1}\right) * V_1 +$ $+\left(\frac{1}{2}\right) * \left(\frac{q_1}{q_1+q_2+q_3}\right) * V_1$ $L_2 = \left(\frac{1}{2}\right) * \left(\frac{q_2 - q_3}{q_1}\right) * V_1 +$ $+\left(\frac{1}{2}\right) * \left(\frac{q_2}{q_1+q_2+q_3}\right) * V_1 + R_1$ $L_3 = \left(\frac{1}{2}\right) * \left(\frac{q_3}{q_1}\right) * V_1 +$ $+\left(\frac{1}{2}\right) * \left(\frac{q_3}{q_1+q_2+q_3}\right) * V_1 +$ $+R_2 + W_3$	Proposed by Lindeijer 1994; a combination of methodological propositions made by Lindeijer 1994 and Brouwer and Lindeijer 1993; see also Schneider 1996:254 This allocation procedure is non-additive and is thus not in line with the 100%-rule.

The parameter q_i stands for 'material quality'.
Note: capital letters stand for the effective environmental interventions of a process.

6.5.3 Three product system with importation of primary material and waste disposal in all product systems of the cascade

The following Figure 6-9 and Table 6-4 describe allocation procedures applicable to a cascade of three systems. Primary material is used in all product systems; each product system generates waste to be disposed of. Life cycle stages that are not relevant for the modelling of recycling are omitted.

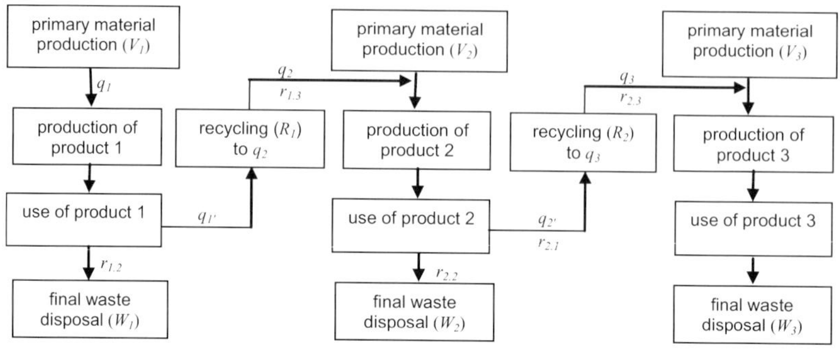

Figure 6-9. Simplified illustration of processes and material flows of an ideal cascade of three product systems with the import of primary material and waste disposal in all product systems
Note: capital letters stand for the effective environmental interventions of a process
The parameter q_i stands for 'material quality'; the parameters r_{i1} stands for relative amount of material going to recycling compared to total input; the parameters r_{i2} stands for relative amount of material going to final waste compared to total input; the parameters r_{i3} stands for relative amount of material delivered to the next product system final waste compared to total input

Table 6-4. Description of allocation procedures for a cascade of three product systems with importation of primary material and waste disposal in all product systems (various references)

	Normative assumptions	Equations	Remarks
S	Each product system is attributed the environmental interventions *directly* caused by that system. The material to be recycled is considered raw material for the following product system. Recycling substitutes primary material production and avoids waste treatment.	$L_1 = V_1 + W_1$ $L_2 = V_2 + R_1 + W_2$ $L_3 = V_3 + R_2 + W_3$	Cut-off procedure for three product systems; after Fava *et al.* 1991:79ff; Ekvall and Tillman 1997:160; Frischknecht 1994:127. Buhé *et al.* 1997 for cascades with two product systems.
S	Each product system is attributed the environmental interventions *directly* caused by that system. Collection and compacting or the collected post-consumer material is considered waste treatment. Reprocessing of recycling material is considered secondary material production and allocated to the product using the secondary material.	$L_1 = V_1 + R_{1,1} + W_1$ $L_2 = V_2 + R_{1,2} + R_{2,1} + W_2$ $L_3 = V_3 + R_{2,3} + W_3$	A modification of the cut-off procedure. Klöpffer 1996 quotes the allocation procedure applied in the LCA-study on packaging materials by the Frauenhofer-Institute for UBA Berlin (Anonymous 1994b:31ff).

continued

Table 6-4 continued

	Normative assumptions	Equations	Remarks
●	Material used as input is held responsible for primary material production proportional to its 'quality' compared to primary material. The use of recycling material and its supply to other uses saves primary material and waste treatment. For the avoidance of primary material production a credit is given to both product systems involved on a half and half basis; for the waste treatment a credit is implicitly assumed in the allocation of the waste treatment processes. Recycling processes are caused by the product systems preceding and following them. Environmental interventions of recycling processes are partitioned on a half-and-half basis. Environmental interventions of waste treatment within the product systems considered are allocated based on the weighting factor 'material quality of the product'. For the material degradation over a product system, the environmental interventions related to waste treatment processes in the following product systems are attributed weighting accordingly.	$L_1 = V_1 - \left(\frac{1}{2}\right) * r_{1,1} * V_1 + \left(\frac{1}{2}\right) * R_1 + r_{1,2} * W_1 + r_{1,1} * \left(\frac{q_1 - q_2}{q_1}\right) * (W_2 + W_3)$ $L_2 = V_2 + \left(\frac{q_2}{q_1}\right) * V_1 - \left(\frac{1}{2}\right) * r_{1,1} * \left(\frac{q_2}{q_1}\right) * V_1 - \left(\frac{1}{2}\right) * r_{2,1} * \left(\frac{q_3}{q_1}\right) * V_3 + \left(\frac{1}{2}\right) * (R_1 + R_2) + r_{2,2} * \left(\frac{q_{2'}}{q_1}\right) * W_2 + r_{2,1} * \left(\frac{q_2 - q_3}{q_1}\right) * W_3$ $L_3 = V_3 + \left(\frac{q3}{q_1}\right) * V_2 - \left(\frac{1}{2}\right) * r_{2,1} * \left(\frac{q_3}{q_1}\right) * V_2 + \left(\frac{1}{2}\right) * R_2 + \left(\frac{q_{3'}}{q_1}\right) * W_3$	Proposed by Kim *et al.* 1997. The partition of the environmental interventions on a half-and-half basis corresponds to the 50/50-procedure. By giving a credit for supplying/using recycling material, the 100%-rule is violated. Within the logical frame proposed by Kim *et al.* (1997), the credit for avoiding primary material production does not have an analogy for waste treatment. Anyhow, a credit for avoiding waste treatment similar to the one for avoiding primary material production would still violate the 100%-rule. This procedure is not additive as the sum of all environmental interventions attributed is bigger the sum of the single environmental interventions.
S	Secondary material is taken from a homogeneous material pool and is recycled into the pool after the use of the product. Net output of secondary material substitutes primary material production.	$L_1 = V_1 + W_1 - r_{1,1} * V_1$ $L_2 = V_2 + R_1 + W_2 - r_{2,2} * V_2$ $L_3 = V_3 + R_2 + W_3$	Material pools procedure according to ISO/TR 14049, chap. 8.3.2

continued

Table 6-4 continued

	Normative assumptions	Equations	Remarks
	Each product system is attributed the environmental interventions *directly* caused by that system. The environm. interventions of recycling are allocated to the system as long as the recycling material has a negative market price. Every additional process (price of the recycling material > 0) is allocated to the system using secondary material.	$L_1 = V_1 + R_{1\,(p<0)} + W_1$ $L_2 = V_2 + R_{1(p>0)} + R_{2\,(p<0)} + W_2$ $L_3 = V_3 + R_{2\,(p>0)} + W_3$	A strict coproduct allocation procedure by Huppes 1994:74; see also Schneider 1996:256.
	Environmental interventions avoided and additionally caused with recycling are shared on the relative mass of the products produced in the cascade	$L_1 = V_1 + W_1 + \left(\frac{m_1}{m_1+m_2+m_3}\right) * \left(\sum_1^n R_i * r_i - V_i * R_{ci} - W_i * r_i\right)$ $L_2 = V_2 + W_2 + \left(\frac{m_2}{m_1+m_2+m_3}\right) * \left(\sum_1^n R_i * r_i - V_i * R_{ci} - W_i * r_i\right)$ $L_3 = V_3 + W_3 + \left(\frac{m_3}{m_1+m_2+m_3}\right) * \left(\sum_1^n R_i * r_i - V_i * R_{ci} - W_i * r_i\right)$ m_i = mass of product *i* n = number of product systems involved in the cascade R_{ci} = recycled metal content of product *i*	Franklin-methodology for three product systems, proposed by Boguski *et al.* 1994:153ff Upgrading is attributed to the system using the material. See also Newell and Field 1998 for the sensitivity of the procedure depending on n and R_{ci}
S	Over a product system, the material is normally 'downcycled'. Output material substitutes primary material production of following product systems, at least partially. Environmental interventions associated with this devaluation (primary material production and eventually final waste disposal) are allocated to the product system causing it.Recycling processes are dealt with depending on the allocation parameter chosen.	$L_1 = V_1 + R_1 + W_1 - r_{1,3} * \left(\frac{q_1 - q_2}{q_1}\right) * V_1$ $L_2 = V_2 + R_2 + W_2 + r_{1,3} * \left(\frac{q_1 - q_2}{q_1}\right) * V_1 - + r_{2,3} * \left(\frac{q_2 - q_3}{q_1}\right) * V_1$ $L_3 = V_3 + W_3 + r_{2,3} * \left(\frac{q_2 - q_3}{q_1}\right) * V_1$	The value-corrected substitution, proy-posed by Werner 2000; Werner and Richter 2000a; see also Huppes 2000a See also Chap. 10

Abbreviations, see Figure 6-9

6.5.4 Some more allocation procedures for marginal LCA

Some allocation procedures are exclusively developed for marginal LCA. These allocation procedures do not fit into the descriptive frame as used above. They are described separately in the following:

For marginal LCA, Weidema[258] favours the avoided-burden procedure for the allocation of coproduction and recycling. This procedure aims to identify substituting processes and technologies within the whole socio-economic system that are affected by a marginal change.

In the case of open-loop recycling, this procedure can possibly lead to extensive cascades of products and substituted processes:

"Even when the decision-maker is not able to influence directly any substitutions elsewhere in the life cycle (*i.e.* when most processes are in the background), the studied substitution at one stage in a life cycle in the foreground (processes whose production volume will be affected directly by the change studied; WF) may still lead indirectly to product substitutions in other life cycle stages (in the background)" (Weidema 1998b:3).

Price elasticities of primary material, secondary material, and possibly other substituting materials have to be considered for the determination of processes that will be substituted by a change related to recycling in the system[259]. This procedure is not based on a material cascade model but on an effect cascade model following substitution processes induced by a change as can be assessed in environmental terms.

For marginal LCA, another procedure for the handling of open-loop recycling has been proposed by Azapagic and Clift (1994/1999b) based on system expansion and linear programming, applied to the whole cascade of product systems as a multifunctional net of physical flows. Environmental interventions of indirect effects are attributed to single marginal changes of the system (see Chap. 6.4).

Ekvall (2000) has proposed another whole-system approach to marginal LCA: price elasticities for the determination of material flows within a pre-defined system. Environmental interventions of indirect effects are attributed to single marginal changes of the system induced by changes in demand or supply.

Ekvall (2000) proposes the 50/50-procedure (see above) as the default method if recycling flows are principally caused by governmental interventions. In this case:

[258] Weidema 1998b; Weidema *et al.* 1999; Weidema 1999

[259] Weidema 1999:43ff

"...we do not know if an increased supply of recovered material from a specific product or a specific location will replace 100% virgin material, 100% recycled material from other sources, or anything in between. ... "One way to deal with the uncertainty is to make the assumption that 50% of an outflow of cascade material from the life cycle investigated replace virgin material and that the remaining 50% replace recycled material from other sources. To be consistent, an increased inflow of cascade material to the life cycle investigated should then in the base case be assumed to result in 50% increased collection and 50% reduced use of cascade material in other products" (Ekvall 2000:104).

The underlying assumption is that price elasticities of virgin material and recycled material are equal[260].

As a final remark, it should be stated that some allocation procedures are developed for the evaluation of recycling (waste management options) and not for the environmental assessment of products, *e.g.,* in Schneider (1994/1996).

6.6 Choice of an allocation procedure for reuse & recycling

In view of the broad variety of proposed approaches to system definition in general and in particular allocation related to reuse & recycling, the question concerning the appropriateness of an allocation procedure arises. This question takes its importance from the fact that the choice of an allocation procedure for the integration of recycling in LCA can have a predominant effect on the outcome of a study[261].

On a long-term basis, data, time and human resource intensity of LCA have to be minimised if LCA should be established within a company's daily work as a tool for environmental product assessment and environmental product development (see Chap. 3.1.2). Furthermore, allocation procedures need designing which depict environmental improvement strategies and the options followed during product design in an adequate manner. Furthermore, no 'perverse' incentives should arise: assumptions underlying an allocation procedure may not contradict the decision-maker's mental models and values in a context-specific decision situation[262].

[260] see also Weidema 1999:43ff

[261] Such comparisons have been made, *e.g.*, by Werner 2000, Werner and Richter 2000b or Lindfors *et al.* 1995, Technical Report No 7, pp. 6ff.

[262] see Chap. 3.2.4 for the questions whose mental models and values shall be depicted in LCA

In view of the complexity of the object, simplified modelling (*e.g.*, applying the cut-off procedure or solving allocation problems related to recycling on process level) implicitly requires the use of subjective mental models or values. On the one hand, system expansion (the whole-system approach) avoids these subjective assumptions at first sight. On the other hand whole-system models can lead to extensive and unmanageable models that – in the sense of von Winterfeldt and Edwards (1986) – do not support a decision-making process[263]. But also whole-system models are shaped by subjective elements. Further mental models and value choices are applied if recommendations for a specific product system have to be made.

A priori, none of the three levels of approaching allocation of recycling can be excluded. The choice of an allocation procedure must be made considering the:

- Object to be assessed, especially its materials and their material and market characteristics in view of the management rules for sustainable development;
- Later use of LCA-results.

A whole-system approach might be suitable for materials, whereby a (short) material cascade can unambiguously be defined. This is especially the case in descriptive LCA if results should 'indirectly' influence decisions during product development by integrating them into recommendations or check-lists (see Chap. 3.1.2).

With metals for instance, allocation procedures on the process or product system level should be considered, as whole-system approaches would possibly result in an undefinable quantity of product systems (see case study on aluminium, Chap. 8ff).

Allocation procedures on the process or product-system level should be considered with priority for the direct (descriptive) evaluation of design alternatives to be as simple, transparent, instant, cheap, quick and 'adequate' as possible. Special emphasis has to be put on the allocation procedure for these applications, depicting material and market characteristics of the materials involved in view of their corresponding management rules for sustainable development. The point is also relevant for the question of how far the range of the decision-maker's responsibility should be extended.

Non-additive allocation procedures should generally be avoided as they *a priori* lead to double-counting or leave environmental interventions unallocated. An exception is the closed-loop procedure as proposed in ISO/EN 14041, chap. 6.5.4 or by Lübkert (1991). In closed-loop situations, the

[263] see Chap. 3.2.1f

(assumed) unlimited number of re-uses makes the share of primary material production and disposal to be allocated to one product system irrelevant.

According to surveys on future tendencies of LCA within product development, an increased direct application of LCA within the design process has to be assumed (see Chap. 3.1.2). Further research is needed to investigate how open-loop recycling of single materials or of a class of similar materials can adequately be modelled in consistency with their corresponding management rules for sustainable management. This issue will be investigated in the following two case studies.

Finally, the requirements on a product system that best support the process of efficient and effective decision-making should be remembered (see Chap. 3.4). The approach chosen when tackling allocation issues related to recycling (*i.e.* on process level, product system level or whole-system level) and the allocation procedure should be selected in accordance with these requirements.

Chapter 7

THE DECISION-MAKER'S MENTAL MODELS AND VALUES IN INVENTORY ANALYSIS

In the life cycle inventory step of LCA, unambiguous, scientifically based, objective attribution of material and energy flow to a product from the complexity of the socio-economic system is pure fiction. This is due to the fundamental epistemological conditions of LCA. It has been demonstrated in the previous chapters that various mental models and values guide the attribution of material and energy flows to a product under study in the life cycle inventory analysis of LCA.

Among the most important mental models influencing modelling in the life cycle inventory analysis are the:

- *Life cycle of a product itself*: primarily, the product system as life cycle model of a product is a mental model itself. The life cycle of a product is a theoretical construct that has to be made concrete during modelling in the life cycle inventory analysis;
- *The technosphere and ecosphere* as demarcations which are predefined by the assessment methods applied in the impact assessment step (see Chap. 5.2 and footnote 188);
- *The internal structure of the technosphere*, *i.e.* knowledge on processes and technologies, guiding the definition of modules, sub-modules and unit processes;
- *Material and market characteristics* of the materials and products involved;
- *The organisational principle of the socio-economic system*. This mental model is relevant whenever changes within the technosphere have to be modelled, *e.g.*, if substitution effects or marginal changes have to be depicted or if 'arbitrary' allocation factors have to be chosen;
- *The range of the decision-maker's responsibility in view of the management rules for sustainable development* defines the life cycle steps and

processes for which the decision-maker feels responsible. This mental model is especially relevant if reuse & recycling are modelled in the life cycle inventory analysis. It is also closely linked to the following point;
- *The role of environmental information for different actors for the achievement of sustainable development*, which possibly guides the choice of a marginal or descriptive LCA, but which is also closely linked to the mental model of the range of the decision-maker's responsibility.

The decision-maker's values involved in the definition of the product system and expressed as preferences are stated in the following decision situations:
- Attribution of material and energy flows to the product system in the sense of 'less is better';
- Level of relevance for cut-off (the percentage) below which 'irrelevant' inputs and outputs (mass, energy or environmental interventions) are cut off;
- Valuation of outputs (or functions) as coproducts of a multifunctional process, which are attributed environmental interventions, in contrast to 'by-products' or 'waste', which go free of environmental interventions;
- Valuation of the functionality of input and output materials when deciding on modelling closed-loop or open-loop recycling. This valuation is strongly influenced by the mental model of the organisational principle of the socio-economic system[264];
- Definition of material-specific management rules for a sustainable use of the materials involved, which is closely linked to the mental models of the material and market characteristics. This point affects the attribution of material and energy flows to the product system in a fundamental way;
- Modelling of changes within the socio-economic system, stating the decision-maker's attitude towards risk expressed in his/her temporal preferences. This is particularly relevant when modelling future reuse & and recycling options[265].

All these mental models and values guide the setting up of the product system as an attribution problem in LCA. These mental models and values have to be properly addressed and depicted in a product system if the results of an LCA should be relevant for a decision-maker. The decision situations encountered during the setting up of a product system, where these mental

[264] Functionality issues are of course also relevant in the goal and scope definition when defining equi-functional functional units.

[265] affects also the definition of service life assumed for different products during goal and scope definition

models and values are required, have been discussed in Chapter 5 and Chapter 6.

Of course, further mental models and values are involved when conducting an LCA during goal and scope definition as well as in the impact assessment step[266]. Further mental models cover, *e.g.*, the structure and behaviour of the ecosphere for modelling environmental (impact) categories.

Further values are stated as preferences in, *e.g.*, the choice of the temporal and financial scope of the LCA-study; the goal definition; the definition of an environmental impact assessment method – including the definition of environmental impact categories (criteria) and their weighting – as well as in the selection of a specific environmental impact assessment method. These mental models and values are related to decisions made outside the life cycle inventory analysis. Nonetheless, they influence modelling decisions in the life cycle inventory analysis in an indirect way.

A variety of methodological propositions have been made for solving the attribution problem and, more specifically, the allocation problem in LCA. An extensive view of methodological propositions related to reuse & recycling has been presented in Chapter 6. At first sight, this variety of options leaves the impression of an arbitrary and possibility of interest-guided biasing of LCA-results. On the other hand, these modelling options allow one to depict the decision-maker's mental models and values in a specific situation in the best-suited way.

These options allow one to suitably model the material and market characteristics of the materials and products involved, taking into account the limited 'descriptive' power of LCA models. Furthermore, they allow one to incorporate case-specific normative aspects of sustainable development related to the materials and products involved that are not covered by current life cycle impact assessment methods. Considering these aspects, LCA must be understood as a case-specific decision support tool and not as a tool delivering objectively right measures of the environmental relevance of products.

The following two sections contain two case studies where the methodological issues related to the modelling of recycling and other end-of-life options are discussed and solved for two different products made from different main materials.

The first case study (Chapters 8 to 13) treats the modelling of the recycling of an aluminium window where the maintenance of the functional properties (prevention of downcycling) over a product life cycle is considered the primary design strategy for a sustainable use of aluminium. The

[266] see also previous footnotes 264 and 265

second case study (Chapters 14 to 19) investigates the modelling of recycling or incineration of a beech wood railway sleeper.

These two materials are chosen because aluminium as a non-renewable material and wood as a renewable material play different roles in the discussion on sustainable development. The theoretical framework and the set of criteria for LCIs serve in problem recognition and reasoning; they can also be used as creative rules for addressing methodological problems related to modelling end-of-life options of the considered products.

IV.

CASE STUDY I: DEVELOPMENT OF THE VALUE-CORRECTED SUBSTITUTION FOR ALUMINIUM WINDOWS

Chapter 8

INTRODUCTION

The first case study deals with the LCA of aluminium window frames (see Chapter 1.1). This study illustrates how recycling of aluminium used in building applications should be integrated into product systems to meet the optimal requirements of a product system and its LCI, as stated in Chapter 3.4.

The value-corrected substitution (VCS) is outlined as an allocation procedure, which has been developed to best meet the requirements of LCA of aluminium products in the building sector[267].

The VCS is compared to traditional allocation procedures like the 'cut-off method' proposed by SETAC and others[268] (see Chap. 11.1), or the closed-loop procedure based on material pools described in ISO/TR 14049:2000, pp. 27ff (see Chap. 11.2).

A third approach, the strict coproduct allocation according to Huppes (1994:84f)[269] and Lindeijer and Huppes (2001) is discussed but not included into the calculations as it is not yet considered fully developed (see Chap. 11.3).

These allocation procedures are applied in an LCA of a fictitious aluminium window in two variations, one containing zinc-parts and the other containing no zinc. As different allocation procedures rely on different accounting principles, the system boundaries also differ for each of the allocation procedures. A sensitivity analysis is conducted on the inclusion and exclusion of dross treatment for each of the allocation procedures to investigate the influence of different system boundaries.

[267] Werner 2000; Werner and Richter 1998a; Werner and Richter 2000a; Werner and Richter 2000b

[268] see, *e.g.*, Fava and Denison 1991; Fava *et al.* 1991; Lindfors *et al.* 1995a

[269] called economic allocation by the authors

The suitability of the allocation procedures is discussed on an argumentative basis[270] referring to a) the set of requirements deduced in Chapter 3.4; b) to the market and material characteristics of aluminium, and c) to the strategies for its sustainable use described first in the following chapters (Chapters 9 and 10). This puts emphasis on the role of the decision-maker's mental models of the material and market characteristics of aluminium plus the decision-maker's value system for the choice of an allocation procedure.

[270] In this sense, reference is made to 'practical rationality' by Habermas 1987, which relies on substantiatability and intersubjective comprehensiveness.

Chapter 9

MATERIAL AND MARKET CHARACTERISTICS OF ALUMINIUM

Taking into account the market and material characteristics as well as the respective management rules for a sustainable use of the material has been claimed essential for meeting 'causality' in the attribution of material and energy flows to the product under study. The following chapters outline the material and market characteristics of aluminium, aluminium recycling and strategies for its sustainable use.

9.1 Characteristics of aluminium as material

Aluminium is the second most frequent metal in the earth's crust[271]. It is commonly gained from bauxite via alumina, an aluminium oxide, from which primary aluminium is gained by electrolysis (Hall-Heroult process).

Aluminium is used world-wide in a variety of applications utilising its low density, high strength-to-weight ratio, high thermal conductivity, high corrosion resistance and its electrical conductivity. Aluminium is also non-toxic, very malleable, and nonmagnetic[272]. Its applications range from the ground transport sector, packaging, aviation, building and construction to electrical equipment, lithographic or tooling plates[273].

Basically, two types of aluminium alloys are commonly distinguished: wrought alloys and casting alloys. As a rule of thumb, casting alloys tend to have higher contents of non-aluminium metals, especially of iron, silicon and/or zinc, whereas wrought alloys tend to have higher manganese or mag-

[271] Askeland 1996:362
[272] Askeland 1996:362f
[273] Woodward 1995:161

nesium contents. There may be exceptions to the above-stated rule for other alloy elements such as chromium, zirconium, and lithium[274]. The material properties of aluminium are also influenced by the crystallographic microstructure, mainly depending on the process parameters during melting and cooling.

As casting alloys tend to be made out of secondary material (scrap) for their higher alloy content, the products of the casting alloy industry ('refiners') are sometimes addressed as secondary aluminium. This may lead to confusion as wrought alloys can also be made out of remelted scrap - mainly home and industrial scrap - but also from sorted old scrap. The products of the wrought alloy producers ('remelters') are sometimes addressed as primary aluminium although using secondary material (scrap) for their production.

Intermediate products of aluminium have the characteristics of investment goods, serving as inputs in further production processes. They are homogenous mass products, dividable at will, which can be split into defined classes and sorts[275]. This makes them tradable on commodity exchanges such as the London Metal Exchange (LME; see also Chap. 10.3).

9.2 Characteristics of the supply side of primary aluminium

The bauxite price is a very small share of the price of aluminium. The rest of the aluminium price is caused by capital, labour and in particular[276] the energy costs.

Short-term supply elasticity is a very low factor in bauxite and aluminium processing facilities as capital, labour costs and the technology are considered almost quasi-fixed[277]. High capital intensity combined with gains of economies of scale has led to an oligopolistic and/or oligopsonic structure of the transglobally active bauxite-aluminium industry with a high degree of vertical integration[278,279].

[274] after Caron 1997

[275] after Engelhardt and Günter 1996:212ff

[276] Carlsen 1980; Legrand 1997:4; Slade 1980:233ff

[277] after Lindquist 1995:361f; see also Bird 1990:140ff

[278] Hojman 1980:290; see also Maizels 1984:25ff

[279] On the other hand, cartelistic efforts by the producer countries of bauxite have tried to counterweight the structural power of the bauxite-aluminium industry by the formation of the International Bauxite Association (IBA) (Hojman 1980). Market interventions to stabilise commodity prices such as cartelistic price-fixing or export quota agreements, buffer stocks, or compensatory financing systems are a quite usual phenomenon in the interna-

In the long run, by adjusting the capital structure the supply curve is optimised in the sense that any given output is produced at the lowest possible cost.

The largest primary aluminium producers have established a producers' price of primary aluminium. Compared to the quotations at the LME, this price is of low significance for the price forming process[280].

9.3 Characteristics of the supply side of secondary material

Aluminium has been recycled in Central Europe since the late twenties of the last century when the availability of recyclable aluminium became a commercial viability. Since then, the secondary sector has grown constantly (with some drawbacks during World War II and during the recent recession of the 1990's), being an integral part of the present day raw material supply[281].

Just as the primary industries mine from stocks of ore and ore quality is a determinant of primary industry cost, one can think of the secondary industries as mining from disperse stocks of scrap, and scrap quality and availability as determinants of secondary production cost. The price for primary aluminium can be considered the upper 'ceiling' for secondary aluminium, as in principle alloyed aluminium can also be produced from primary aluminium. Temporally, the difference in prices between primary aluminium and secondary aluminium can fluctuate and is sometimes close to the limit, whereby substitution of secondary aluminium occurs[282,283].

Carlsen (1980) has determined the cost of energy, the scrap prices relative to primary aluminium and the level of industrial production to be the

---

tional trade of primary commodities. On one side, these market interventions are favoured by third world countries heavily depending on export earnings, and on technical and structural input by the mining companies (Siebert 1984). On the other side, they are favoured by the demand side to guarantee availability and price stability of a commodity market to prevent inflation (Hermann *et al.* 1990). For many commodities, these market interventions have not been very effective on the long run (Hermann *et al.* 1990:283ff). See Hallett and Ramanujam 1990 for the role of futures markets as stabilisers of commodity earnings.

280 Werner 1990:72

281 after Krone 2000:7/19

282 see the year 1995 in Figure 10-3, where the prices for secondary aluminium equal the ones of primary aluminium

283 after Krone 2000:28f

main factors responsible for the recycling coefficients[284] of aluminium (for the U.S.).

Secondary producers are usually small- to medium-sized firms, who function as price takers in a highly competitive, atomistic market. The secondary-processing industry tends to be labour-intensive in comparison to the capital-intensive operations, which typify the primary processing industry. As the main part of aluminium is consumed in industrialised countries, the secondary aluminium industry is therefore mainly located in these countries[285].

Secondary (recycled) aluminium alloys are composed by remelters on demand out of varying types of scrap (and primary aluminium) in varying quantities depending on the alloys required by the market.

The supply sides of primary and secondary aluminium have to be considered highly insulated for technical reasons as different processing technologies are used.

> "This insulation has been weakening in recent years but it still holds in general. Thus it is possible to consider the primary aluminium industry on its own, without the complication of destabilising feedback which scrap recycling can produce" (Bird 1990:135)[286].

9.4 Demand of aluminium

Short-term demand elasticity of aluminium is low. The small share of the material cost compared to the total costs of an aluminium product does not lead to material substitution due to small price changes of aluminium[287]. Labour and aluminium-processing machinery have to be considered semi-fixed.

For instance, Banks (1979:83) does not expect the substitution of aluminium by copper due to small changes of the aluminium price. The long-term substitution of copper by aluminium will mainly depend on the depletion

[284] the ratio of recycled industrial and old aluminium scrap and total aluminium consumption

[285] after Krone 2000:11ff

[286] The influence of the primary aluminium industry on the secondary industry has been the topic of a famous verdict on ALCOA in the U.S. in 1945. As the supply of old scrap is determined by the use of primary material, a monopolistic supplier might influence the scrap market to maximise its monopolistic rents. The court judged affirmative in the case of ALCOA. As a consequence, ALCOA was divided into three different companies. This verdict has given birth to a long contradictory discussion among economists on the possibilities of a commodity monopolist to take influence on the secondary commodity market (see, *e.g.*, Grant 1999; Hollander and Lasserre 1988; Martin 1982; Swan 1980).

[287] Banks 1979:83

rate of high quality ore for copper and the development of the electricity prices for aluminium as the main price forming parameters of the two metals[288].

Long-term demand elasticity has to be considered higher. Consumption is strongly influenced by changes in industrial production, as a large share of aluminium is used in building construction and electrical machinery. The demand for these items is very sensitive to movements in the business cycle[289].

> "In other words, the key factor in the adjustment of aluminium consumption to changes in price is the size of the stock of aluminium-using machinery and the rate of its depreciation" (Banks 1979:91).

Nonetheless, technical and functional properties of copper, adaptation costs, lack of experience with aluminium, etc. have so far impeded the expected substitution of copper by aluminium. This has been the case despite the fact that the price ratio per weight, per conductivity and per volume has been mostly in favour of aluminium over the last 50 years[290].

A last point has to be mentioned relating to the demand function of aluminium. With regard to the material characteristics, it cannot be determined if an alloy is produced from post-consumer scrap, process scrap or primary aluminium, or a mixture of them all. As a consequence of non-distinguishability, demand functions of an alloy made from primary or from secondary aluminium must be equal.

9.5 Principles of a more sustainable use of aluminium

9.5.1 Paradigm: value preservation

Regarding the redirection of the current material flows within the technosphere into more sustainable paths, the closing of material cycles and the best possible value-preserving application of materials – *i.e.*, the least possible downcycling to prevent entropic losses – become key strategies for non-renewable materials.

Based on theoretical considerations, Messner (1999) formulates management rules for the sustainable use of non-renewable resources, referring particularly to entropic losses (see also Chap. 3.3.3):

[288] Slade 1980:233f
[289] Banks 1979:84; Carlsen 1980; Krone 2000:20
[290] Messner 1999:526ff/631ff

"*Third Management Rule: Non-renewable resources* are to be used in a way which maintains energetic potential and guarantees the long-term availability of high-quality material resources for a spectrum of material applications as broad as possible, especially for all fields of applications to satisfy basic needs, without exceeding the maximum tolerable level of nature for emissions or mechanical interventions. Therefore: (...)

- "The thermodynamic potentials of subsequent uses of non-renewable material shall be used and corresponding future options shall be kept open bearing in mind the more restrictive economic limitation of material resources;
- "Non-renewable material shall be kept in circular flows and its dispersion shall be avoided as far as possible. With the intention of reaching high thermodynamic efficiency in the circular flow economy, material flows are to be aspired for where the change of entropy of the material is kept as low as possible over subsequent uses. In this sense, the principle of low entropic differences in material cycles is valid; (...)

"*Fifth management rule*: In a *combined use of renewable and non-renewable resources* shocks in resource availability, quality losses in the resource use and forms of inefficient resource use are to be avoided in a long run. As, in effect, renewable resources will constitute the only resource basis for future generations, the following guidelines have to be considered for the realisation of an inter-temporal continuous transformation path in resource use:

- "The use of high-quality resources shall be reduced to a minimum level as soon as possible to ensure a transformation without major troubles towards an economy based on renewable resources. The principle of resource conservation is valid;
- "The economic processes of an economy are to be organised and co-ordinated in a way that the lowest possible material and energy losses, a lowest possible affection of natural resources and natural areas and a lowest possible dispersion of material to reach the production targets occur. In this sense, the principle of thermodynamic efficiency in the production is valid;
- "The demand side has to be integrated also into the transformation process. Therefore, thermodynamically efficient forms of use of products and services as well as a generally resource conserving lifestyle has to be aspired for. In this sense, the principle of sustainable consumption is valid;
- "Material-related and technological innovations are essential components for the transformation process in the resource use. Having in mind the environmental risks associated with the inter-temporal

> trends for higher entropy differences of innovations, a concept for material recycling should be in place for each product innovation at market entry, according to the principle of low entropic differences" (Messner 1999:400ff).

The importance of closing material cycles and the material value-preserving application of materials has also been recognised by political decision-makers. For instance, the Dutch government has marked off various environmental policy lines, one of which deals with the concept of Sustainable Construction.

> "One of the main principles of Sustainable Construction is Integral Substance Chain Management (ISCM). ISCM implies closing the various raw materials chains in such a way that a minimum amount of the materials is dumped or incinerated and a maximum amount of the released materials is reused, preferably in the same field of application..." (Stuip 1998:199).

This means regarding raw materials and waste products:

> "that in various life cycles all substances have to be applied at the highest possible level of quality" (Stuip 1998:199).

Preserving the material value as a strategy towards sustainability has already been incorporated in concepts like 'design for recycling' by the aluminium industry:

> "For example the value of wrought aluminium alloys cannot be recovered if melted with lower-grade aluminium castings. This loss of value offers incentive for designers to optimise the compatibility of alloys used in automobiles and for industry to develop cost-effective alloy separation technologies that facilitate the highest value post-consumption recovery" (Wernick and Themelis 1998:486).

Krone (2000) considers it absolutely necessary from an economic and environmental point of view to use the energy embodied in aluminium during primary material production in a perpetual way[291].

Material value preservation of the material is also chosen as the starting point for the development of a uni-alloyed aluminium can as beverage packaging[292].

Material value preservation as an economic incentive has led the secondary sector to become more and more technically sophisticated concerning

[291] Krone 2000:18

[292] see Sillekens *et al.* 1997

collection systems (distinguishing process scrap, pure old scrap and old scrap), scrap separation and processing technologies[293]. In the development of scrap separation technologies for instance, emphasis is now put on separating cast and wrought alloys by image analysis, atom-emission-spectroscopy or roentgen-fluorescence-spectroscopy for the same purpose of material value preservation[294].

Two similar models have been proposed to address the issue of entropic losses ('downcycling') from a resource perspective: the material cascade model by Sirkin and ten Houten (1994) and the model of different material pools, described, *e.g.*, by Tillman (1994).

9.5.2 Approach I: the material cascade model

The closing of material flows as well as the best possible value preservation of a material over a product system is illustrated by the model of a material cascade by Sirkin and ten Houten (1994) (Figure 9-1).

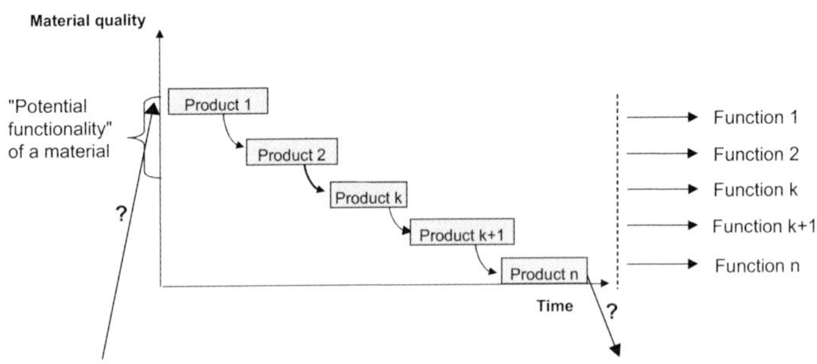

Figure 9-1. Material cascade: the material is recycled after its use and serves as 'downcycled' input material for a new product in another product system (after Sirkin and ten Houten 1994:230; reprinted with permission from Elsevier)

The material cascade model starts from the point that a material is used first in the technically most demanding application for an effective and efficient use of a resource. During the use phase and during recycling, the material's inherent properties are changed. These changes imply that the material can only be used in a less demanding application, or that virgin material must be added to reach the same functionality and performance of the material. Examples for reduced applicability are the shortening of fibre length

[293] Krone 2000:18

[294] see, *e.g.*, Krone 2000:556ff; Nijhof and Rem 1999; Pötzschke 1993:39; Rink 1993:121; Sattler 1992

during pulping, the inclusion of dirt or chemical modifications to the material. The reduction of the possible applicability, *i.e.* lower functionality, is a value reduction of the material and is commonly called 'downcycling'[295].

The value reduction occurring over a product system can now be interpreted as resource consumption. In this case resource consumption is not understood as the extraction of a resource *in situ* but as the loss of potential functionality of the material within the socio-economic system (see 'asset approach' in Chap. 5.8.3).

Following this reasoning, product systems causing a devaluation of the material are partially held responsible for processes enabling material use, such as resource extraction or final disposal of the material (*e.g.*, landfill, incineration).

9.5.3 Approach II: the material pools model

Another model has been proposed for the illustration of material devaluation: the material pools model (Figure 9-2)[296].

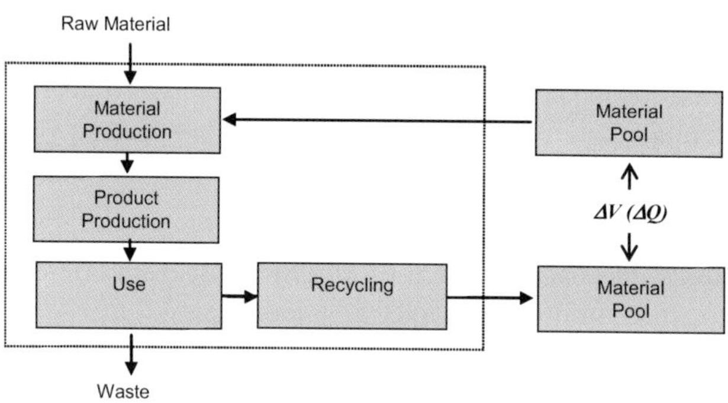

Figure 9-2. System with external material pools (after Tillman *et al.* 1994:25, modified; reprinted with permission from Elsevier)

> "The concept of material pools has been introduced implicitly in methods where recycled or recyclable materials cross system boundaries as non-elementary flows. The material in the pool is given an intrinsic 'environ-

[295] The reduction of the market price is the crucial aspect for aluminium concerning 'downcycling', "as each metal or alloy offers unique combinations of useful physical, chemical, and structural properties that are made available by its particular composition combined with the proper choice of a processing method" (Caron 1997:383).

[296] see also Karlsson 1995:92ff; ISO/TR 14'049, chap. 8.3.2

mental value' expressed in terms of environmental loadings. Material pools with different qualities have different 'environmental values'. Taking material from a material pool costs 'environmental value' and delivering material to a pool results in a credit. ...It may be given the value of the difference between virgin production and reprocessing to the original quality. This leads to an allocation of the loadings from the virgin production in proportion to the quality degradation caused by each product in the cascade" (Tillman *et al.* 1994:25).

Of course, there is no reason for *a priori* excluding final waste disposal from the above-mentioned model of 'environmental value' consisting of the associated environmental impacts. The reasoning on the external material pools model is quite similar to the cascade model, although the terminology is different.

9.5.4 Defining material value

The most striking difference in terminology of the two approaches is the addressing of the potential functionality of a material as 'material quality' in the material pools model. The term 'material quality' is quite usual in LCA literature. However, in engineering science and quality control[297], quality has a quite different, broader meaning and is strictly oriented towards the fulfilment of customers' requirements. In quality management, quality is defined as:

"(The) totality of characteristics of an entity (1.1) that bear on its ability to satisfy stated and implied needs. (...)
NOTE 3: Needs are usually translated into characteristics with specified criteria... Needs may include, for example, aspects of performance, usability, dependability (2.5) (availability, reliability, maintainability), safety, environment, ...economic and aesthetics..." (ISO/EN 8402:1995, chap. 2.1).

Alternately, according to the standards covering quality management:

"(The) ability of a set of inherent characteristics (2.5.1) of a product (2.4.2), system (2.1.1) or process (2.4.1) to fulfil requirements (2.1.2) of customers (2.3.5) and other interested parties (2.3.7).
NOTE: The term "quality" may be used with adjectives such as poor, good or excellent" (ISO/DIS 9000:1999, chap. 2.1.1)

[297] see, *e.g.*, Taguchi 1986:13ff

The term 'material value' is used instead to prevent confusions with the definition of quality used in quality management. Material value is defined as follows:

> *The material value is an expression of the potential functionality of a material considering all potential applications within the socio-economic system. The higher the material value, the higher functional requirements are met.*

This definition is in line with one of the important neo-classical resource economists who claims:

> "Most routine natural resources are desirable for what they do, not for what they are. It is their capacity to provide usable goods and services that we value" (Solow 1993:171, quoted after Radke 1999:149).

Up to now, several allocation procedures using 'material value' ('material quality' respectively) for allocation have been published so far[298]. They are among the procedures outlined in Chapter 6. The fundamental question for all the methods is how to describe material value.

There are several options for parameters describing material value:

- Physical, chemical, mechanical parameters;
- Prices.

Functionality of a material in one application can hardly be described by one single physical, chemical or mechanical parameter. Furthermore, the material can be used in a variety of different applications along the material cascade where the relevant physical or chemical parameters responsible for functionality can be totally different, depending on the application (see y-axis in Figure 9-1).

Entropy has been proposed also as allocation factor[299]. But entropy is no current parameter within the technical world and hardly determinable in a specific case. Entropy can thus not be recommended as basis for an allocation procedure[300].

Considering the above, the only applicable allocation factor making reference to material value is a value in economic terms.

In this case study, an allocation procedure is described and discussed which incorporates the model of the material cascade and which considers particular market prices - the quotations of the London Metal Exchange

[298] Ekvall and Tillman 1997; Karlsson 1994; Kim *et al.* 1997; Lindeijer 1994; Tillman *et al.* 1994 among others

[299] see, *e.g.*, Huppes and Schneider 1994

[300] see also Lindfors *et al.* 1995a:90

(LME) - as allocation factors: the value-corrected substitution. It will have to be discussed if the LME quotations - in fact the difference between two LME-quotations - reflect differences in 'material value' as defined above. Furthermore, the premises related to the use of economic data for allocation as outlined in Chap. 5.7.2 will have to be discussed (see Chap. 10.3.4).

9.6 Requirements of allocation procedures for reuse & recycling of aluminium

The following requirements are *a priori* established for the comparison of different allocation procedures for reuse & recycling of aluminium. An allocation procedure for reuse & recycling of aluminium should:

- *Concentrate on the product system under study*, and thus not be based on system expansion requiring the re-definition of the functional unit (see Chapters 5.6.4 and 6.4);
- *Focus of an LCA on the recycling characteristics* of a product and on the recycling processes itself as primary design strategies for the environmental improvement of the aluminium industry.

The allocation procedure should respect the decision-maker's mental model of the organisational principle of the socio-economic system. It should provide recommendations that take the material and market characteristics of aluminium into account as well as the applicable management rules for the environmental dimension of sustainable development (see Chap. 3.3.3).

Specifically, the allocation procedure should:

- *Not be based on a recycled metal content approach*, as the recycled metal content is not meaningfully determinable for aluminium;
- *Consider changes of the material's inherent properties* of aluminium as source of material devaluation;
- *Contain no 'jumps' of the allocation factor*, *i.e.* provide a steady allocation function in dependency of the changes of the materials inherent properties.

Moreover, the resulting product system and the respective LCI should be in line with the requirements established in Chapter 3.4.

Last but not least, an allocation procedure for reuse & recycling of aluminium should be in conformity to relevant standards, mainly the series of standards ISO 14040.

Chapter 10

DEVELOPMENT OF THE VALUE-CORRECTED SUBSTITUTION FOR ALUMINIUM WINDOWS

The value-corrected substitution has been developed in close co-operation with representatives of the aluminium industry and with exponents of the LCA community over the last few years. The value-corrected substitution adopts a principle described in Wenzel *et al.* (1996) and operationalises it for aluminium. The value-corrected substitution for aluminium has been published in Werner and Richter (2000b)[301]. See also Chapter 1.1 for a retrospective.

10.1 Principles

The value-corrected substitution starts from the assumption that material leaving a product system will substitute processes like primary material production processes of further systems, at least partly. This is in analogy to the closed-loop procedure, ISO/EN 14041, chap. 6.5.4, point 1, by which allocation is 'avoided' by the substitution assumption.

As the inherent properties of a material are normally changed in the sense of 'downcycling', the substitution is not complete but only partial. The devaluation consists of a reduced potential applicability and functionality of the material. Normally, no physical or chemical parameter can be found that reflects the functionality of a material, *i.e.* the material value (see Chap. 9.5.4), over the whole material cascade. Instead, market prices can be used for the description of the material devaluation over a product system. A fun-

[301] During the peer review process, G. Huppes (CML, The Netherlands) suggested addressing this approach formerly called economic allocation properly as value-corrected substitution (see Huppes 2000a; Werner and Richter 2000a).

damental question for the applicability of the value-corrected substitution is, how well do the market prices reflect material value as defined in Chap. 9.5.4?

The material and energy flows of processes associated with this devaluation – in general terms the primary material production and final disposal – are allocated to the system causing the devaluation.

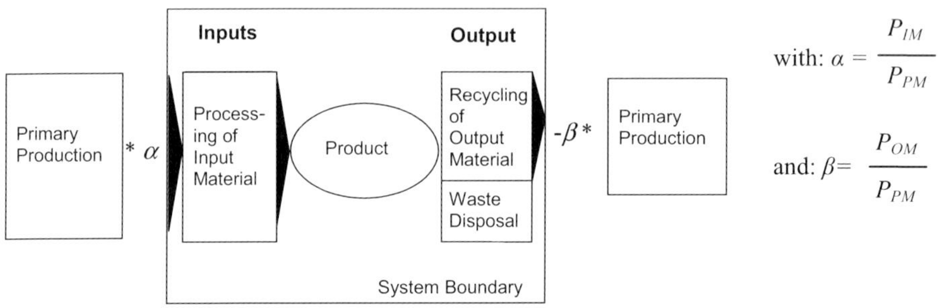

Figure 10-1. Principle of the value-corrected substitution
p_{OM} : price of output material; p_{IM} : price of input material; p_{PM}: price of primary material

Figure 10-1 shows the principle of the value-corrected substitution. The process requiring allocation – in this case, *e.g.,* a primary material production process – provides two functions:

- Provision of material for the product system under study;
- Provision of material to be used in further applications.

It is argued that, in principle, recycled material as output of the product system under study substitutes material from primary material production in other product systems. In analogy to the closed-loop procedure according to EN/ISO 14041, chap. 6.5.4, this substitution is considered by subtracting the material and energy flows that will be substituted.

In most cases, material's inherent properties are changed over a product system in the sense of 'downcycling'. Therefore, substitution is not complete but only partial. The substituted (subtracted) process is multiplied by a value-correction factor β, reflecting the relationship between the material value of the output compared to the material value of primary material.

The same reasoning has to be applied at the input side: The environmental interventions from primary material production are attributed to the input material by weighting it with the value-correction factor α which reflects the relationship between the material value of the input compared to the material value of primary material.

The difference of the two value-correction factors (α - β) indicates the devaluation of the material over the product system. The value-correction factor β can also be seen as the value preservation factor of the output material compared to primary material.

$$E_{pp}(P_n) = \alpha * E_{pp} - \beta * E_{pp} = (\alpha - \beta) * E_{pp} \tag{10.1}$$

$$\text{with } \alpha = \frac{p_{IM}}{p_{PM}} \quad \text{and} \quad \beta = \frac{p_{OM}}{p_{PM}}$$

In principle, similar considerations can also be made for allocating final disposal to the product system under study. For reasons given in Chapter 10.5 final disposal of 'used-up' aluminium is disregarded in this study[302].

The value-corrected substitution is based on the following assumptions:

- The relative difference of the prices between different grades of a material (or primary material and recovered material) reflects differing material's inherent properties and is a consequence of differing (reduced) applicability; for example: a higher content of iron and zinc reduces the potential applicability of aluminium and causes a lower price than (purer) primary aluminium;
- As a precondition, the prices used (depending on the setting of system boundaries) must reflect the definition of 'material value' in Chap. 9.5.4 as much as possible;
- The devaluation of the material as well as material losses during the use phase, collection and recycling efficiency cause a relative 'resource consumption' of a material over a product life cycle;
- The environmental interventions of primary material production processes (and final disposal at the end of the material cascade) are allocated to the system under study according to the relative 'resource consumption';
- Recycling processes after the use phase of the product are attributed to the system under study (see Chap. 10.2 for reasoning).

10.2 Operationalisation for the recycling of aluminium

While applying the value-corrected substitution, special attention has to be paid to the setting of system boundaries and to the attribution of the recycling processes. Three possible options for system boundaries and for the

[302] It has been argued that, under the material and market conditions of aluminium, it is not 'final disposal' of 'used-up' aluminium but future purification technologies that have to be attributed in accordance with the devaluation of aluminium over a product system. This argument is discussed in Chapter 13.4.

attribution of the recycling processes are evaluated for a fictitious aluminium product system. The process substituted differs for each option, as only similar processing stages can be reasonably considered in a substitution[303]. A crucial aspect for the implementation of the VCS for aluminium is the question, how well do the prices of the material reflect the definition of material value as defined in Chap. 9.5.4?

Option VCS/1: System boundaries on the demolition site: There are several factors to be considered in the classification of scrap on the demolition site, besides the elementary metallic composition of the material. In general, scrap prices are fixed relative to the quotations of the relevant international market – in the case of aluminium to the LME[304]. In practice, they vary considerably due to the homogeneity (large vs. small pieces), impurities (dirt, plastics etc.), humidity, amount available at one site, transport distance, scrap availability, etc.[305]. Furthermore, they can vary due to fluctuations of the exchange rate of the local currency compared to the currency of the reference market[306].

The system boundary setting implies knowledge of the recycled metal content of the aluminium used for the product because system boundaries have to be set consistently at the input and output side of the product system. Due to the characteristics of the secondary aluminium market (scrap from different processing stages of aluminium), this cannot reasonably be determined for a specific alloy nor for a specific product.

The equivalent 'processing stage' of primary material for scrap on the demolition site would be bauxite *in situ*. As no technical process is linked to a resource *in situ* (except probably exploration of reserves) no substitution of technical processes occurs on this level of system boundaries.

For these reasons this option is not considered any further.

Option VCS/2: system boundaries after sorting before remelting: The prices for sorted scrap will largely depend on non-material related parameters such as the size of the pieces, their homogeneity, moisture content, the quantity available, transport distance, etc.

The equivalent processing stage of primary material for scrap after recollection and sorting before remelting would possibly be aluminium hydroxide production (the processing stage before electrolysis). Hence, the recollection and sorting would substitute the exploitation of bauxite and the processing to aluminium hydroxide. Although a possible option for system

[303] see, *e.g.*, Borg *et al.* 2001
[304] Maag 2000; see also Pötzschke 1993:35
[305] Henstock 1980:243ff; see also Bower 1977:10; Krone 2000:92ff; Münster 1992:13
[306] Münster 1992:12

boundaries, aluminium hydroxide and sorted scrap remain two different products requiring different treatment to reach the processing stage where the (technical) functionality will be reflected in their prices.

For the modelling, these system boundaries imply knowing the recycled metal content of the aluminium used for the product at the input side. For the characteristics of the secondary aluminium market, this cannot reasonably be done (see above).

For these reasons this option cannot be recommended for aluminium.

Option VCS/3: System boundaries after remelting (*intermediate goods*): This system boundaries correspond to the results of the calculation procedure for the closed-loop procedure based on a material pool as outlined in ISO/TR 14049, chap. 8.3.3 (see also Chap. 11.2).

The equivalent processing stage of primary material as ingot is the remelted recycling material as ingot. Under given market conditions, it can be assumed that ingots from recycled material substitutes ingots made from primary material and its respective processes (see Chap. 13.2).

Two alternatives are possible (see following pages):

- Alternative 'A' is suitable if prices for the recycled material as intermediate products are available or if the value of the incoming materials equals the one leaving the system. The value-corrected substitution or the 'closed-loop' procedure (in the latter case) can easily be justified. For aluminium, the prices of intermediate products used as inputs are compared to the ones obtained after the recycling processes. No additional primary material (*PME**) is added;
- Alternative 'B': The (hypothetical) addition of primary material is necessary if the recycled material has to be 'upgraded' with primary material to theoretically comply with material specifications, for which prices are available (*e.g.,* LME prices for the alloyed aluminium contract)[307]. For aluminium, primary material is added to up-grade the recycled material depending on the material production process (refining) (see, *e.g.*, Uchida *et al.* 1999:985). The inclusion of the primary material production process (*PME**) may also be necessary if material has to be upgraded (theoretically) to reach a certain material quality for which prices are available (in brackets).

For both alternatives system boundaries are set after 'Material Production'.

The result of this approach will show very clearly the results of measures aiming at maintaining the material's inherent properties.

[307] see also Borg *et al.* 2001

In the case of aluminium, the third option VCS/3 is favoured, as allocation can be based on highly transparent LME prices (for further details, see Chap. 10.3). Furthermore, the VCS is following and refining a methodological proposition made in ISO/TR 14049 by introducing the aspect of material value.

The third option is illustrated in Figure 10-2 and Table 10-1.

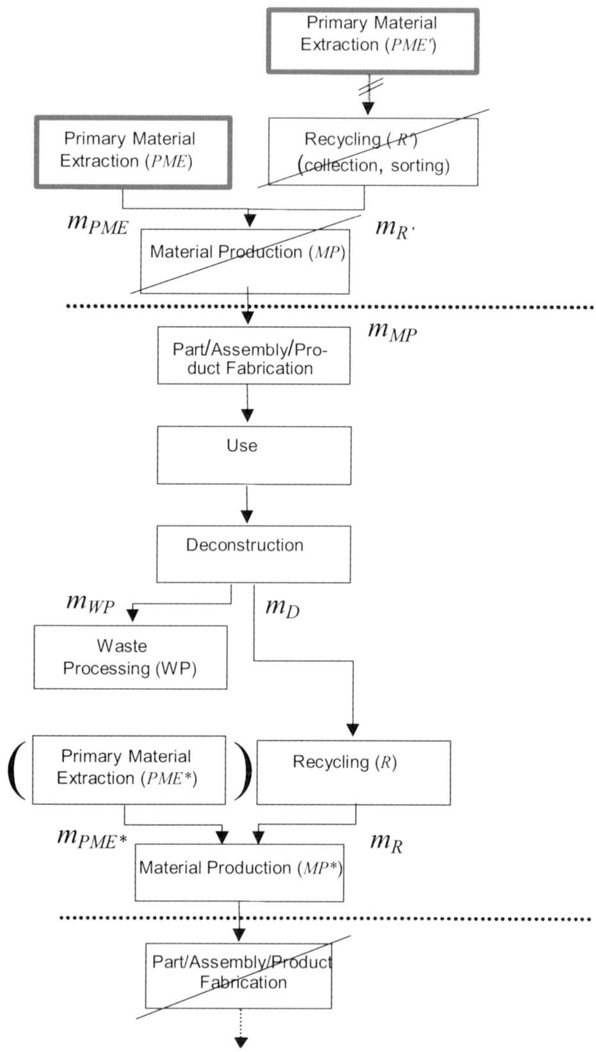

Figure 10-2. Value-corrected substitution applied to an aluminium window

Table 10-1. Value-corrected substitution applied to an aluminium window

Allocation of specific processes	Allocation factor	
Primary Material Extraction (*PME'*)	$\alpha * m_{R'}$	
Recycling (*R'*)		whereas:
Primary Material Extraction (*PME*)	m_{PME}	m_n : mass flows related to process *n*
Material Production (*MP*)		$\alpha = p_{\text{ingot input material}} / p_{\text{ingot primary material}}$
Waste Processing (*WP*)	m_{WP}	$\beta = p_{\text{ingot output material}} / p_{\text{ingot primary material}}$
Recycling (*R*)	m_D	$PME' \sim PME$
Primary Material Extraction (*PME**)	$-\beta * m_R + m_{PME*}$	
Material Production (*MP**)	$m_{R*} + m_{PME*}$	

10.3 LME quotations as reference points for the price formation of aluminium

10.3.1 Primary aluminium and alloyed (secondary) aluminium contracts

> "Many – perhaps most – of the international commodities markets are complex packages of different trading channels and different market structures. Nonetheless, in many cases it is possible to identify a dominant pattern that acts as the 'price leader' in the market. For commodities with competitive conditions, even in only a minor trading channel, the price so determined would tend to be used as a reference point for long-term contracts. This is the case, for example, in the copper market (outside the United States) where the London Metal Exchange (LME) price is normally the current reference price, even though the LME is only a marginal market" (Maizels 1984:31).

The same is true for aluminium, which is quoted on the two major commodity exchanges[308], the London Metal Exchange (LME)[309] and the New York Commodity Exchange (COMEX). The first of these two, the LME is considered the most important in terms of turnover, physical deliveries and its influence on the pricing of metals in general.

[308] A commodity exchange is an organised, regular market event, where fungible, representable goods, which are not present at the venue, are traded according to an agreed contractual routine. The prices established by supply and demand, are officially determined (after Engelhardt and Günter 1996:214).

[309] London Metal Exchange, see http:www.lme.co.uk

Two aluminium quotations exist at the LME: the primary aluminium contract and the alloyed (secondary) aluminium contract. The LME quotations of the primary aluminium contract are presently the worldwide reference[310]. The quotations of the alloyed aluminium contract are of less importance; the turnover has been low up to now. However, the quotations of the alloyed aluminium contract are referred to occasionally[311].

Like most primary commodities, aluminium prices have been highly volatile in recent years. Since the LME quotation of the primary aluminium contract was established in 1978, it has fluctuated significantly. The same is true for the quotation of the alloyed aluminium contract, which was introduced to the LME in 1991. However, statistical analysis carried out at the World Bank support the notion

> "...of the aluminium market being 'efficient'. The volatility of the aluminium price is due mostly to fluctuations in industrial production, as well as changes in aluminium production capacity" (Anonymous 1994c:121 quoting Hobeika 1990).

Frank (1990:72) states that a stock exchange as a punctual market without any spatial, temporal or product-related preferences is reacting very sensitively to incremental discrepancies in supply and demand[312]. Thus, stock exchanges as international markets are highly transparent. Bird (1990:145ff) calls the LME quotations a 'justified price'; he deduces that discrepancies between market prices and cost-justified prices of primary aluminium are due to changes in the inventories of the LME, at least over short periods.

10.3.2 Influence of speculation on price formation

In recent years, the steep increase in transactions by banks, financial institutions and investment funds on the non-ferrous exchanges, especially on the LME, has led to what is commonly known as 'market financialisation'[313]. After developing new and increasingly sophisticated risk-manage-

[310] Frank 1990:70; Krone 2000:30; Legrand 1997:8; for the importance of the LME as forward-market see also Anonymous 2000a

[311] according to Krone 2000:32f

[312] In this respect, it is no coincidence that Léon Walras (1834-1910) introduced an auctioneer for the explanation of the general conditions of economic equilibrium (see Anonymous 2000d).

[313] The linkage of the commodity price with the financial market has been known as the Hotelling-rule, published in Hotelling 1931. The Hotelling-rule states that in a perfectly informed market in equilibrium, "the resource owner must expect the net price of an ore (*in situ*; WF) to be increasing exponentially at a rate equal to the rate of interest (Solow 1974:2; see also Robinson 1980:279; Siebert 1983). Thus, the fundamental principle of the

ment instruments for financial portfolios, the financial institutions have turned to commodities and their derivates, mainly in the course of the last decade. It enabled them to pursue their strategy of assets diversification with a view to minimising the risks and maximise their portfolios yield[314]. The increasing level of intervention by financial institutions, funds and banks is evidenced by the fact that the volume of transactions on the LME has multiplied by five from 1989 to 1995[315].

On the influence of speculation on the price forming process of a commodity Bank (1080) states:

> "...some evidence seems to exist, which indicates that price movements in speculative markets follow a random walk, which among other things means that current (short-term; WF) price changes cannot be predicted by previous price changes. This does not deny the validity of the underlying long run supply-demand trend as the ultimate arbiter of the average market price..." Banks (1980:488) [316].

This goes in line with Solow (1974) who comments, based on theoretical considerations on the influence of speculation on commodity prices

> "...that in tranquil conditions resource markets are likely to track their equilibrium paths moderately well" (Solow 1974:7).

Besides the above-mentioned hedging and speculating[317], which normally does not include the transaction of materials, one should also be aware of the possibility of speculating in physical items[318].

10.3.3 Statistical analysis of LME quotations

As mentioned above, two aluminium contracts are actually traded at the LME, a primary aluminium contract and an alloyed aluminium contract with

---

economics of exhaustible resources is (...) simultaneously a condition of flow equilibrium in the market for the ore (or for traded intermediate products; WF) and of asset equilibrium in the market for deposits" (Solow 1974:5).

314 Commodity funds provide a opportunity for hedging as their performance is often delayed in time compared to other stocks or bonds (see Birch 2000).

315 Legrand 1997:9

316 see also Kantor 1979: "If markets are efficient, the price changes on those markets should follow an approximately random walk. If price changes are unrelated to previous prices, then all the forces known to affect prices are assumed to have been incorporated in ruling market prices. Efficient markets are equilibrium prices conditional on all information available when prices are established" (Kantor 1979:1430)

317 see Gibson-Jarvie 1976:119ff; Hull 1997:10f/31ff

318 Banks 1980:480; see also Solow 1974:6

three slightly different specifications[319]. In Table 10-2, the characteristics of the two commodities are summarised.

For each of the two traded contracts, one official price in US $/ton is established for both the spot market and the futures market (three-month) during the daily morning sessions at the LME. Another unofficial price is determined at the end of the afternoon session.

The official monthly average 'bid/ask) LME Primary AL' and average '(bid/ask) LME Secondary AL' prices in US $/ton can be found in Appendix A1 for the spot price and for the 3-month futures price, dated from January 1993 to May 1999. Their trends are illustrated in Figure 10-3.

Several statistical preconditions for the appropriateness of LME quotations as a basis for allocation can be formulated:

- The price ratio of the alloyed (secondary) aluminium contract to the primary aluminium contract (p_A/p_P) should be stable over the time period. The arithmetic mean, its standard deviation $\bar{s}$ and the median are calculated as outlined in Chapter 10.3.3. The standard deviation $\bar{s}$ can be used for describing an error interval, covering about two third of the data($\pm\ \bar{s}$ in % of the arithmetic mean)[320]. Monthly, quarterly, half-year and annual data is analysed in order evaluate the random distributed price effects of speculation;

Table 10-2. Elementary specifications of primary and secondary (alloyed) aluminium contracts traded at the LME (various sources)

Specifications for traded primary aluminium (in% mass)[321]												
Standard	Abbreviation										others	
		Si	Fe	Cu	Mn	Mg	Ni	Zn	Pb	Sn	each	total
AA P1020A		<0.1	<0.2									
Specifications for traded aluminium alloy (in% mass)												
Standard	Abbreviation										others	
		Si	Fe	Cu	Mn	Mg	Ni	Zn	Pb	Sn	each	total
DIN 1725/5[322]: 226	GBD-AlSi9Cu3	8.0-11.0	1.0 (0.8)	2.0-3.5	0.1-0.4	0.1-0.5	0.3	1.2	0.2	0.1	0.05	0.15
AA[323] A380.1 ASTM B179		7.5-9.5	1.0	3.0-4.0	0.5	0.1	0.5	2.9		0.35		0.5
JIS H2118-1976, Class 12 D12S	AlSi11Cu2.5Fe	9.6-12.0	<1.3	1.5-3.5	<0.5	<0.3	<0.5	<1.0		<0.3		

* up to 0.3% impurities in total including Fe and Si

[319] Krone 2000:32
[320] see Sachs 1974:51
[321] taken from http://www.lme.co.uk/AlSpecs.htm (07/2000)
[322] see DIN 1725:1986
[323] see Anonymous 1977:59

- The price time series should highly correlate, determinable by the Pearson correlation[324];
- The choice of using the spot price relation or the futures price relation should be made based on lower standard deviation and higher Pearson correlation.

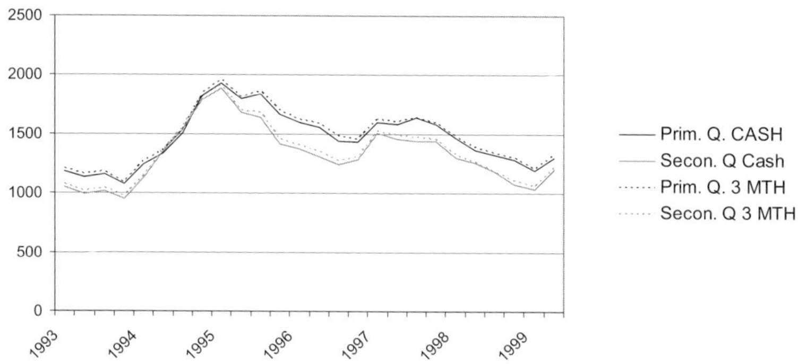

Figure 10-3. Average monthly (BID/ASK) LME prices of the primary and alloyed contracts as spot ('cash') and futures prices ('3 mth') 1993/1-1999/5 per ton (Lysheim 1999)

The results of the statistical analysis are shown in Table 10-3.

Table 10-3. Statistical analysis of the (BID/ASK) LME prices of the primary and alloyed contracts as spot ('cash') and futures prices ('3 mth') 1993/1-1999/5 per ton (own calculations)

Cash prices	monthly	quarterly	half year	annual
$n =$	77	26	13	7
arithmetic mean p_A/p_p	0.9060	0.9061	0.9059	0.9052
median	0.8951	0.8938	0.8887	0.8887
standard deviation ($\bar{s}$)	0.0515	0.0490	0.0440	0.03916
$\pm\ \bar{s}$ in %	5.68%	5.40%	4.85%	4.33%
Pearson correlation	0.9543	0.9581	0.9637	0.9714

Futures prices (*3 mth*)	monthly	quarterly	half year	annual
$n =$	77	26	13	7
arithmetic mean p_A/p_p	0.9074	0.9075	0.9074	0.9069
median	0.8978	0.8977	0.8926	0.8926
standard deviation ($\bar{s}$)	0.0463	0.0449	0.0416	0.03724
$\pm\ \bar{s}$ in %	5.11%	4.94%	4.59%	4.11%
Pearson correlation	0.9629	0.9650	0.9678	0.9741

[324] see Sachs 1974:315ff

Having in mind Figure 10-3 it is not surprising to see that:

- The arithmetic mean of the futures price ratio is generally slightly higher[325];
- The arithmetic means and the medians are almost stable over all steps of aggregation;
- The medians are slightly lower than the arithmetic means[326];
- The standard deviation decreases for higher aggregated data for both the spot and the futures price ratios as aggregation 'smoothens' time series;
- The correlation of the price ratio is generally high (> 0.95) for both the spot and the futures price, slightly increasing for higher aggregated data.

From these results and considering the characteristics of the aluminium market, the conclusion can be drawn that the price ratio of the LME quotations for primary and secondary aluminium are a stable and adequate measure for the value of these two commodities within the actual socio-economic system. The short-term influence of speculative transactions on the price formation cannot be proven any more, given the degree of aggregation of the available data.

Allocation for recycling in LCA can reasonably be based on this parameter if a relation can be established of the recycled aluminium to one of the two commodities. For practical purposes and for further calculation, the price ratio of the secondary (alloyed) grade versus primary aluminium is considered to be:

$$p_A/p_p = 0.9 +/- 0.0405 \quad (sd = 4.5\%) \tag{10.2}$$

This value with a variation of +/- 4.5 % covers about two third of the price relations between 1993/1 to 1999/5.

10.3.4 LME-quotations as allocation factors in LCA

The third option of setting system boundaries favoured in Chapter 10.2 enables the use of LME quotations as market prices for allocation purposes if using the value-corrected substitution. As has been mentioned previously, a crucial question for its applicability is the question how well LME quotations reflect 'material value' as an expression of the potential functionality of a material considering all potential applications within the socio-economic system (Chap. 9.5.4).

From the above, the following points can be concluded:

[325] because of the Hotelling Rule (Hotelling 1931)

[326] Due to the period in 1995, where shortage of alloyed material lead to higher prices of the secondary grade, while structural adjustments took their time. Thus, the assumed Gaussian distribution of the price ratio is therefore slightly asymmetric.

- The LME as a punctual market without any spatial, temporal or product-related preferences is reacting very sensitively to incremental discrepancies in world wide supply and demand. The prices can be considered close to equilibrium prices;
- Country-specific, product-related or case-specific aspects, such as available quantities, location, inhomogeneities, dirt, specific transport distances, etc. do not distort the quotations;
- As international markets, stock exchanges are highly transparent;
- Fluctuating exchange rates, differing import or export taxes for primary and secondary material, subsidies, dividends, or inflation does not directly distort the quotations. Indirect distortions, such as regulation-induced market effects in particular countries, will affect both quotations in the same way;
- The price-ratio of the two aluminium contracts can be considered rather stable, at least for the time the alloyed aluminium contract has existed (since 1992);
- Speculation-induced distortion of prices can be considered random distributed and thus be neglected for long-term price considerations;
- Sensitive (competitive) market conditions can be assumed at the LME for the primary aluminium contract as well as for the alloyed (secondary) aluminium contract. As alloyed material can in principle be substituted by primary material (by adding alloy elements) competitive market conditions at the LME can also be assumed for the primary and secondary (alloyed) grade markets in general. This is supposed to be the main reason for the high correlation of the time series of for the primary and secondary (alloyed) grade at the LME.

This means that LME quotations reflect the world wide marginal demands of the aluminium alloys covered by the two contracts without product-related, spatial, or temporal preferences on a highly transparent market. Thus, LME quotations are reasonably good (or the best available) parameters for 'material value' as defined in Chap. 9.5.4.

10.4 Time dependency in the value-corrected substitution

As the socio-economic system changes over time, assumptions made for the future in LCA become more uncertain the further the assumptions are based in the future. Especially for long-term applications of materials, uncertainties related to assumptions on future technologies have to be considered.

For the value-corrected substitution, time effects play an important role concerning the:

- Technology that will be substituted in the future and its respective environmental interventions $E_{pp}(\tau)$;
- Time dependency of the price ratio β;
- Recycling processes that will be used for the treatment of the post-consumer material;
- Percentage of the material collected for recycling and the yield of the recycling processes;
- Service life τ assumed.

All of them depend on the service life τ of the product. This time dependency is illustrated adding (τ) into Eq. (9.1):

$$E_{pp}(P_n) = \alpha * E_{pp} - \beta(\tau) * E_{pp}(\tau) \tag{10.3}$$

$$\text{with } \alpha = \frac{p_{IM}}{p_{PM}} \quad \text{and } \beta(\tau) = \frac{p(\tau)_{OM}}{p(\tau)_{PM}}$$

with:
τ : service life of the product
E_{PP} : environmental interventions related to primary material production
P_n : product n
p_{OM} : price of output material
p_{IM} : price of input material
p_{PM} : price of primary material

Scenario development can help to deal with time-related aspects[327]. Anyhow, 'best guess' scenarios should be complemented with an 'actual situation' scenario in a sensitivity analysis.

The price ratio of secondary and primary material in the future can be estimated by analysing the mean of the price ratio[328] of the time period back $(t_0 - \tau)$ equal to the time period over which the LCA is to be applied or extrapolated $(t_0 + \tau)$. This is a standard procedure for price predictions used in the financial analysis of assets applied to the allocation problem[329].

$$\beta(\tau) = \frac{p(\tau)_{OM}}{p(\tau)_{PM}} \sim \frac{1}{n} \sum_{t = t_0 - \tau}^{t_0} \frac{p(t)_{OM}}{p(t)_{PM}} = \left[\overline{\frac{p(t)_{OM}}{p(t)_{PM}}}\right] \tag{10.4}$$

with:
n : number of data pairs available during period $t_0 - \tau$ and t_0
t_0 : actual period
τ : time of post use recycling
β : value-correction factor
p_{OM} : price of output material
p_{PM} : price of primary material

[327] see, *e.g.*, Coates 1994; Gordon 1994

[328] It can be considered reasonable to take the median of the price ratio instead of the mean of the price ratio.

[329] see Hull 1997:233

The corresponding standard deviation can be used as an error interval $\pm\sqrt{s^2(\beta(\tau))}$ covering about two third of the data points[330], based on:

$$s^2(\beta(\tau)) \sim \frac{1}{n-1} \sum_{t=t_0-\tau}^{t_0} \left[\frac{p(t)_{OM}}{p(t)_{PM}} - \overline{\left(\frac{p(t)_{OM}}{p(t)_{PM}} \right)} \right]^2 \tag{10.5}$$

with:

n : number of data pairs available during period t_0-τ and t_0
t_0 : actual period
τ : time of post use recycling
β : value-correction factor
p_{OM} : price of output material
p_{PM} : price of primary material

As a further statistical precondition of the applicability of Eq. (9.4), the time series of the two prices taken for allocation must be highly correlated (Pearson correlation[331]). Instead of using the mean for the estimation of the future price ratio $\beta(\tau)$, more sophisticated analytical tools of time series analysis or scenario development could be applied to predict the future price ratio based on price trend analysis. However, causality of the correlation must be discussed.

10.5 'Final disposal' of aluminium

To be in line with ISO/EN 14041, chap. 6.5.4, point a), end-of-life processes (final disposal) have to be considered possible processes to be allocated to subsequent product systems. A closer look at the global aluminium market is necessary to decide on how to deal with final disposal at the end of the material cascade.

Whereas the annual global quantity of aluminium manufactured from bauxite entering the technosphere totals some 20 million tons, a mere 2 million tons are dumped or otherwise disposed of each year. Consequently, the amount of aluminium in the technosphere – currently around 500 million tons – is increasing by an annual rate of approx. 18 million tons[332]. Forecasts of the World Bank estimate an annual worldwide growth of primary aluminium consumption of 2.3% until 2005[333]. By comparison, approx. 12 million t of process scrap and 4 million t of post-consumer scrap are recycled each year.

[330] Sachs 1974:51
[331] see Sachs 1974:315ff
[332] Buxmann 1997
[333] Anonymous 1994c:111ff

For the consideration of 'end-of-life' disposal of aluminium this has several implications:

- Aluminium as a metal is composed of atoms, not of molecules that might be destroyed or 'used up'. Theoretically, this allows nearly indefinite recycling[334] of aluminium and metals in general[335]. But even in uni-alloyed applications and with closed-loop recycling, the accumulation of non-aluminium elements deriving from surface treatment of processing can reduce the recyclability of aluminium, mainly for products with short service life[336];
- The amount of aluminium available worldwide for technical applications is increasing annually by an average of 3.6% per year[337]. The high demand of aluminium also causes that many scrap fractions have positive market price leading to an almost complete recycling. This is particularly true for aluminium in building applications;
- Compared to the use of primary aluminium and recycled process scrap, the amount of post-consumer scrap continues to play a minor role. This is especially true for construction products with long service life, where much more aluminium is entering into constructions than is actually taken out[338]. The situation is different for packaging materials (cans) or litho plates with much shorter service life and a corresponding throughput[339]. Nonetheless, the recycling of post-consumer scrap is actually of minor importance in quantitative terms compared to the global flows of primary aluminium and recycled process scrap;
- On a small scale, purification technologies (electrolytic separation, partial solidification, exposition to chlorine gas, etc.) have been developed for the removal of specific alloy elements and are used in special applications[340]. For cost reasons, these technologies are not applied on large scale to high-alloy aluminium. In the refinement process, high-alloy scrap is usually mixed with various other scraps and primary aluminium to add certain alloy elements to a required alloy;

[334] not considering material losses and oxidation

[335] Krone 2000:18; Melo 1999:92; Wernick and Themelis 1998:466

[336] Krone 2000:41

[337] Buxmann 1997

[338] At Mücksch GmbH (1998), the ratio of production scrap to post-consumer scrap for aluminium windows is approx. 0.9/0.1 (Werner and Richter 1998a).

[339] Currently the recycling rate of aluminium beverage cans is at 91% of the amount sold in the year 2000 (Anonymous 2001b).

[340] see Wernick and Themelis 1998:471/485 and references given there; Krone 2000:41f/370ff

- Another hypothetical option for the 'final disposal' of high-alloy aluminium scrap would be its incineration for electricity generation, allowing for the recovery of the energy invested in primary material production[341]. The costs of these technologies have prevented the application or further development of this technology so far.

Considering the material and market characteristics, there exist good reasons for assuming that aluminium will not be disposed of as 'used up' material as a consequence of its too high non-aluminium alloy content. Therefore, 'final disposal' of aluminium is not taken into account in this case study.

Aluminium losses from the technosphere are assumed to occur due to dissipative use and due to incomplete re-collection and losses during recycling. The paradigm of value preservation (Chap. 9.5.1) will help to:

- Prevent these losses, giving incentive to the most complete recycling and avoiding dissipative uses;
- Provide high-value scrap to the recycling facilities as a precondition for the most environmentally efficient use of aluminium in a closed-loop economy.

[341] see Partington 1992

Chapter 11

SELECTED ALLOCATION PROCEDURES FOR COMPARISON

In this chapter, the allocation procedures used for comparison are documented and discussed. It starts with two allocation procedures using input recycling percentage and output recycling percentage respectively:

- The cut-off procedure, as applied in the studies on window frames[342] for comparison;
- The closed-loop procedure based on a material pool model, as applied in ISO/TR 14049, chap. 8.3.2 on aluminium packaging material.

A third procedure, the strict coproduct allocation according to Huppes (1994:84f) and Lindeijer and Huppes (2001) is discussed but not included into calculations for not yet being fully developed[343].

These allocation procedures are selected because of:

- Having been used in the former LCA study on window frames[344] (the cut-off procedure);
- Being proposed in ISO/TR 14049:2000 for aluminium products (the closed-loop procedure based on a material pool model);
- Being favoured by the peer reviewer of Werner (2000) and being proposed in the up-date CML methodology document[345] (the strict coproduct allocation).

All these allocation procedures can also be seen as reflecting the consequences of:

[342] Richter *et al.* 1996; Werner and Richter 1998a
[343] see also Vogtländer *et al.* 2001
[344] Richter *et al.* 1996
[345] Lindeijer and Huppes 2001; Guinée *et al.* 2004

- Different mental models of the material and market characteristics of aluminium;
- Different paradigms on the sustainable use of aluminium and thus different information requirements of the actors involved;
- Different attitudes towards risk in view of the selection of present or future data required for modelling (see also Chap. 5.8.5).

11.1 Cut-off procedure

11.1.1 Principles

The cut-off procedure is one of the most traditional allocation procedures for reuse & recycling. It has been proposed by SETAC in the early years of life-cycle assessment[346](see also Chap. 5.8.3). Its underlying principle is summarised by:

> "Each product should only be assigned the environmental impacts directly caused by that product" Ekvall and Tillman (1997:160).

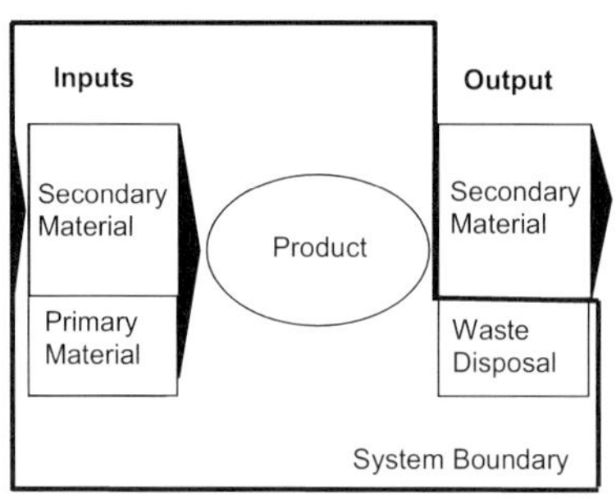

Figure 11-1. System boundaries setting in the cut-off procedure

The environmental burdens are allocated according to the following principles:

- Primary material as input: the environmental burden of the raw material extraction and processing is entirely included into the system studied;
- Recycled material as input: the environmental effects of collecting, processing and reintegrating the material are included into the system;
- Recyclable materials as output: any environmental effects of collecting and recycling are excluded (as burden to the next product system);

346 Fava *et al.* 1991:79ff; see also Boguski *et al.* 1994:151ff; Ekvall and Tillman 1997:160; Heijungs *et al.* 1992, guide, pp. 29; Klöpffer 1996:27ff; Rydberg 1995:236

- Non-recyclable wastes: the effects of disposing non-recyclable wastes are included to the system.

The procedure is illustrated in Figure 11-1:

The cut-off procedure has frequently been used for its low data requirements, the simple allocation rules and for the system boundaries, which are easy to communicate. In the end, the cut-off procedure solves the allocation problem related to recycling on process level by the setting of system boundaries as no processes are divided for allocation. Material's inherent properties or material value of the input and output materials are disregarded.

11.1.2 Operationalisation for the recycling of aluminium

In the Figure 11.2 and Table 11.1, the cut-off procedure is applied to a fictitious product system of an aluminium window.

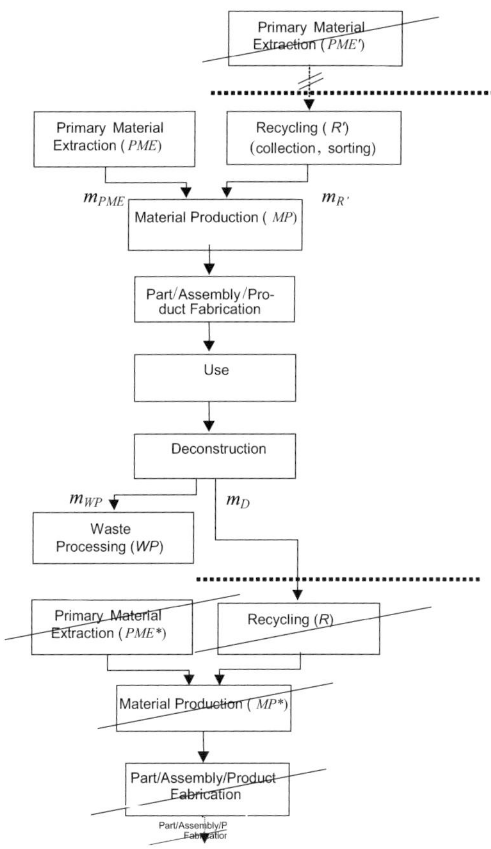

Figure 11-2. Cut-off procedure applied to an aluminium window

Table 11-1. Cut-off procedure applied to an aluminium window

Allocation of specific processes	Allocation factor	
Primary Material Extraction (*PME'*)		
Recycling (*R'*)	$m_{R'}$	
Primary Material Extraction (*PME*)	m_{PME}	whereas:
Material Production(*MP*)	$m_{PMP}+m_{R'}$	m_n mass flows related to process *n*
Waste Processing (*WP*)	m_{WP}	
Recycling (*R*)		
Primary Material Extraction (*PME**)		
Material Production (*MP**)		

11.2 Closed-loop procedure with material pools model

In the Technical Report ISO/TR 14049, chap. 8.3.2, an example is outlined how open-loop recycling can be handled with a closed-loop procedure based on material pools.

11.2.1 Principles

A material pools model is applied. It is assumed that the product system delivers secondary raw material into that pool and is supplied with secondary material by the pool (see also Figure 9-2). As the amount of secondary raw material delivered to the pool is larger than the amount taken out of the pool, the product system has a net output of scrap. It is argued that the net output of scrap increases the amount of available secondary aluminium that displaces primary aluminium metals in other systems. Thus, the recycling process is attributed to the system under study and a closed-loop procedure is applied for the net aluminium output assuming that the output substitutes part of the primary material input of the system. With this procedure, the difference between producing aluminium from secondary raw material and producing the same product from primary material is calculated and allocated to the system under study.

Note that the aluminium pool is considered to be of a homogenous composition equivalent to primary aluminium. This is in contrast to the materials pools model presented in Chapter 9.5.3.

The procedure is illustrated in Figure 11-3:

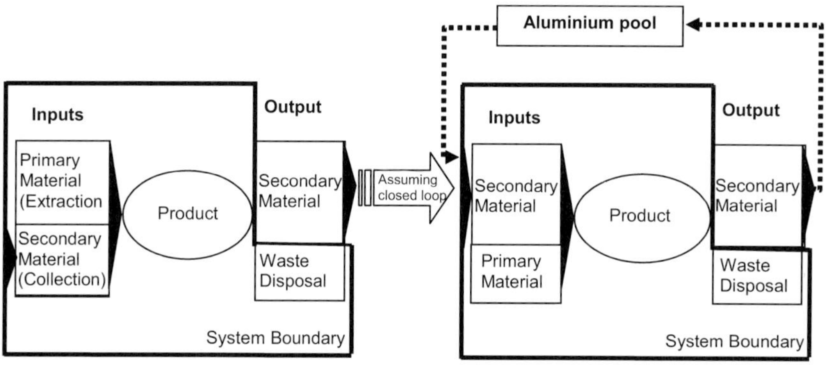

Figure 11-3. Transforming an open-loop system into a closed-loop system with the aid of the introduction of a virtual material pool assuming that recycled material substitutes primary material

11.2.2 Operationalisation for the recycling of aluminium

In Figure 11-3 and Table 11-2, the closed-loop procedure according to ISO/TR 14049 is applied to a fictitious product system of an aluminium window.

No clear rule can be established for the system boundary setting, as the recycling processes are considered to be the same for input and output secondary material. In the calculation below, the recycling processes of the output are considered to substitute the recycling processes of the input and future primary material production based on masses. Allocation is based on the recycled metal content determined by the recycling yield.

Table 11-2. Closed-loop procedure according to ISO/TR 14049 applied to an aluminium window

Allocation of specific processes	Allocation factor	
Primary Material Extraction (*PME'*)		
Recycling (*R'*)	$m_{R''} - m_{R'} = 0$	
Primary Material Extraction (*PME*)	m_{PME}	whereas:
Material Production (*MP*)	$m_{PMP} + m_{R'}$	m_n mass flows related to process n
Waste Processing (*WP*)	m_{WP}	$m_{R'} = m_{R'}$
Recycling (*R*)	$m_D - m_{R''}$	
Primary Material Extraction (*PME**)	$-m_{D'}$	
Material Production (*MP**)		

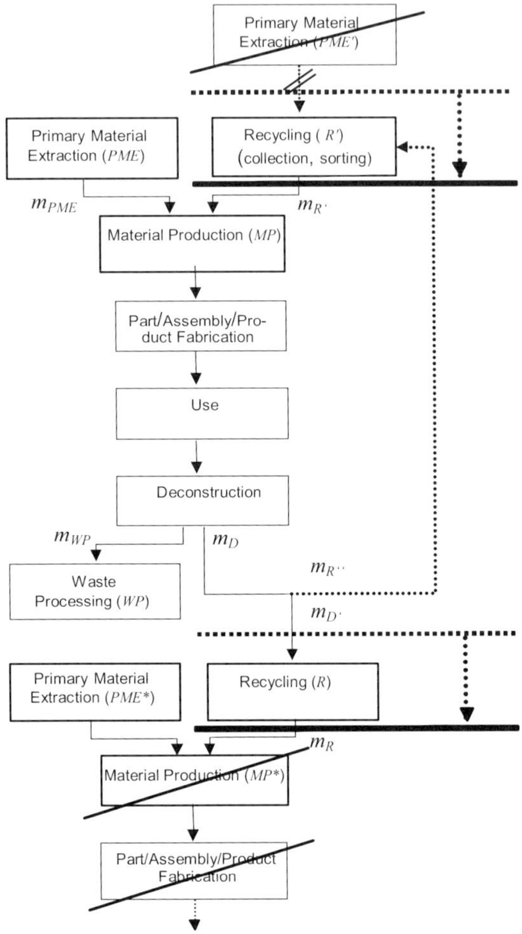

Figure 11-4. Closed-loop procedure according to ISO/TR 14049 applied to an aluminium window

11.3 Strict coproduct allocation

11.3.1 Principles

The principles of the strict coproduct allocation according to Huppes (1994), Lindeijer and Huppes (2001) and Guinée *et al.* (2004) are described extensively in Chapter 6.2. The strict coproduct allocation solves the attribution problem on a strict process-by-process basis. In such cases, the allocation problem is reduced to allocating the recycling processes; primary material production processes are not allocated to subsequent product systems.

The allocation of the environmental interventions of the recycling process can be made using allocation procedures developed for multifunctional (multi-output) processes. For the allocation of recycling processes, the two functions to be weighted for the determination of the allocation factor are the function 'waste disposal' and 'provision of secondary material'. Only economic allocation factors are applicable as these two functions are inseparably joined[347].

11.3.2 Operationalisation for the recycling of aluminium

Huppes (1994), Lindeijer and Huppes (2001) and Guinée *et al.* (2004) propose to consider the recycling process as a waste treatment process until the negative market price of the material crosses the zero-point and becomes positive; all the processing with positive market price of the material is considered secondary material extraction and is attributed to the following product system[348].

Aluminium window profiles and aluminium facade elements currently obtain positive market prices *in situ* (depending on the quantities available and on the circumstances of the demolition)[349]. Correspondingly, not only the recycling and upgrading processes are allocated to the following system but also part of the use phase which is considered a multifunctional process providing the function of the window profile and the secondary material[350] (Figure 11-5).

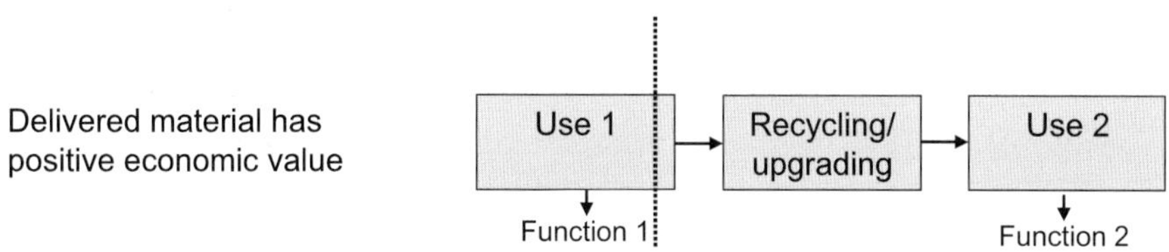

Figure 11-5. Setting of system boundaries and allocation for aluminium window profiles according to the strict coproduct allocation (after Lindeijer and Huppes 2001:689; reprinted with permission from the authors)

In the case of window profiles, the environmental interventions of the use phase are mainly related to the compensation of heat losses through the pro-

[347] see Huppes 1992:64
[348] Huppes 1994:84f and Lindeijer and Huppes 2001:309ff
[349] Buxmann 1998b
[350] reconfirmed by Huppes 2000b, personal communication; also Guinée *et al.* 2004

file by a heating system[351]. According to the strict coproduct allocation, these environmental interventions are allocated partly to the window profile under study and partly to the material leaving the system. In this case, the strict coproduct allocation provides a result that severely interferes with the common understanding of causalities within the socio-economic system. For this reason, the strict coproduct allocation procedure is *a priori* excluded from this case study[352].

The above argument possibly does not hold for all aluminium products. For instance, consumers freely dispose of aluminium beverage cans into collection containers. In this case, the market price of the material reaches a zero-point and recycling would be allocated to the following product system. In this case, the strict coproduct allocation could reasonably be applied; it would provide the same numerical result as the cut-off procedure (see Chap. 11.1).

[351] see, *e.g.*, Richter *et al.* 1996
[352] see also Vogtländer *et al.* 2001

Chapter 12

APPLICATION IN AN LCA OF ALUMINIUM WINDOWS

12.1 Goal and scope definition

The allocation procedures described in Chapters 10 and 11 are applied to a fictitious product system of a window frame. This comparison is made only for methodological considerations. It aims at testing and evaluating the suitability of allocation procedures with emphasis on the value-corrected substitution. Therefore, the main focus is laid on discussing the consequences of different allocation procedures on the LCA-results rather than analysing the causes of environmental burdens on the process level.

Data for the window frame is taken from Richter *et al.* (1996), representing a Zn-containing aluminium frame. For the goal of the case study, a fictitious Zn-free variant is included into the study.

The product system of the window is modelled from a life cycle perspective, including raw material extraction, material and window production, use phase, demolition and recycling (from 'cradle to cradle'). It corresponds with the product system and data evaluated in Werner and Richter (1998a). Additionally, data for thermal insulation glazing with some data update on energy conversion systems and on recycling processes is included.

As the comparison focuses on methodological considerations and not on product comparison, the requirements of data quality are not of primary concern. However, representative data is used (see Chap. 12.5).

The case study uses Central European data of the mid-nineties. Anyhow, the methodological issues focused on are of general interest, independently of the geographical scope. Depending on the allocation procedure, statements are made on recycling processes after the use phase of the window after approx. 30 years of service life. The consideration of changes of the system under study over time has to be discussed.

The studies, on which this case study is based[353], have been subject to internal and external critical peer reviews.

12.2 Functional unit

The aluminium window, on which the different allocation procedures are applied, corresponds to the window evaluated in Werner and Richter (1998a).

A detailed description of the two-winged window under study is made in Richter *et al.* (1996); the thermal insulation glazing is described in Richter and Brunner (1996). As a general overview, some data on the functional unit is presented below (Figure 12-1):

Frame materials: aluminium folding-casement with steel fittings, zinc corner cleats, glass fibre reinforced PA-bridges and EPDM sealing
Dimensions: 1650 x 1300
u-value frame: 1.9 W/m2K
u-value glazing: 1.6 W/m2K
Assumed service life: 30 years

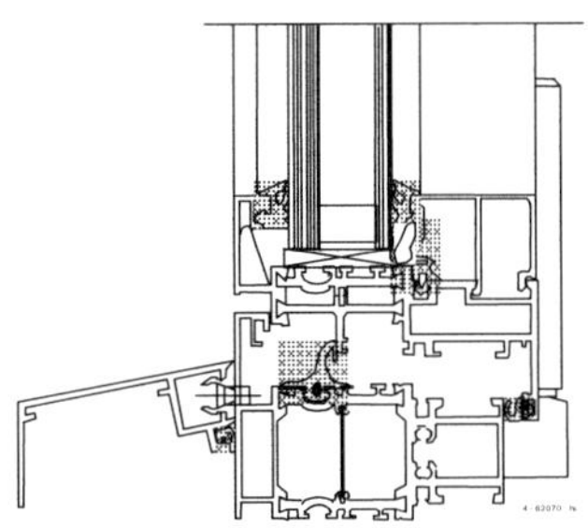

Figure 12-1. Cross-section of the window studied and some specific data

In a scenario, a hypothetical construction is considered where equal masses of the aluminium alloy of the frame substitute for the zinc corner cleats and the brass parts. The alloy composition of the frame material AlMgSi0.5 is shown in Table 12-10.

12.3 Scenarios considered and system boundaries

Some allocation procedures require specific assumptions. Several scenarios are developed to test the influence of these assumptions for other allocation procedures:

- A zinc-free Al-window and a zinc- and brass containing Al(Zn)-window are considered;

[353] Werner 2000; Werner and Richter 1998a

- As dross treatment can be included in the value-corrected substitution (and probably would not be included in the cut-off procedure) scenarios are developed including and excluding dross treatment ('d.t.' for including dross treatment and "no d.t." for not including dross treatment);
- Three allocation procedures are applied and evaluated. They are theoretically described in Chapter 10 and Chapter 11:
  - the *cut-off procedure*: as in Richter *et al.* (1996) a recycled aluminium content of 35% as a general recycling quota for aluminium is assumed;
  - the *closed-loop procedure after ISO/TR 14049* (chap. 8.3.2) with a system-specific recycling yield of 90% (see Table 12-2);
  - the *value-corrected substitution*, starting from frame material equivalent to primary material (VCS).

  The resulting scenarios are (Table 12-1):

Table 12-1. Life cycle scenarios calculated for the Al-window and the Al(Zn)-window

Al-window	Al(Zn)-window
Al; cut-off; d.t.	Al(Zn); cut-off; d.t.
Al; cut-off; no d.t.	Al(Zn); cut-off; no d.t.
Al; closed-loop; d.t.	Al(Zn); closed-loop; d.t.
Al; closed-loop; no d.t.	Al(Zn); closed-loop; no d.t.
Al; VCS; d.t.	Al(Zn); VCS; d.t.
Al; VCS; no d.t.	Al(Zn); VCS; no d.t.

System boundaries are set in correspondence with the basic assumptions underlying each of the allocation procedures.

12.4 Model of the reference flow

The product systems are modelled with several modules representing different stages of the life cycle of the window. The general structure is taken from Richter *et al.* (1996) and Werner and Richter (1998a). Note that depending on the allocation procedure and due to different system boundaries, the modules (*) contain different data for each procedure. All the modules developed for this case study can be found in Werner (2000).

*Frame material**

The window frame consists of AlMgSi05 with glass fibre reinforced PA 6.6 bridges. Most of the aluminium is extruded; a small part of it is hot rolled.Depending on the allocation procedure the aluminium is calculated as follows:

- Cut-off procedure: 35% of the material is assumed to be recycled material, using post-consumer recycling data (see below); 65% is accounted as primary aluminium;
- Closed-loop procedure: 90% of the material is recycled material, using post-consumer recycling data (see below); 10 % is accounted as primary aluminium;
- Value-corrected substitution: all the aluminium input is accounted as primary aluminium. A credit is given for the material value of the recycled material at the end of the life cycle in accordance with the value-corrected substitution.

*Fittings**

Two variants of fittings are considered:

- Al(Zn) fittings consisting of aluminium, zinc pressure die-casting, hot-dip galvanised unalloyed steel and brass;
- Al-fittings consisting of aluminium and hot-dip galvanised unalloyed steel. The zinc and brass parts are replaced by an equal amount of aluminium.

Aluminium for the fittings is assumed to be the same as for the frame material. To aluminium, the same calculations are applied as outlined for the frame material.

Surface treatment Al

The surface of the aluminium window is chromated and finished with a polyester powder lacquer using electricity and natural gas burnt in an industrial LowNO$_x$ burner >100kW.

Window production

For the window production, electricity, glue (epoxy resin), PE(HD) for packaging and propanol for cleaning is used.

Sealing

The sealing is made out of vulcanised EPDM profiles.

Thermal insulation glazing (LCA)

A life cycle inventory of a thermal insulation glazing (k-value=1.6 W/m^{2}K, g-value=0.65 W/m^{2}K) is integrated into the calculation[354].

[354] see Richter and Brunner 1996 for details

Use phase

The Department of Applied Physics in Buildings at EMPA Duebendorf have calculated heat flows through the frame and the glazing. A standard use according to the standard SIA 380/1, which is based on heat grade days of Zurich, is assumed for the modelling of the use phase. A LowNO_x gas heating compensates the heat losses.

*Recycling of process scrap**

The recycling of process scrap is modelled into the frame material in the cut-off and the closed-loop scenarios, assuming material losses of 3% during remelting[355]. For the value-corrected substitution procedure, the recycling of process scrap is integrated as a module of its own[356]. The different treatments are caused by the moved system boundaries; the same material flows are depicted with all procedures.

*Recycling of post-consumer windows**

The data of the recycling processes is based on mass balances and taken from Werner and Richter (1998a). The assumptions are:

- All aluminium windows are entering recycling processes, no land filling of Al-windows;
- The windows enter the shredder without prior treatment such as cutting or stripping;
- Zn-containing windows enter sink/float separation, Zn-free Al-windows enter eddy current separation;
- The treatment of Al-granulate gained from salt slag recycling stemming from the remelting of Al-granulate from dross recycling is not considered any further (< 0.3% of total Al) (see also Figure 12-2);
- For the eddy current separation, a separation efficiency of 99% for Fe and 95% for Zn and brass is assumed, whereas 5% of the total Al remains in these fractions;
- For the sink/float separation, no window-specific data is available concerning the separation efficiency. Additionally, no data has been found to quantify the Fe- input to the aluminium fraction due to incomplete washing off of the ferrosilicon after processing[357]. Based on information from scrap processors in Germany, it is assumed that the output material

[355] a strict closed-loop is assumed for process scrap
[356] quantities and material value are modelled explicitly
[357] see Krone 2000

has the elementary composition of a casting alloy corresponding to DIN S226/231[358].
The data for an Al- and an Al(Zn)-window is shown in Table 12-2.
The underlying aluminium flows are shown in Figure 12-2.

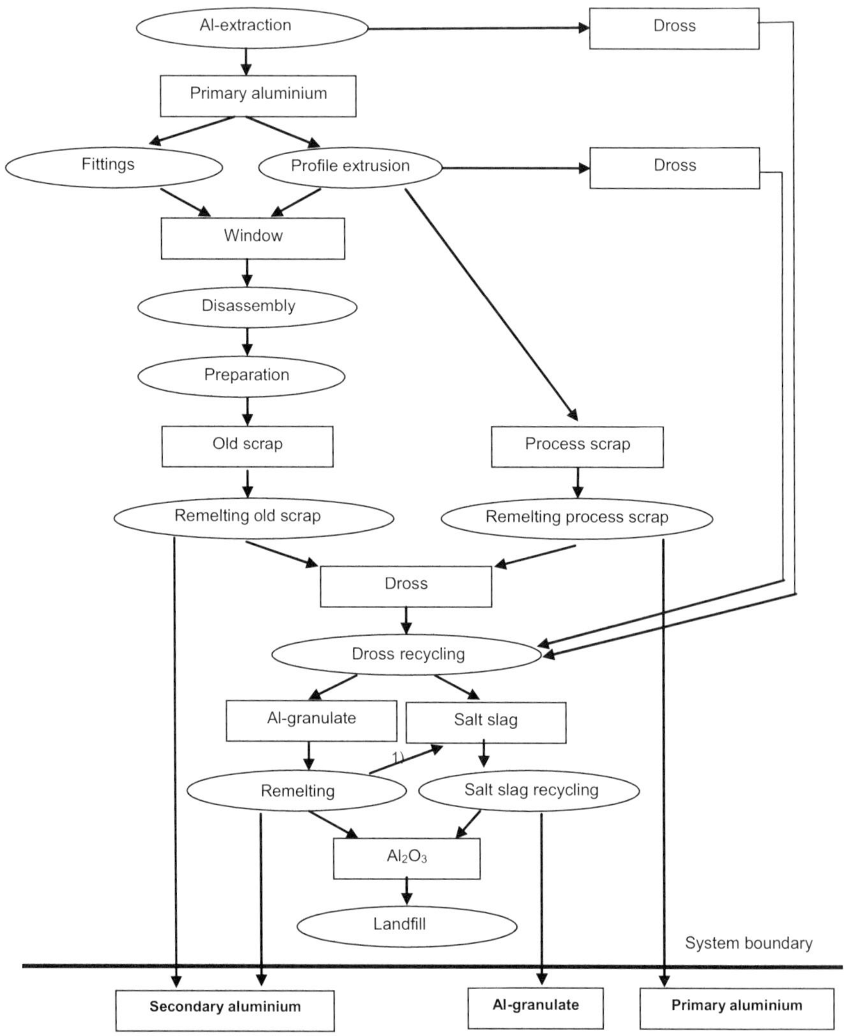

Figure 12-2. Material flow diagram for aluminium including the allocation factors
[1)] No recursion is calculated for Al-granulate stemming from the salt slag treatment originating from the remelting from Al-granulate.

[358] No elementary analysis of the recycled material was made (Werner and Richter 1998a:17).

Table 12-2. Material flows of the post-consumer recycling processes for the Al-window and the Al(Zn)-window (after Walter 1998, Anonymous 1996)

		Al(Zn)-window		Al-window		
	Output	%	kg	%	kg	Processing, remarks
Demolition	Scrap	100%	39.78	100%	39.79	Shredding; containing 29.03 kg and 31.25 kg aluminium respectively
	Waste	0%	0.00	0%	0.00	Landfill
	Total		39.78		39.79	
Shredding	Al/Zn-fraction	94%	37.28			Sink/float separation
	Fe(Al)-fraction	3.9%	1.67			5% Al/99% Fe
	Al/Fe-dust	1.1%	0.44			Landfill
	Loss	1.0%	0.40			Loss
	Total		39.78			
Shredding	Al-fraction			93%	37.17	Eddy current separation
	Fe(Al)-fraction				1.78	99% Fe/5% Al: Fe-recycling
	Al/Fe-dust			1.1%	0.44	Landfill
	Loss			1.0%	0.40	Loss
	Total				39.79	
Sink/float separation	Al-fraction	71%	26.37			Remelting
	Fe/Zn-fraction	6%	2.22			Recycling
	Non-metals	22%	8.32			Landfill
	Loss	1%	0.37			Loss
	Total		37.28			
Eddy current separation	Al-fraction			77%	28.48	Remelting
	Non-metals			22%	8. 32	Landfill
	Loss			1%	0.37	Loss
	Total				37.17	
Remelting of old scrap	Secondary Al	94%	24.79	94%	26.77	
	Al into dross	6%	1.58	6%	1.71	Dross treatment
	Total		26.37		28.48	
Dross recycling	Salt slag	6%	0.13	6%	0.14	Real weight including oxygen: 0.694 kg and 0.748
	Al-granulate	94%	2.02	94%	2.17	Real weight including oxygen: 2.25 kg and 2.43 kg
	Total		2.15		2.31	Real weight including oxygen: 2.94 kg and 3.17 kg
Salt slag recycling	Al-granulate	33%	0.04	33%	0.05	Into rotary furnace (considered in module)
	Al_2O_3	67%	0.09	67%	0.09	Landfill
	Total		0.13		0.14	
Rotary furnace	Secondary Al	80%	1.61	80%	1.74	
	Salt slag	3%	0.06	3%	0.07	
	Al_2O_3	17%	0.34	17%	0.37	Landfill
	Total		2.02		2.18	

This data is converted into inventory data, linking it with EAA data as follows in the Table 12-3 and Table 12-4.

Table 12-3. Total mass flows for the window recycling for the Al(Zn)-window, based on mass flow calculations from Table 12-2.

Al(Zn)-window	total kg	incl. dross treatment per kg scrap input kg	incl. dross treatment per kg Al output kg	no dross treatment per kg scrap input kg	no dross treatment per kg Al output kg
Shredding	39.790	1.000	1.507	1.000	1.605
Eddy current-separation	-	-	-	-	-
Sink/float-separation	37.280	0.937	1.412	0.937	1.504
Compacting/refining	26.370	0.663	0.999	0.663	1.064
Dross treatment	2.150	0.054	0.081	0	0
Landfill	8.760	0.220	0.332	0.220	0.353
Recycled aluminium yield	28.160	0.666	1.000	0.623	1.000

Table 12-4. Total mass flows for the window recycling for the Al-window, based on mass flow calculations from Table 12-2.

Al-window	total kg	incl. dross treatment per kg scrap input kg	incl. dross treatment per kg Al output kg	no dross treatment per kg scrap input kg	no dross treatment per kg Al output kg
Shredding	39.790	1.000	1.396	1.000	1.486
Eddy current-separation	37.170	0.934	1.304	0.934	1.388
Sink/float-separation	-	-	-	-	-
Compacting/refining	28.480	0.716	0.999	0.716	1.064
Dross treatment	2.310	0.058	0.081	0	0
Landfill	8.760	0.220	0.307	0.220	0.327
Recycled aluminium yield	30.310	0.712	1.000	0.673	1.000

The aluminium mass balances of the two window variants are presented in Table 12-5 and Table 12-6. The values for the value-corrected substitution are determined for both options including dross treatment and disregarding dross treatment. The allocation factors for the two window variants are determined in Table 12-9 and Table 12-10.

*Transport**

The assumptions about vehicles, transport distances and loaded capacity are taken from Richter *et al.* (1996). They include the transport of the intermediate products to the production site, the delivery of the window to the construction site and the transport for both the process scrap and post-consumer scrap to recycling facilities.

Table 12-7 and Table 12-8 show the assumptions made to model the transport related to recycling for the two window variants and for the inclusion or exclusion of the dross treatment. For the recycling of the process scrap, transportation with a truck 28 t over 100 km is assumed. As the process scrap is transported in a 'closed-loop' (back and forth) with 3% material

Table 12-5. Aluminium mass balance for the Al(Zn)-window and determination of the values of the VCS, depending on the inclusion or exclusion of dross recycling

	kg		kg	VCS; incl. d.t. alloc. factor	kg	VCS; no d.t. alloc. factor	kg
Input aluminium		*Output aluminium*					
Frame material/fittings (Al(Zn))	29.03	Landfill from demolition	0.00	0	0.00	0	0.00
Process scrap	1.15	Secondary Al from old scrap	24.79	0.9	22.31	0.9	22.31
Dross from Al production	0.18	Al in process scrap	1.11	1	1.11	1	1.11
Dross from profile production	0.35	Secondary Al from dross treat.	1.61	0.9	1.45	0	0.00
Total Al(Zn)	30.71	Loss	0.77	0	0.00	0	0.00
		Al in salt slag/ Al-granulate	0.11	0	0.00	0	0.00
		Landfill from recycling	0.87	0	0.00	0	0.00
		Total Al	29.26		24.88		23.43
Input other materials		*Output other materials*					
Fe/other metals/non-metals	10.75	Fe/other metals/ non-metals	12.20	including Al loss due to incomplete separation			

Table 12-6. Aluminium mass balances for the Al-window and determination of the values for the VCS, depending on the inclusion or exclusion of dross recycling

	kg		kg	VCS; incl. d.t. alloc. factor	kg	VCS; no d.t. alloc. factor	kg
Input aluminium		*Output aluminium*					
Frame material/fittings (Al)	31.25	Landfill from demolition	0.00	0	0.00	0	0.00
Process scrap	1.15	Secondary Al from old scrap	26.77	1	26.77	1	26.77
Dross from Al production	0.19	Al in process scrap	1.11	1	1.11	1	1.11
Dross from profile production	0.37	Secondary Al from dross treat.	1.74	1	1.74	0.0	0.00
Total Al	32.96	Loss	0.77	0	0.00	0	0.00
		Al in salt slag/ Al-granulate	0.12	0	0.00	0	0.00
		Landfill from recycling	0.90	0	0.00	0	0.00
		Total Al	31.41		29.62		27.88
Input other materials		*Output other materials*					
Fe/other metals/non-metals	8.54	Fe/non-metals	10.10	incl. Al loss due to incomplete separation			

losses, the factor 2.03 is used to calculate the tkm used for the process scrap transportation[359].

Table 12-7. Assumptions made to model the transport during the post-consumer recycling step for the Al(Zn)-window

Al(Zn)-Window		Al(Zn) incl. dross treat.		Al(Zn) no dross treat.	
		km	kg	tkm	tkm
Demolition-shredder/EC-separator	Truck 16 t		39.78		
	Truck 40 t		39.78		
Demolition-shredder	Truck 16 t	50	39.78	1.989	1.99
	Truck 40 t	500	39.78	19.89	19.89
Shredder - sink/float-separation	Truck 40 t	100	37.28	3.73	3.73
Separation – remelting	Train	700	26.37	18.46	18.46
Remelting - dross treatment	Train	2000	2.94	5.88	
Landfill	Truck 16 t	50	8.76	0.44	0.44
Total	Truck 16 t			2.437	2.43
	Truck 40 t			23.62	23.62
	Train			24.34	18.46

Table 12-8. Assumptions made to model the transport during the post-consumer recycling step for the Al-window

Al-Window		Al incl. dross treat.		Al no dross treat.	
		km	kg	tkm	tkm
Demolition-shredder/EC-separator	Truck 16 t	50	39.78	1.99	1.99
	Truck 40 t	500	39.78	19.89	19.89
Demolition-shredder	Truck 16 t	50			
	Truck 40 t	500			
Shredder - sink/float-separation	Truck 40 t	100			
Separation – remelting	Train	700	28.48	19.94	19.94
Remelting - dross treatment	Train	2000	3.17	6.34	
Landfill	Truck 16 t	50	8.76	0.44	0.44
Total	Truck 16 t			2.437	2.43
	Truck 40 t			23.62	23.62
	Train			24.34	18.46

*Value-corrected substitution**

A material credit is given for partly substituting primary aluminium production based on the yield of the recycling processes and on the value of the recycled material. Final disposal of aluminium is not considered in the value-corrected substitution for reasons given in Chapter 10.5.

[359] tkm = 2.03 x process scrap reused in window x distance window producer to remelter

Table 12-9. Alloy composition of the Al(Zn)-window scraps compared to input alloy

Al(Zn)-window	Quantity	Al	Fe	Zn	Mg	Cu	Si
Input							
Frame (AlMgSi0.5) (kg) [1)]	28.690	28.320	0.057	0.029	0.136	0.014	0.129
Fittings (AlMgSi0.5) (kg) [1)]	0.340	0.340	0.001	3E-04	0.002	2E-04	0.002
Composition AlMgSi0.5[1)]			0.20%	0.10%	0.48%	0.05%	0.45%
Brass (kg)	0.040			0.017		0.023	
Steel unalloyed (kg)	0.220		0.22				
Zn(Al) pressure die-casting (kg)	2.180	0.109		2.03		0.044	
Total (kg)	31.470	28.770	0.278	2.073	0.138	0.081	0.131
Composition in % total weight		91.42%	0.88%	6.59%	0.44%	0.26%	0.42%
Ouput							
Composition after recycling (kg)	27.790	n.k	n.k	n.k	n.k	n.k	n.k
Composition in % total weight		n.k	n.k	n.k	n.k	n.k	n.k
Assumed composition							
Secondary alloy DIN S226 [2)]		rest	1.00%	1.20%	0.50%	3.50%	11.00%

[1)] after DIN 1725/1
[2)] maximal values

Table 12-10. Alloy composition of the Al-window scraps, compared to input alloy

Al-window	Quantity	Al	Fe	Zn	Mg	Cu	Si
Input							
Frame (AlMgSi0.5) (kg) [1)]	28.690	28.320	0.057	0.029	0.136	0.014	0.129
Fittings (AlMgSi0.5) (kg) [1)]	2.560	2.530	0.005	0.003	0.012	0.001	0.012
Brass (kg)	0						
Steel unalloyed (kg)	0.220		0.22				
Zn(Al) pressure die-casting (kg)	0						
Total (kg)	31.470	30.850	0.283	0.031	0.148	0.016	0.141
Composition after recycling (kg)	29.690	29.310	0.062	0.030	0.141	0.015	0.134
Output							
Comparison Al-input vs. recycled Al					n.k	n.k	n.k
Composition Al Input (AlMgSi0.5) in % [1)]			0.20%	0.10%	0.48%	0.05%	0.45%
Composition recycled Al fraction in %		98.72%	0.21%	0.10%	0.47%	0.05%	0.45%

[1)] after DIN 1725/1

In the Tables 12-9 and 12-10, mass balances for the aluminium alloy elements are established for both window variants and compared to the grades traded at the LME. The elementary composition of the Al scrap has not been changed over the product life cycle compared to the alloy used to produce the window profile. Therefore, a value correction factor of 1 is applied, resulting in a closed-loop procedure based on equivalent in- and outputs.

As separation efficiency of the sink/float process for different scrap fractions is not know, it is assumed that the Al(Zn)-window scrap corresponds to the grade DIN S226. A value correction of 0.9 is applied to the Al(Zn)-window. This means that in addition to the 10% material lost, another 10% (of the 90%) are subtracted for the value loss. This factor corresponds to the long-term price ratio of primary aluminium contract and the alloyed aluminium contract (the grade DIN S226 as one of them) at the LME.

12.5 Data quality

The data used for the inventory corresponds to average Central-European technology in the second half of the 1990's (1997).

For the window frame, data is taken from:

a) One production site of aluminium window frames in Switzerland[360];
b) Average EAA-inventories for aluminium specific data[361]; EAA-data of recycling processes refers to automotive scrap;
c) ESU-inventories for data of transport, energy and disposal systems[362];
d) Two recycling sites in Germany for the yield of the recycling processes[363].

For electricity production the UCPTE-mix is applied except the aluminium production processes, where the specific electricity-mix of the European aluminium smelters is used[364]. Data used for the modelling of the main processes of the aluminium window frame can be considered to be the best available.

Data for the thermal insulation glazing has been collected from several glazing producers in Switzerland[365]. Data quality for the glazing can be described as good – at least for the glazing production and its preceding processes.

360 see Richter *et al.* 1996
361 Anonymous 1996c and unpublished EAA-data
362 Frischknecht *et al.* 1996
363 Werner and Richter 1998a
364 Anonymous 1996c
365 see Richter and Brunner 1996

Data quality can be considered adequate for the goal and scope of the study.

12.6 Inventory analysis

A complete life cycle inventory can be ordered from the author.

12.7 Impact assessment

In line with the goal of this study, updated Eco-Indicator 95[366] is chosen for the assessment of the LCI. Eco-Indicator 95 is an assessment method that is based on effect-oriented classification as the CML-methodology[367]. But Eco-indicator 95 goes some steps further by normalising the impact categories with the average impact of one European person, evaluating and aggregating them based on critical flows with regard to three safeguard subjects: fatalities, human health impairment and eco-system impairment. As a sum, a single score – Eco-indicator 95 points – results.

Table 12-11. Impact assessment of the Al-window (Eco-indicator 95 points)

Al-windows Allocation procedure incl. dross treatment/no	cut-off incl. d.t.	cut-off no d.t.	closed-loop incl. d.t.	closed-loop no d.t.	VCS incl. d.t.	VCS no d.t.
Value-corrected substitution					-2.182	-2.055
Frame material	1.681	1.684	0.407	0.408	2.585	2.585
Surface treatment	0.032	0.032	0.032	0.032	0.032	0.032
Window production	0.012	0.012	0.012	0.012	0.012	0.012
Fittings Al	0.212	0.212	0.102	0.102	0.263	0.263
Sealing	0.032	0.032	0.032	0.032	0.032	0.032
Thermal insulation glazing (LCA)	0.732	0.732	0.732	0.732	0.732	0.732
Use phase	0.164	0.164	0.164	0.164	0.164	0.164
Transport window	0.037	0.037	0.057	0.057	0.057	0.055
Recycling process scrap	1)	1)	1)	1)	0.002	0.002
Recycling old scrap					0.047	0.045
Total	2.901	2.904	1.537	1.538	1.743	1.867

1) considered closed-loop; added to the amount calculated as recycled metal content for the frame material

366 Goedkoop 1995; Förster *et al.* 1998

367 see Heijungs *et al.* 1992; anyhow, Eco-indicator 95 uses some different impact categories

Table 12-12. Impact assessment of the Al(Zn)-window (Eco-indicator 95 points)

Al(Zn)-windows Allocation procedure incl. dross treatment/no	cut-off incl. d.t.	cut-off no d.t.	closed-loop incl. d.t.	closed-loop no d.t.	VCS incl. d.t.	VCS no d.t.
Value-corrected substitution					-2.028	-1.910
Frame material	1.681	1.684	0.407	0.408	2.585	2.585
Surface treatment	0.012	0.012	0.012	0.012	0.012	0.012
Al-window production	0.032	0.032	0.032	0.032	0.032	0.032
Fittings Al(Zn)	2.830	2.830	2.816	2.816	2.843	2.843
Sealing	0.032	0.032	0.032	0.032	0.032	0.032
Thermal insulation glazing (LCA)	0.732	0.732	0.732	0.732	0.732	0.732
Use phase	0.164	0.164	0.164	0.164	0.164	0.164
Transport Al-window	0.038	0.038	0.061	0.061	0.060	0.058
Recycling process scrap	1)	1)	1)	1)	0.002	0.002
Recycling old scrap					0.050	0.047
Total	5.521	5.524	4.254	4.256	4.484	4.596

1) considered closed-loop; added to the amount calculated as recycled metal content for the frame material

As an effect-oriented, single score impact assessment method, it allows one to focus on the influence of the different allocation procedures to be discussed in this case study. The results are shown in Table 12-11 and Table 12-12.

Figures 12-3 to 12-11 illustrate the results of different comparisons (net impacts): for graphical reasons, the credit given for the 'value-corrected substitution' (the negative part in the figures) is subtracted from the 'frame material'. The two absolute amounts add up to the total impacts of the 'frame material' at the input side[368].

Contributions to the Global Warming Potential determined after Heijungs *et al.* (1992) and Houghton *et al.* (1996) are presented as additional information in Appendix A2.

[368] 'Value-corrected substitution' includes not only material recovered from 'frame materials' but also from 'fittings'.

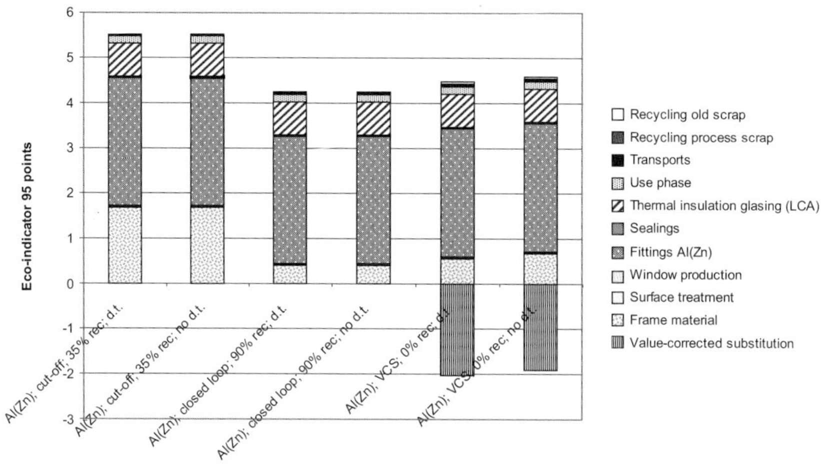

Figure 12-3. Impact assessment of the Al(Zn)-window, overview (net impacts)
For graphical reasons, the credit given for the 'value-corrected substitution' (the negative part in the Figure 12-3) is subtracted from the 'frame material'. The two absolute amounts add up to the total impacts of the 'frame material' at the input side.

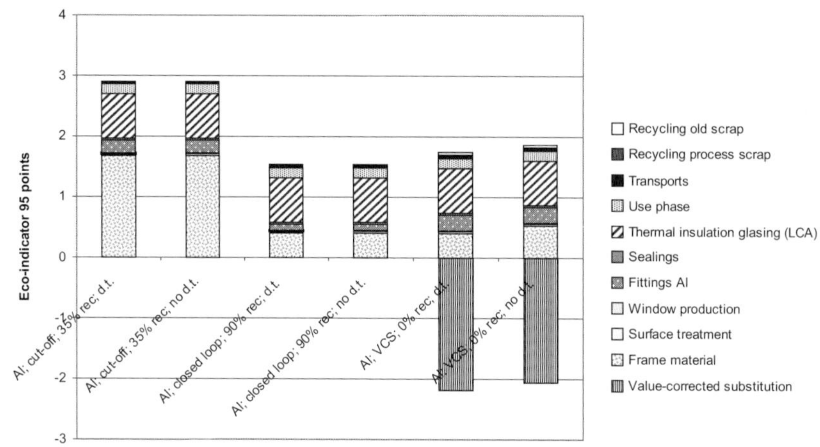

Figure 12-4. Impact assessment of the Al-window, overview (net impacts)
For graphical reasons, the credit given for the 'value-corrected substitution' (the negative part in the Figure 12-4) is subtracted from the 'frame material'. The two absolute amounts add up to the total impacts of the 'frame material' at the input side.

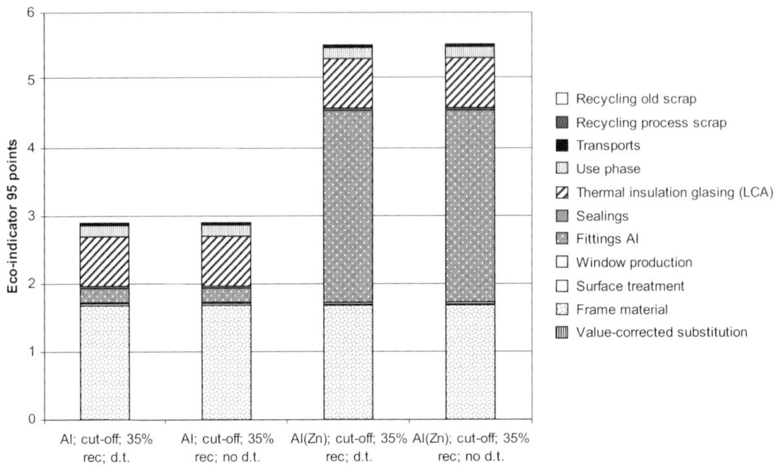

Figure 12-5. Impact assessment applying the cut-off allocation procedure

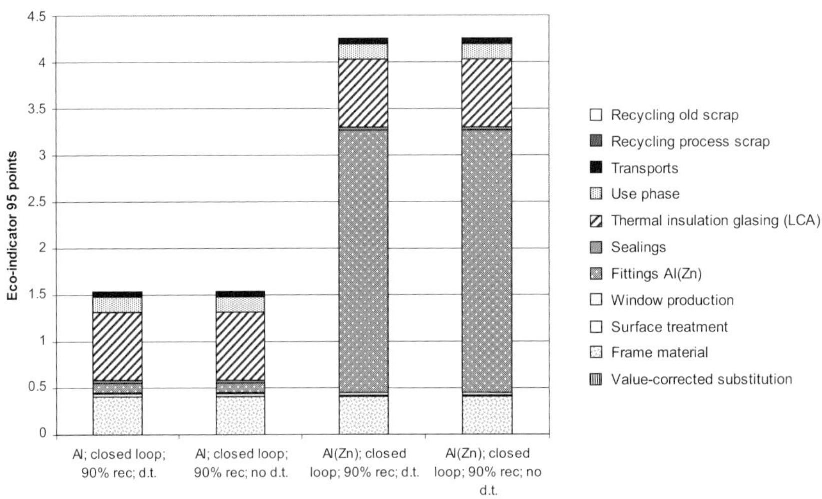

Figure 12-6. Impact assessment applying the closed-loop allocation procedure

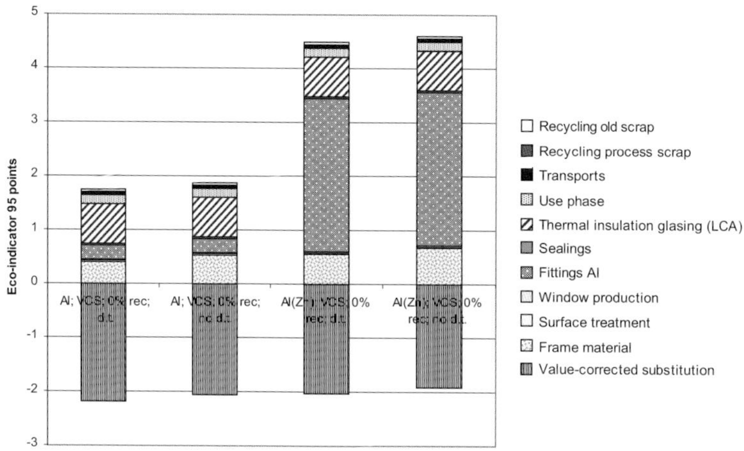

Figure 12-7. Impact assessment applying the value-corrected substitution procedure
For graphical reasons, the credit given for the 'value-corrected substitution' (the negative part in the Figure 12-7) is subtracted from the 'frame material'. The two absolute amounts add up to the total impacts of the 'frame material' at the input side.

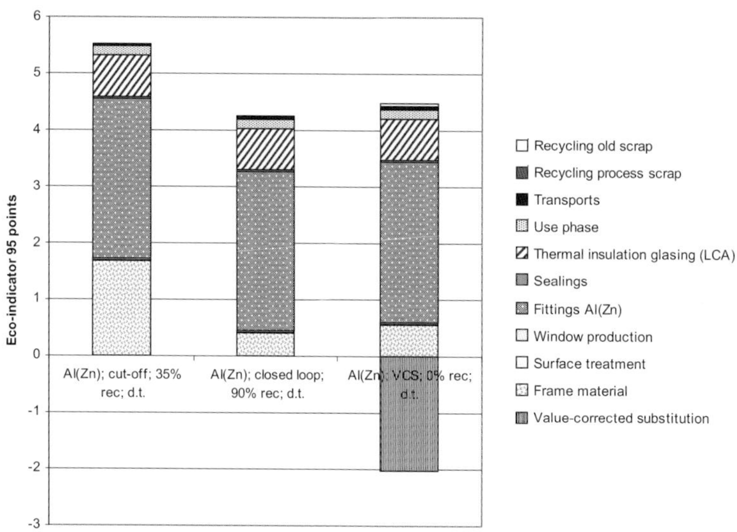

Figure 12-8. Impact assessment for the Al(Zn)-window applying different allocation procedures, considering dross recycling
For graphical reasons, the credit given for the 'value-corrected substitution' (the negative part in the Figure 12-8) is subtracted from the 'frame material'. The two absolute amounts add up to the total impacts of the 'frame material' at the input side.

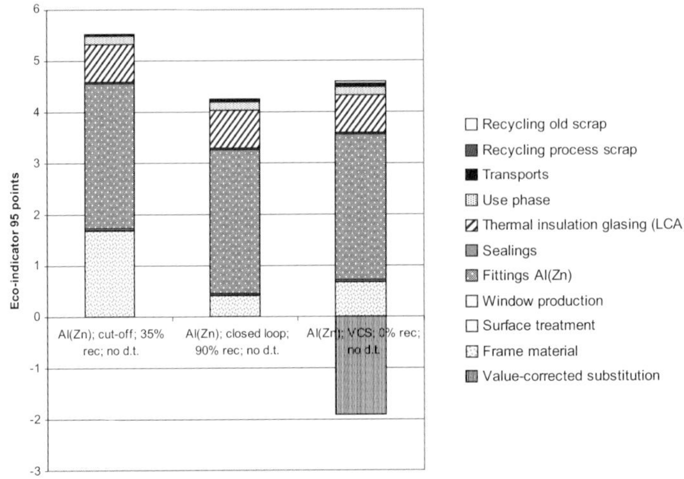

Figure 12-9. Impact assessment for the Al(Zn)-window applying different allocation procedures, not considering dross recycling
For graphical reasons, the credit given for the 'value-corrected substitution' (the negative part in the Figure 12-9) is subtracted from the 'frame material'. The two absolute amounts add up to the total impacts of the 'frame material' at the input side.

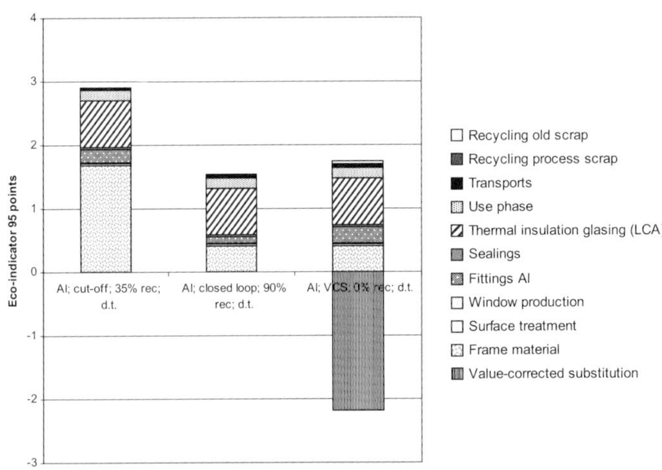

Figure 12-10. Impact assessment for the Al-window applying different allocation procedures, considering dross recycling
For graphical reasons, the credit given for the 'value-corrected substitution' (the negative part in the Figure 12-10) is subtracted from the 'frame material'. The two absolute amounts add up to the total impacts of the 'frame material' at the input side.

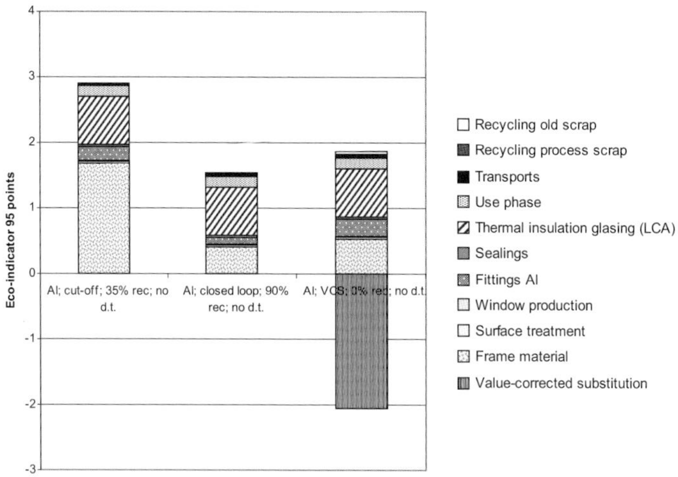

Figure 12-11. Impact assessment for the Al-window applying different allocation procedures, not considering dross recycling
For graphical reasons, the credit given for the 'value-corrected substitution' (the negative part in the Figure 12-11) is subtracted from the 'frame material'. The two absolute amounts add up to the total impacts of the 'frame material' at the input side.

12.8 Interpretation

12.8.1 General remark

In line with the goal of the case study, the following interpretation of the impact assessment focuses on methodological considerations related to the application of different allocation procedures. It has abstained from analysing the results on the process level to determine the main influences. Complementary information on the main impacts on the process level can be gained from Richter *et al.* (1996a/1996b) and Richter (1999).

12.8.2 Interpretation focusing on methodological considerations

Compared to secondary aluminium production, primary aluminium production causes high environmental impacts. One of the most relevant environmental aspects is the greater energy consumption and the respective airborne emissions[369]. This is clearly reflected in the results:

- The higher the amount of recycled metal attributed to the product system the lower the environmental impacts (cut-off with 35% recycled material attributed > closed-loop with 90% recycled material attributed);
- The value-corrected substitution leads to slightly higher environmental impacts than the closed-loop procedure.

From a numerical point of view, the results can be summarised as follows:

cut-off » VCS > closed-loop

The Al(Zn)-window scores approximately twice as high as the Al-construction. This is due to high amounts of zinc released into air (10 g/kg Zn) during zinc extraction for the fitting production[370]. Zinc – as with all metal emissions – are weighted strongly in the Eco-indicator 95-method. The dominating effect of metals in this impact assessment method in LCA-results is a well-known phenomenon.

The scores are much closer if, *e.g.,* the greenhouse potential is used for comparison (see Appendix A2). However, the ranking of the two window variants remains the same as discussed in the following.

Apart from the influence of the fitting production, the results do not differ between the zinc-free and the zinc-containing construction for methods not including material value (closed-loop, cut-off).

The value-corrected substitution leads up to around 10% lower results for the Zn-free construction (besides the effect of the Zn-emissions) for both the Eco-indicator 95 and the greenhouse gas potential. This is due to the fact that the window profiles of the zinc-free Al-window can be recycled without devaluation; the profiles of the Al(Zn) have to be devalued about 10% and carry the respective environmental burdens from primary material production. Additionally, the Al(Zn)-window has to carry additional environmental burdens from the sink-float processes and from the corresponding aluminium losses:

Al(Zn)-window > Al-window

[369] Secondary aluminium production needs about 5% of the energy needed for primary aluminium production (Krone 2000:13).

[370] Data taken from Frischknecht *et al.* 1996

The inclusion of the dross treatment into the recycling processes leads to slightly lower environmental impacts than not considering the treatment. As the environmental impact of the dross treatment is relatively low, the differences between including and excluding the dross treatment almost does not (<0.1%) affect the results for allocation procedures that do not take into account the value of the recovered material (closed-loop, cut-off).

For the value-corrected substitution, the value of the aluminium gained from the dross treatment offsets the environmental impacts of the transport and processing.

excluding dross treatment > including dross treatment

12.8.3 Data quality and sensitivity

For the goal of the case study, data quality can be considered sufficient, as the best available representative data has been used. The data extrapolation for the definition of a Zn-free version of the aluminium window is based on the best available data; insecurities related to the extrapolation do not affect the methodological conclusions as a goal of this case study.

For the goal of the study, system changes over time inherent to reuse & recycling are of primary concern for sensitivity analysis. This aspect is treated in the following chapter.

Chapter 13

METHODOLOGICAL CONCLUSIONS

The following methodological conclusions can be drawn, also taking into account the requirements of an allocation procedure for reuse & recycling of aluminium (see Chap. 9.6).

13.1 Cut-off procedure

The cut-off procedure is easy to use and therefore widely applied. The easy-to-use advantage of this procedure contrasts with some disadvantages for aluminium products, *e.g.,* in building applications.

As a precondition for the applicability of the cut-off procedure, the content of recycled material has to be determinable in a reasonable way. Due to the specific situation in aluminium production, the recycling content of a specific intermediate product or alloy cannot be determined in a sensible way: a specific alloy grade is usually achieved by mixing primary material and/or a variety of different old and new scraps with different metal compositions to get the desired metal content of the alloy[371].

Average data on recycled metals compared to total aluminium consumption is used to determine the recycled metal content instead. However, data is used for calculation that is not directly linked to the specific system under study. As a consequence, the cut-off procedure does not provide actor-based and site-specific results.

The cut-off procedure favours the use of secondary material for materials like aluminium for which primary material production processes cause considerably higher environmental interventions than the recycling processes.

[371] see Rink 1993:117ff; Rombach 1998:421; Rombach 1999:119

This is due to the fact that no environmental burden of primary material production of the secondary material is considered.

The resulting recommendation to use more recycled material is only valid in expandable markets, where the supply of recycled material is elastic. Boguski *et al.* (1996) point out:

> "The increase of industrial scrap or trim by one company simply takes the scrap or trim away from another company because the supply of scrap or trim is not expandable. From a narrow, individual company point of view, recycling of industrial scrap is good and may reduce the energy and wastes allocated to its operations, but from the larger point of view, no environmental value accrues to society. One possible exception that increased demand for industrial scrap may force marginal users to replace industrial scrap with post-consumer scrap" (Boguski *et al.* 1996:20f).

Thus the cut-off allocation procedure does not consider material and market characteristics of aluminium in a suitable way. Therefor, the procedure does not support the primary environmental improvement strategies applied in the environmentally conscious product design of aluminium products:

- No emphasis at all is put on the value of the secondary material at the end of the life cycle and therefore no incentive is given for the preservation of the value of the material. As a consequence, efforts to optimise the recyclability of a product by applying strategies like design for recycling will not be depicted in an LCA;
- Furthermore, only small incentives are given to produce recyclable goods; the incentive consists in preventing environmental impacts of waste treatment processes[372].

Thus, this allocation procedure makes LCA unsuitable for one of its most important applications: decision support during environmental product design of aluminium products.

As primary material production and final waste disposal may have to be allocated over several subsequent product systems (ISO/EN 14041, chap. 6.5.4, point a), the choice of the cut-off procedure has to be substantiated, referring to the shortcomings mentioned above.

In view of the criteria for LCIs established in Chap. 3, the following points have to be mentioned:

- The cut-off procedure is very easy-to-use and provides a consistent allocation procedure. As such, the resulting model is complete within the

[372] see also Tillman *et al.* 1994:28

logic of the cut-off procedure; it is operational, decomposable, non-redundant and minimal. The resulting LCI is not necessarily comparable to LCIs of products made of other materials. Applying the same allocation procedure to different materials does not necessarily provide consistent and adequate – and thus comparable – models, as management rules for the sustainable use of different materials can differ (see Chap. 3.3.5). This has to be reflected in LCIs, *e.g.*, by applying different allocation procedures;

- The cut-off procedure treats allocation of recycling in a simple, manageable, transparent way and allows quick recalculations. On the other hand, it does not provide an 'adequate' system description (see below);
- The cut-off procedure is not actor-based in the case of windows or facade elements of aluminium. It recommends environmental improvement options that lay beyond the action space of aluminium window producers (and even scrap dealers) because it focuses on the recycled metal content of the input material. On the other hand, recycling properties of the products – which are of primary interest because they can directly be influenced by the decision-maker – lie outside the model. Furthermore, the cut-off procedure does not allow one to consider different grades of aluminium; downcycling cannot be taken into account. It must be concluded that material and market characteristics of the aluminium in windows and facades are not properly addressed by the cut-off procedure;
- For the same reason, the cut-off procedure does not provide a very case-specific model. On the other hand, it does not require the use of 'uncertain data' of future recycling processes. For this reason, the modellers and decision-makers with an averse attitude towards risk could prefer the cut-off procedure;
- The cut-off procedure solves the allocation problem related to recycling by system boundary setting. The decision-maker's mental model of the organisational principle is of minor importance for this allocation problem (no choice of an allocation factor). Still, the conception remains that treating recollection and recycling processes as 'secondary raw material extraction' in LCA leads to a good basis for decision-making;
- As outlined above, the cut-off procedure provides improvement options that are not in line with the sustainable management rules for aluminium, referring to the material and energy flows within the technosphere. Following improvement options indicated by the cut-off procedure, over-all environmental impacts rise (see also Chap. 1.1);
- The question concerning environmental efficiency and eco-efficiency of the resulting model cannot be answered in general terms but only in a specific application and decision situation.

To conclude, the cut-off procedure cannot be recommended for LCA of aluminium products.

13.2 Closed-loop procedure based on a material pool model

The closed-loop procedure based on a material pool model does not make a distinction between different aluminium grades, as material value is not considered. This implies that environmental improvement strategies applied to aluminium products – such as maintaining material value by developing single alloy products or fostering a higher separation effectiveness of the recycling processes – are not adequately depicted in an LCA.

The equivalency of primary and secondary material as a crucial assumption and precondition of the closed-loop approach will always be a debatable approximation of reality where downcycling commonly occurs.

As a mathematical consequence of the closed-loop procedure, the procedure calculates the recycled output to be the input of recycled material. Thus, product-specific recycling data is used for the allocation procedure.

The closed-loop procedure assumes that recycled material substitutes primary material. But recycled material does not only compete with primary material but also with other materials. The recycled material could even not replace any material at all if used for functions which would otherwise not be generated. The mix of materials actually replaced by an increased outflow of recycled material depends on how the market for the recycled material reacts to the increased supply[373]. Due to the much higher production costs of primary aluminium compared to secondary aluminium and due to the high cross-substitution elasticity of aluminium scrap compared to primary aluminium prices[374], the assumption that secondary material replaces primary material seems justifiable.

Time dependency of future recycling and waste disposal processes need to be taken into account. 'Best guess' scenarios should be complemented with an 'actual situation' scenario in a sensitivity analysis.

By applying the closed-loop procedure based on a material pool, incentive is given to produce recyclable products. No incentive is given to use recycled material; no emphasis is put on the value preservation of a material as equivalency of input and output materials is *a priori* assumed[375]. Hence,

[373] after Ekvall and Finnveden 1999:10

[374] Carlsen 1980:46f

[375] see also Tillman *et al.* 1994:28

the primary environmental improvement strategies applied in product design of aluminium products are only partially depicted.

In view of the criteria for LCIs established in Chap. 3, the following points have to be mentioned:

- The procedure is very easy-to-use and provides a consistent allocation procedure. As such the resulting model is complete within the logic of the closed-loop procedure based on material pools. Furthermore, the resulting model is operational, decomposable, non-redundant and minimal. The resulting LCI is not necessarily comparable to LCIs of products made of other materials (see above);
- The closed-loop procedure treats allocation of recycling in a simple, manageable, transparent way and allows quick recalculations. On the other hand, it does not provide an 'adequate' system description (see below);
- The closed-loop procedure is not actor-based in the case of windows or facade elements of aluminium, although more so than the cut-off procedure. The scope for action lies on the recycling properties of the aluminium product; still, the procedure only considers mass flows; qualitative issues, *i.e.* downcycling, are disregarded. It must be concluded that material and market characteristics of the aluminium in windows and facades are not completely addressed by the closed-loop procedure based on material pools;
- The closed-loop procedure based on material pools provides a more case-specific model than the cut-off procedure. On the other hand, it requires the use of data of future recycling processes, which could be considered as less site- and case-specific. For this reason, the modellers and decision-makers with a risk-calculating or risk-taking attitude towards risk (*e.g.*, depending on service life) could prefer the closed-loop procedure to the cut-off procedure;
- The closed-loop procedure based on material pools solves the allocation problem related to recycling by system boundary setting. The decision-maker's mental model of the organisational principle is of minor importance for this allocation problem (no choice of an allocation factor). Still, the conception remains that focussing on post-consumer recycling in LCA leads to a good basis for decision-making;
- As outlined above, the closed-loop procedure provides improvement options that only consider quantitative aspects of the sustainable management rules for aluminium referring to the material and energy flows within the technosphere. Qualitative, entropic aspects, *i.e.* downcycling, are disregarded;

- The question about environmental efficiency and eco-efficiency of the resulting model cannot be answered in general terms but only in a specific application and decision situation.

The closed-loop procedure based on the materials-pool model can partly be recommended for LCA of aluminium products if equivalency of the input and output material can be assumed. Furthermore, it can be applied in cases in which – for any reason – price-based allocation procedures are rejected. In any case, the closed-loop procedure based on the materials-pool model is preferable to the cut-off procedure for LCA of aluminium products in building applications.

13.3 Strict coproduct allocation

If the material to be recycled achieves positive market prices (which is the case for many aluminium products, especially in building applications) the use phase is considered a multifunctional process, enabling the function of the product in use and providing the secondary material for subsequent applications.

Aluminium window profiles and aluminium facade elements currently obtain positive market prices *in situ* (depending on the quantities available and on the circumstances of the demolition). Correspondingly, not only the recycling and upgrading processes are allocated to the following system but also part of the use phase, which is considered a multifunctional process providing the function of the window profile and the secondary material. In the case of window profiles, the environmental interventions of the use phase are mainly related to the compensation of heat losses through the profile by a heating system. According to the strict coproduct allocation, these environmental interventions are allocated partly to the window profile under study and partly to the material leaving the system. As a consequence, the strict coproduct allocation procedure can conflict with common mental models of causalities within the socio-economic system.

The above argument possibly does not hold for all aluminium products. For instance, consumers freely dispose of aluminium beverage cans into collection containers. In this case, the market price of the material reaches a zero-point and recycling would be allocated to the following product system. In this case, the strict coproduct allocation could reasonably be applied; under the above assumptions, it would provide the same numerical result as the cut-off procedure.

Still, the crucial point of the strict coproduct allocation procedure applied to recycling processes is the definition of the zero-point. Economic data on the price development of the material in recycling processes is hardly obtainable, even worse if future processes have to be modelled. The price development of the recycling material during processing depends on a variety of very case-specific factors, such as quantities, location and transport distances, case-specific demolition and recycling techniques, etc. In addition, scrap prices are much more volatile than the prices for ingots as semi-fabricats. Compared to the volatility of scrap prices, the ratio of the LME quotations is quite stable.

One could argue that recycling processes are of minor environmental importance compared to primary material production processes or environmental impacts of the use phase, and that for this reason, an assumption on the zero-point – and thus the allocation of the recycling process – is of minor importance for the over-all result of the LCA. This is true as long as the zero-point lies after the use phase. As outlined above, this argument does not hold for aluminium facades and large quantities of aluminium windows, where the zero-point can lie within the use phase.

In view of the criteria for LCIs established in Chap. 3, the following points have to be mentioned:

- The strict-coproduct allocation procedure is rather difficult to use because specific economic and process-related data on future processes is required. As such the resulting model is consistent, complete within the logic of the strict coproduct allocation procedure. Furthermore, the resulting model is operational, decomposable, non-redundant and minimal. The resulting LCI is not necessarily comparable to LCIs of products made of other materials (see above);
- The strict coproduct allocation procedure treats allocation of recycling in a conceptually simple, manageable, transparent way and allows quick recalculations. On the other hand, it does not provide an 'adequate' system description if market prices of materials *in situ* are positive (see below);
- The strict coproduct allocation procedure is actor-based and case-specific in the sense that system boundaries (and as such the environmental impacts attributed to the product) can be influenced by decisions during process design. Quantitative (mass) and qualitative (entropic or economic aspects) issues are depicted. On the other hand, data required on the recycling processes and materials is case-specific; assumptions on future processes may be necessary, with far-reaching consequences on the result;
- Concerning risk, the uncertainties about the future recycling processes, materials and their economics are of primary concern for this procedure; the assessment of the related risk is case-specific;
- The strict coproduct allocation uses market prices for system boundary setting and allocation. It is based on the assumption on the organisational principle of the socio-economic system that a decision-maker (*e.g.,* a

product designer) is responsible for a material until it is reprocessed and available to further users at zero cost;

- The strict coproduct allocation in principle provides improvement options that consider qualitative and quantitative aspects of the sustainable management rules for aluminium referring to the material and energy flows within the technosphere. On the other hand, the data on future processes and their economics, enabling these considerations, is highly case-specific and thus highly dependent on assumptions;
- The question about environmental efficiency and eco-efficiency of the resulting model cannot be answered in general terms but only in a specific application and decision situation.

The strict coproduct allocation procedure cannot generally be applied in LCAs of aluminium products.

13.4 Value-corrected substitution

As preconditions for the applicability of the value-corrected substitution, the following points must be made:

- A primary material grade (and price) must clearly be definable as a reference point for the value-corrected substitution;
- Further (secondary) material grades must clearly be definable;
- The price ratio of primary material and further material grades must be determinable for similar (re-) processing stages;
- The price ratio of primary material and further material grades must be stable in a developed recycling material market;
- Primary material production processes (and final disposal processes) can be assumed equal for each type of products produced from the resource;
- The assumption that recycled material substitutes primary material production processes must be plausible;
- The value-corrected substitution is applicable to homogenous goods.

Due to current market conditions and material characteristics, these preconditions are given for aluminium as is outlined below. Furthermore, the influence of the decision based on the LCA must be insignificant for the structure of the socio-economic system and for the price ratio of different grades of a material. For the applicability of the value-corrected substitution for the handling of reuse & recycling of aluminium products, the following methodological considerations can be made.

Applicability in the light of ISO/EN 14041

The value-corrected substitution with the value-correction factors α and β provide a common, simple, manageable and transparent framework for addressing open-loop recycling of aluminium products, whether the material's inherent properties are changed ('open-loop recycling' according to ISO/EN 14041, chap. 6.5.4) or unchanged ('closed-loop recycling'). If the material is not devalued over the product system the value-correction factors have the same value, which means complete substitution capability of in- and outputs. The value-correction factor $\beta < \alpha$ if the material is devalued over the life cycle of the product. Thus, the value-corrected substitution provides a steady allocation function, as claimed a requirement in Chapter 9.6.

The environmental burdens of the primary material production processes can be reasonably allocated to the product system under study, as stipulated by ISO/EN 14041, chap. 6.5.4, point a. Final waste disposal of 'used-up' aluminium at the end of the material cascade is not taken into account because of the material and market characteristics of aluminium.

It has been argued in the peer review process that instead of final disposal future purification technologies should be allocated partially to a product system causing devaluation. It is replied that specific purification technologies can already be integrated into the modelling of post-consumer recycling processes. Value-losses of the recycled post-consumer material which go beyond the specifications of the alloyed aluminium LME contract are hypothetically compensated by adding primary material until these specifications are met.

Substitution of recycled material for recycled material

The same line of reasoning on the substitution of recycled aluminium for primary aluminium holds for the value-corrected substitution, as presented for the closed-loop procedure (Chap. 13.2).

Suitability of the LME quotations as reference points

A precondition of the applicability of the value-corrected substitution is a high statistical correlation between the prices used to calculate the value-correction factors α and β. A causal correlation of the two prices is necessary.

The London Metal Exchange (LME) is a punctual, highly transparent world wide market without any spatial, temporal or product-related preferences that acts very sensitively to incremental discrepancies in supply and demand. Thus, LME quotations provide reasonable (or the best available) parameters for 'material value' as defined in Chap. 9.5.4. LME prices thus provide a usable basis and a stable price ratio for allocation in LCA. Accordingly, system boundaries are set on the level of intermediate products (ingots).

Temporal aspects

Time dependency of the process or system to be substituted, the yield of the collection and recycling processes, and the time dependency of the price ratio utilised as allocation factors have to be taken into account. 'Best guess' scenarios in the future should be complemented with a 'present situation' scenario in a sensitivity analysis.

The value-corrected substitution uses the ratio of future LME quotations as a basis for the determination of the value-correction factor for the recycled post-consumer material. It has been argued that the ratio of current primary aluminium quotations and (discounted) future alloyed aluminium quotations could be alternatively used. But VCS depicts value relations at the input side as well as at the output-side and the (hypothetical) substitutions occurring at both times.

Allocating the environmental impacts based on the sum of the actual price of a material and the future price of the recovered material ignores the fact that aluminium can be recycled more often than just twice[376].

As future price ratios are difficult to forecast, the price ratios of the last ten years are taken as approximation for future price relations. Besides, estimated future price relations and their impacts on the result can be studied in a sensitivity analysis.

The fluctuation of the price ratio of the LME quotations influence the credit given for the quantity and material value of the recycled material by ± 4.5 %, covering about 2/3 of the price ratios since the existence of the alloyed aluminium contract.

Restrictions and methodological difficulties

The determination of the material's inherent properties – the (theoretical) alloy composition after post-consumer recycling – represents a methodological difficulty. Elementary analysis, expert opinions or mass flow calculations can provide useful solutions.

Alloys with more valuable alloy components than aluminium could set limits to this allocation procedure. As the LME quotation for primary aluminium is used as the reference point for the value-corrected substitution, the (complete) recycling of a more valuable alloy would mathematically result in an over-compensation of primary material production. In such a case, the application of the closed-loop procedure might be a better option.

Latest material and processing developments for aluminium such as unconventional aluminium alloys with additives, aluminium foams, metal matrix composites reinforced with particles or fibres, or thixoforming (process scrap) lead to reduced recyclability[377]. This does not *a priori* affect the

[376] This is also a critique on ISO/EN 14'041, chap. 6.5.4, where "scrap value in relation to primary material" is proposed as basis for allocation.

[377] after Krone 2000:498ff/501

value-corrected substitution although re-evaluation might be considered when these materials and processes are applied in larger scale possibly affecting the structure of the aluminium market.

Consequences for the use of aluminium

By applying the value-corrected substitution, emphasis is put on complete material re-collection and re-processing as well as on maintaining the material's inherent properties (or its value respectively) over the life cycle of a product. This allows one to support decisions of environmental product design of aluminium products towards closing material cycles while preserving best material's inherent properties of aluminium. This creates the incentive that aluminium is only used in applications that allow for an efficient and value-preserving recycling. Furthermore, it creates the precondition that high-value aluminium scrap is generated in a way that allows for the substitution of primary material.

Hence, the value-corrected substitution provides a framework for open-loop recycling that respects material and market characteristics of aluminium and can lead decisions towards sustainable material use.

For products with long service life, the environmental impacts of primary material production are mathematically distributed over a long time period. As a consequence, the primary material production processes occurring today are attributed to uses in the future. The aluminium exploited today is analytically subsidised by environmental credits that future generations will have to pay for as environmental burdens. This can lead to an increased environmental competitiveness of aluminium and could lead to an increased exploitation of primary aluminium. This point is accentuated in the case of aluminium whose market is currently expanding.

This point has been stated as an argument against the value-corrected substitution[378]. It has to be replied that the compression of time is a modelling characteristic of LCA and as such not inherent to a specific allocation procedure. In an a-temporal model (or in a model with no temporal preferences), one big impact of primary material production and several small impacts of recycling can be less than several intermediate impacts of repeated primary material production[379]. Nonetheless, path-dependencies are assumed that do not necessarily turn out to be true. But this is not an argument against the value-corrected substitution but against the use of descriptive instead of marginal LCA as a basis for decision-making.

[378] Frischknecht 2001

[379] The above reasoning is only valid for descriptive LCA and does not hold for decision-oriented marginal LCA.

It is beyond the possibilities of LCA to answer the question: Is too much bauxite *currently* being exploited from an environmental or resource-consumption perspective? The stakeholders of the aluminium industry are called upon in this aspect to work out and commit to a credible strategy for the sustainable use of the resource aluminium.

Anyhow, for products with long service life, it can be suitable to use a second allocation procedure such as the closed-loop procedure based on a material pool in complement with a sensitivity analysis.

Evaluation based on the criteria for LCIs established in Chapter 3

In view of the criteria for LCIs established in Chapter 3, the following points have to be mentioned:

- The value-corrected substitution procedure is rather difficult to use because specific material and process-related data on future processes is required. As such, the resulting model is consistent, complete within the logic of the value-corrected substitution procedure. Furthermore, the resulting model is operational, decomposable, non-redundant and minimal. The resulting LCI is not necessarily comparable to LCIs of products made of other materials (see above);
- The strict coproduct allocation procedure treats allocation of recycling in a conceptually simple, manageable, transparent way and allows quick recalculations. It provides a consistent framework for closed-loop and open-loop recycling;
- The strict coproduct allocation procedure is actor-based and case-specific in the sense that consequences of design decisions for the recycling properties of a product can adequately be depicted as quantitative (mass) and qualitative (entropic or economic aspects) issues are depicted. On the other hand, the determination of the hypothetical alloy content of the recycled material appears to be a methodological difficulty. Mass-balances and expert opinions can solve this problem;
- Related to risk, the uncertainties about the future recycling processes and materials and their economics are of primary concern for this procedure; the assessment of the related risk is case-specific: it depends on the time frame considered as well as on the market conditions (developed market, expanding market, etc.);
- The value-corrected substitution provides improvement options that consider qualitative and quantitative aspects of the sustainable management rules for aluminium referring to the material and energy flows within the technosphere;
- The question about environmental efficiency and the eco-efficiency of the resulting model cannot be answered in general terms but only in a specific application and decision situation.

It has been demonstrated that value-corrected substitution is a suitable and adequate allocation procedure for LCA of aluminium products, especially in building applications. It fulfils the requirements of an allocation procedure for reuse & recycling of aluminium declared in Chapter 9.6. Furthermore, the resulting product system fulfils the requirements stated in Chapter 3.4.

Anyhow, calculations should be complemented with another allocation procedure such as the closed-loop procedure based on a material pool model in a sensitivity analysis (after ISO/EN 14041, chap. 6.5.2).

V.

CASE STUDY II: MODELLING END-OF-LIFE OPTIONS FOR BEECH WOOD RAILWAY SLEEPERS

Chapter 14

INTRODUCTION

The second case study treats the modelling of end-of-life options for beech wood railway sleepers in LCA. It aim is to recommend how to model different end-of-life options while respecting the material and market characteristics of wood – post-consumer wood in particular – and providing recommendations in line with the management rules for the sustainable use of wood.

The case study treats different allocation problems encountered throughout the wood processing chain, as the modelling of end-of-life options depends on methodological decisions made during the modelling of previous life stages. From a methodological point of view, the allocation problems encountered consist in allocating subsequent, (inherently linked) joint coproduction processes and finally modelling end-of-life processes to obtain a model consistent with:

- The decision-maker's mental model of the material and market characteristics of wooden railway sleepers;
- His/her values (preferences) on the sustainable use of wood.

The case study is structured as follows. Chapter 15 gives an overview of the material and market characteristics of wood from its production in the forest throughout the wood processing chain to the utilisation options of post-consumer wood. Sustainable wood flow management is then discussed (Chap. 16). In these two Chapters, management rules are derived for the sustainable utilisation of wood and in particular post-consumer wood (Chap. 16.6).

Chapter 17 illustrates the specific attribution and allocation problems related to the wood processing chain. Furthermore, the proposed treatments of allocation problems in two wood-specific methodological LCA guidelines are summarised, focusing on reuse & recycling.

Chapter 18 contains the example of the life cycle of beech wood railway sleeper.

The following issues are treated:

- Allocation of forestry processes for the production of railway sleepers;
- Allocation of the transport from forestry to the production site;
- Allocation of the sleeper production processes;
- Allocation of different end-of-life scenarios;
- Modelling a partial life cycle including all the process steps investigated.

A systematic evaluation of different allocation procedures is made based on the general requirements of a product system and its LCI (Chap. 3.4) and wood-specific aspects (Chapters 15 and 16, particularly Chap. 16.6).

Chapter 15

MATERIAL AND MARKET CHARACTERISTICS OF WOOD

15.1 Wood as material

Wood is the world's most important renewable material and regenerative fuel[380]. Like stone, wood has been utilised as material and fuel since prehistoric times. Therefore, it is deeply rooted in our entire history of civilisation. Today, wood is utilised in many forms, for example, sawn wood, particleboard, plywood, fibreboard and veneer. Its main applications are construction material for housing and civil engineering, furniture, railway ties (sleepers), poles and reinforcement for mining. Even today wood is still one of the leading construction materials.

Wood is also the most important raw material for the pulp and paper industry and plays a significant role in the packaging sector.

Wood can be used as solid wood, particles, and fibres or even on a molecular basis. The material characteristics and the twofold nature of wood as a material and fuel allow for an almost 100% utilisation of this resource.

Wood is anisotropic and inhomogeneous; its properties vary from tree to tree and even within the same log. Anisotropy is reduced in wood panels and engineered wood products by suitable technologies. Even though wood is one of the oldest materials used by humankind, numerous new developments have taken place in recent years, which have broadened the application of wood and wood products. Especially noteworthy are new impregnation techniques and combinations with other materials such as plastics and metals. The main advantages of wood lie in its relatively low price, the favour-

[380] Bowyer 1995; Schulz 1993; Sutton 1993; Eldag 1980:105; Frühwald *et al.* 2001

able strength-to-weight ratio, its vapour permeability, its appealing, individual appearance, and in the ease in which it is worked with proven techniques[381].

As a biogenic material wood is subject to biological and photochemical degradation and, in consequence, outdoor applications may require (regular) constructive and/or chemical wood protection. Its hygroscopic nature leads to changes in dimension ('swelling and shrinking') depending on the moisture content of the environment. This can lead to deformation and shakes if wood is not applied in a carefully considered way.

Reuse & recycling of wood as a material must be considered as downcycling (except reuse in a very strict sense): diameters of wood pieces and fibre length decrease while 'unwanted' contaminants (*e.g.,* from wood treatment or coating) increase with each processing step[382]. The more often wood is reprocessed, the more limited are its potential applications. Original properties can only be restored with the investment of (non-renewable) energy and material[383].

15.2 Some characteristics of wood industry

Wood as a naturally grown material, irregular in shape and structure, is not, *a priori*, suitable for industrial processing or use in technical applications. As a consequence, the management and processing of wood generates a variety of coproducts throughout the wood processing chain from its cultivation in managed forests, its extraction, sawing, processing to intermediate and finished products, on to its recycling, incineration or final disposal. Coproducts generated are, *e.g.*, residues from thinning, bark, sawdust, shavings, chips and fibres, side-cuts, wood waste and waste of intermediate products from wood and wood-based industries[384].

If coproducts cannot be used internally, *e.g.,* for the generation of thermal energy for wood drying, they can be sold to other industrial sectors. This has in fact generated a considerable network of material flows between different wood processing industries that use coproducts from other sectors as material input or fuel (*e.g.,* chips from saw mills are used in the particleboard industry).

Even after the use phase of wood products, the material characteristics of wood (theoretically) still allow for a variety of options for the reuse of wood

[381] after Schulz 1972:1416
[382] see, *e.g.*, Willeitner and Bucki 1994:5.9
[383] see, *e.g.*, Fraanje 1997:22f
[384] see, *e.g.*, Jungmeier *et al.* 2001b

as a material or energy carrier. These options are further discussed in Chapter 15.3.

Similar considerations on the generation and use of coproducts throughout the material chain are also valid for the pulp and paper chain. In contrast to the wood processing industry, the (material) recycling of waste paper is a widely established industry.

15.3 Utilisation options of post-consumer wood

Due to the material characteristics of wood and its twofold nature as material and energy carrier, a variety of options are open for the reuse & recycling of post-consumer wood[385]. They are mainly restricted by the size of the post-consumer wood, particles or fibres as well as their homogeneity, and their content of contaminants (*e.g.*, preservatives, coatings, adhesives, overlays, etc.).

The guideline VDI-2243 by Verein Deutscher Ingenieure[386] identifies four distinct options of reuse & recycling of materials and products, depending on the area of application and the technical effort required in

Table 15-1. Reuse & recycling options at the end of the life cycle of wood products (after Richter 2000:10)

Principle	Technical effort for reprocessing	Material's inherent properties	Area of application product 1 -> prod. 2	Example
Reuse in same application	small (repair, renovation)	unchanged	same	Returnable pallets, Second hand furniture dealers
Recycling into different application	reprocessing necessary	slightly changed	same	Solid wood beam -> laminated wood beam Particleboard -> particleboard
Reuse in different application	small	unchanged	different	Railway sleepers, utility poles -> landscape architecture
Recycling into different application	reprocessing necessary	changed	mostly different	Post-consumer wood -> particleboard -> MDF -> thermal energy

[385] see, *e.g.*, Jungmeier *et al.* 2001a

[386] Anonymous 1991

reprocessing[387]. Reuse & recycling options at the end of the life cycle of wood products can be classified accordingly (Table 15-1).

The first two options are considered 'closed-loop' recycling according to ISO/EN 14041, while the second two are classified as 'open-loop' recycling. However, the classification of a specific case to one of the four options is subjective to a certain extent. Should the production of particleboard from a solid wood board from furniture be classified as reuse or recycling? Should a laminated wood beam made from a solid wood beam be classified as reuse?

The classification becomes even trickier if thermal utilisation has to be classified. When is the incineration of post-consumer wood considered under the aspect of waste treatment; or when would it be classified as thermal recycling, for instance, in the case of post-consumer wood combustion in municipal waste incinerators with energy recovery?[388]

15.3.1 Options for material re-utilisation of post-consumer wood

Reuse of wood products

Direct reuse of wood on an industrial scale is not usually possible. Because of differing shapes, quality and impurities labour-intensive, expensive reprocessing is required[389]. Nevertheless, the reuse of wooden pallets[390] and packaging elements as well as the reuse of standardised building materials such as laths, beams or boards on a small scale, decentralised single cases are common practice[391].

Another exception, also small in quantities, consists in the reuse of valuable timber, antique and handcrafted furniture, panels, wall panels, parquet floorings as well as, occasionally, old doors and windows. In some cases, salvaged wood products can achieve premium prices, especially if the quality of the product – particularly those made from old growth timber and rare species of wood – surpasses all new products available. Antique shops, second hand furniture dealers and 'spare parts exchanges' play a considerable role here[392].

[387] for the application of VDI-2243 to wooden products, see, *e.g.*, Richter 2000; Willeitner and Bucki 1994

[388] Swiss legislation is actually of limited use as various differing definitions of 'post-consumer wood ('Altholz') can be found (for details, see Scheller 2001:12f).

[389] after Stahel *et al.* 1987:31f

[390] see Hekkert *et al.* 2000b:15

[391] Stahel *et al.* 1987:31f; also Orpin 1996; Plume 1996 for small-scale reprocessing of old beams

[392] Buser 1998

Two examples of industrial reprocessing have been reported from The Netherlands. In one case, floorboards are made out of old floorjoists. On average the floorjoists measure 0.75 x 0.20-0.25 m and are at about 4.5 m long. After the removal of the 75-100 year-old beams with special equipment, the joists are freed of nails by hand and controlled with a metal-detector. They are then planed and, if necessary, the ends that were in contact with the wall are cut off. The beam is sawn into 2.5 cm thick floorboards and finished with tongue and groove joints. About 75% of the old beams are used for floorboard. The rest is used in other applications such as wood with smaller dimensions, pile caps or for fibreboard[393]. Another Dutch company makes floorjoists or window frames out of second hand pine wood, depending on the size of the wood[394].

The reuse of the wood used to construct the wooden Swiss pavilion at Expo 2000 in Hannover can be considered another singular example for the reuse of post-consumer wood. Already planned during design, after the exposition closed the wood (sufficient for about 100 detached houses) was sold to various countries for the production of further products[395].

Recycling in wood-based panels

Wood-based boards such as particleboards or fibreboards (*e.g.*, medium-density fibreboards, MDF) are sold in large quantities and have low unit prices. Various processes for the reintegration of particles and fibres from used boards have been developed[396].

Panel board manufacturers have defined quality specifications for post-consumer wood fractions suitable for wood-based panel production[397]. Further quality management systems ensure the quality of their products whether they contain recycled material or not[398].

Even the possibility of using chromated copper arsenate (CCA-) treated wood in particleboard and flakeboard production has been investigated, mainly in the U.S.[399]. However, at present there are no known commercial markets for recycled CCA-treated wood products, mainly due to environ-

[393] after Fraanje 1997:25f

[394] after Fraanje 1997:25f

[395] after Clénin 2001

[396] see, *e.g.*, the wet 'WKI-processing' based on the hydrolysis (see *e.g.*, Michanickl and Boehme 1995; Wittke 1998) or the production of cut small particleboard plates in a dry procedure (see *e.g.*, Möller and Herrlich 1994); for the discussion of different procedures, see also Dupré 1986:143 or Roeffael 1997:7ff

[397] Anonymous 1996d; Anonymous 1999b; Anonymous 2000b; Krooss *et al.* 1998; Schrägle 2001

[398] see, *e.g.*, Harbeke 1998:70

[399] see, *e.g.*, Felton 1996; Smith and Shiau 1998; Mengeloglu and Gardner 2000

mental concerns related to residual chemicals left in the fibres and due to health and safety concerns of the mill workers[400].

Quantitatively, the use in the wood-based board industry is the most important (and most promising) path for the material utilisation of post-consumer wood[401].

Recycling in wood fibre reinforced plastic composites ('plastic lumber')

The integration of post-consumer wood fibres into synthetic material is one of two options of wood-based composite materials. Wood fibre reinforced composites allow for the recycling of plastics and of post-consumer wood. Wood fibre reinforced composites are currently used in automotive industry and for submarine building structures. Content and leakage of harmful pollutants are a major concern related to these materials[402]. The ability to recover thermal energy is maintained.

Recycling in concrete-bond wood fibre boards

Wood (and waste paper) can also be combined with concrete to produce composite products. The wood aggregate is mixed with Portland cement to form lightweight, fireproof building materials, components and insulation material. The organic fibre makes up 91 percent (by volume) of the total composition[403]. For these products, recovery of the embodied energy content of wood is strongly inhibited.

Recycling in pressed wood fibre pallets

Pallets can be pressed from low-grade wood fibres, mostly bark and residues from thinning. The fibres are moulded into a pressed wood pallet with the use of synthetic organic resins[404].

Recycling as soil amendment, animal bedding or mulching material as peat substitute

The production of soil amendment, animal bedding or mulching material as a peat substitute offers a universal utilisation option for residues from wood processing or for post-consumer wood. The requirements of technical properties are low for this option, although high standards that ensure the

[400] after Smith and Shiau 1998:45f
[401] see, *e.g.*, Marutzky 1997:2004
[402] see, *e.g.*, Various 1997; Various 1992; Hettinga 1996
[403] after Anonymous 1996g:31
[404] Hekkert *et al.* 2000b:16

lack of harmful chemicals have to be complied with. This is not, however, a lucrative option[405].

Recycling as chemical basis materials

The extraction of the main chemical parts of wood, the cracking of the wood substance into monomer fractions as well as the extraction and use of accessory components (*e.g.*, methanol) is theoretically feasible. For cost reasons, the production of chemical basis material is currently only done on a very small scale. Furthermore, the actual use of wood as a provider of chemical raw materials is limited to some marginal areas of industrial production and has very limited practical importance[406].

Paper recycling

For the scope of the study, paper recycling[407], energetic uses of residues from pulp and paper processing[408] and further uses of waste paper as insulation material or secondary material for fibreboards[409] are disregarded.

15.3.2 Thermal utilisation

Thermal uses range from industrial scale combustion to decentralised (in most cases illegal) combustion in open fireplaces of private houses. Among these options are:

- Combustion in municipal waste incinerators with energy recovery;
- Residual wood and post-consumer wood combustion plants, mostly associated with the wood-processing industry, but also independent operators;
- Combustion in cement kilns which, due to the high combustion temperature and the specific characteristics of clinker, allows for the environmentally sound combustion of preservative-treated post-consumer wood (like creosote-treated railway sleepers)[410];

[405] see, *e.g.*, Willeitner and Bucki 1994:5.8; Anonymous 1999a

[406] after Willeitner and Bucki 1994:5.8; Anonymous 1995a:1; Kusian 2001

[407] see, *e.g.*, Anonymous 1996g; Baumgarten 1986; Fallscheer 1997; de Fries 1986; Grace *et al.* 1978; Hekkert *et al.* 2000a; Huttunen and Pirttila 1998; Jönson 1998; Kibat 1986; Kinkley 1984; Leach *et al.* 1997; Neumann 1986; Plätzer *et al.* 1996; Skog and Nicholson 1998; Turner *et al.* 1977

[408] see, *e.g.*, Langer 1986

[409] see, *e.g.*, Roeffael 1997:11f; Dix *et al.* 1997

[410] see, *e.g.*, Stahel *et al.* 1987:46; Anonymous 1998a:18

- Incineration in open fireplaces of private houses, which is related to high quantities of highly harmful emissions[411].

The thermal use of wood allows for the generation of thermal energy (*e.g.*, as vapour) and/or electricity[412].

15.3.3 Further end-of-life options

End-of-life options from which no further utility is derived are, *e.g.*, the incineration in municipal waste incinerators without energy recovery; the thermal treatment of wood as hazardous waste without energy recovery, open burning, land filling[413] and dumping.

15.3.4 Economics of wood residuals and post-consumer wood

Following economic preconditions for the material utilisation of post-consumer wood have to be taken into account:

- Wood and wood products to be recycled have to be available in sufficient quantities; this concerns mainly questions of logistics, transport, storing, and processing;
- Wood and wood products to be recycled have to be steadily available on a mid- to long-term basis to attract investment;
- Wood and wood products to be recycled have to be available at prices that make the use of these materials economically viable;
- There must be an acceptance and demand for products partly or entirely produced from post-consumer wood;
- As a consequence of reuse & recycling of post-consumer wood, no foreseeable disposal problems may arise[414].

Cost and price of post-consumer wood are determined by the following factors (after Harbeke 1998:69):

- State of reprocessing;
- Quality, especially the contents of potentially harmful substances;
- Quantities to be delivered;
- Demand from wood products industry and thermal energy producers;
- Purchaser's and supplier's strategic goals;
- Quotas for the processing of post-consumer wood;
- Regional post-consumer wood supply.

[411] see, *e.g.*, Vock 2000:4; Nussbaumer 1994:9.8
[412] see Nussbaumer *et al.* 1997
[413] although landfill gases from anaerobic decomposition can be used for energy recovery
[414] after Willeitner and Bucki 1994:5.5

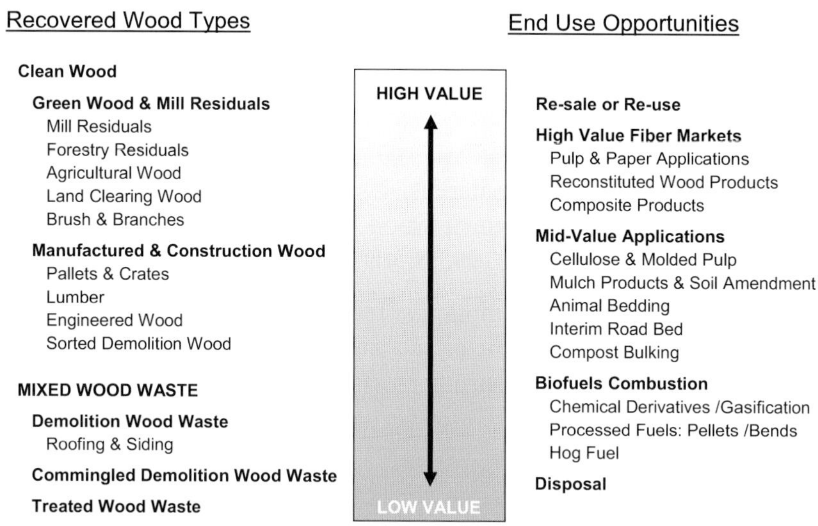

Figure 15-1. Residual wood and post-consumer wood fractions, possibilities of end use and their respective relative market price (after Horne-Brine 1996:37; reprinted with the permission from the Forest Products Society)

The possible content of harmful organic or inorganic chemicals coming from wood preservatives, adhesives, coatings, etc. is a major restriction and of great concern in the reuse & recycling of post-consumer wood (see below).

Figure 15-1 gives an overview of the relative value of wood residues and post-consumer wood fractions and the relative revaluation potential of the different end-of-life options discussed above.

15.4 Post-consumer wood market in Switzerland

A variety of definitions for post-consumer wood ('Altholz') can be found in Swiss legislation and governmental strategy papers. This variety is a consequence of difficulties encountered while defining this material category and the result of particular interests related to post-consumer wood[415].

For this study, the term 'post-consumer wood' is used according to the following definition:

[415] see Scheller 2001:12ff for details

"All parts from wood-based materials such as solid wood, plywood, particleboard, fibreboards, MDF etc. that leave the use process at the consumer after a shorter or longer service life, are considered post-consumer wood. The wooden parts can have been subject to various treatment processes during production: during service life, they can have been combined with different materials. Usually, the treatment and use processes, to which post-consumer wood was subject to, are only rudimentarily known" (Vock *et al.* 1993:8; translated from German by WF).

While data on wood consumption is readily available[416], data on the post-consumer wood market in Switzerland must be considered sketchy[417]. Current estimations on the post-consumer wood sources and its sinks are shown in Table 15-2.

Table 15-2. Estimated current sinks and sources of post-consumer wood in Switzerland (after Vock 2000:4; with data from Anonymous 1998d; Anonymous 1999d)

Sources	t/year	Sinks	t/year
Building parts	300'000	Municipal waste incinerators (estim.)	220'000
Furniture and wooden goods	235'000	Land filling (estimation)	30'000
Packaging	165'000	Firing systems (BFE-statistics 99)	80'000
		Exports (Exports-statistics 99)	160'000
		Illegal disposal (estimation)	210'000
Total	700'000	Total	700'000

Two relevant changes in legislation have strongly influenced the post-consumer wood flows in Switzerland:

- Based on the Technical Ordinance on Waste, land filling of combustible waste fractions has been forbidden since the 1st of January 2000 (TVA, Art. 53a);
- Exports to foreign countries of post-consumer wood have to be declared[418].

Material use of post-consumer wood on industrial scale has been insignificant in Switzerland[419]. Swiss particleboard producers have so far abstained from using post-consumer wood in their products, most probably for reasons of image of their products[420].

416 Wiegand *et al.* 1996; Anonymous 2000b

417 see *e.g.*, Scheller 2001:33ff; Vock 2000:4

418 Anonymous 1998b

419 In Germany the average contents of post-consumer wood lies between 5-20% whereas in Italy, for example, wood-based boards are produced from 100% post-consumer wood (Harbeke 1998:69).

420 after Hofer 2000a:17

Instead, Swiss post-consumer wood is exported mainly to Italy where particleboards and fibreboards are produced from up to 100% post-consumer wood. These exports will probably increase because it is a considerably cheaper option than the disposal of post-consumer wood in municipal waste incinerators or post-consumer wood incinerators in Switzerland[421].

Political programs such as Energie 2000 and its follow up Energie Schweiz have been set up to foster the use of wood as an energy carrier.

Despite governmental support, the use of wood as an energy carrier has been stable in recent years[422]. Nevertheless, the use of post-consumer wood as a source of energy in Switzerland is losing importance; large post-consumer wood incineration facilities like the concrete plant in Rekingen have been closed down for economic reasons. Similarly, some owners of post-consumer wood incineration plants consider switching from post-consumer wood to alternative waste fuels[423]. On the other hand, capacities of existing municipal waste incineration plants are almost exhausted[424]; it is doubtful whether these plants will be able to absorb the increased amounts of post-consumer wood in the future.

[421] Prices for export are estimated around 100 CHF/t, compared to an average of 223 CHF/t for the treatment in municipal waste incinerators or post-consumer wood incinerators (Hofer 2000a:17).

[422] Anonymous 2000b:34f

[423] after Hofer 2000a:17

[424] see, *e.g.*, Anonymous 2001a

Chapter 16

SUSTAINABLE WOOD FLOW MANAGEMENT

16.1 Forestry

The term 'Nachhaltigkeit' (sustainability; WF) is historically closely linked to forestry and is by no means an invention of recent times. Conceptually, it was developed in the early 18th century, when population growth and industrialisation had increased pressure on the resource wood to an extent threatening further supply.[425]

The concept of sustainable forest management entered German and Swiss forest legislation at the turn of 19th century, with the main focus of maintaining forested areas[426]. Today, sustainable management of renewable resources is defined in a much broader sense. One of the most widely accepted definitions has been set up 1993 by the Ministerial Conference for the Conservation of Forests in Europe in Helsinki:

> "The stewardship and use of forests and forest lands in a way, and at a rate, that maintains their biodiversity, productivity, regeneration capacity, vitality and their potential to fulfil, now and in the future, relevant ecological, economic and social functions, at local, national and global lev-

[425] It was Hans Carl Carlowitz who claimed in his 'Sylvicultura Oeconomica. Die Naturmässige Anweisung zur Wilden Baum-Zucht', published 1713, "...dass man mit dem Holtz pfleglich umgehe, and wie eine sothane Conservation und Anbau des Holtzes anzustellen, dass es eine continuirliche, beständige und nachhaltende Nutzung geben" (Carlowitz (1713), quoted after Grober 1999).

[426] for details, see Schuler 2000:498ff

els, and that does not cause damage to other ecosystems" (Anonymous 1998c).

This definition is very much in line with quantitative and qualitative aspects of the first and seventh management rule for sustainable resource management by Messner (1999), quoted in Chapter 3.3.3:

"*First Management Rule*: For the long-term preservation of useable biomass and of the potential of biomass from *renewable natural resources*, it has to be guaranteed:

- That fertile soils and watersheds apt for the production of biomass are kept available;
- That natural production and life support functions remain intact and are not affected by mechanical interventions or emissions from the production of biomass;
- That the consumptive use of renewable resources does not exceed the growth rate of usable biomass minus the resources needed for the maintenance of the renewable resource basis; and
- That the genetic potential of renewable resources is protected and is assigned an adequate habitat. (…)

"…*Seventh management rule*: The pressure on nature caused by *mechanical interventions* may not lead to a degradation of subordinate regional or global life support functions. To secure and preserve the natural functions and the biodiversity, unmodified and regionally interconnected nature protection areas shall be assigned. The claim for nature protection areas might be possible in single cases if compensation is made in form of an ecologically equivalent compensation area. Due to ecological uncertainties in determining the size and structure of nature protection areas, safety margins have to be considered" (Messner 1999:400ff; translated from German by WF).

In Europe, the local potential of wood as a resource is currently not exploited to its full extent[427]. In Switzerland, the average standing volume of wood per hectare has to be considered high, with an increasing tendency. Moreover, forested area in Switzerland has been increasing constantly over the last fifty years. As a consequence of current under-exploitation, Swiss forests contain a disproportionally high share of elderly trees[428]. The use of

[427] Anonymous 1999c
[428] all data taken from Anonymous 2000b

wood could be increased by an average of 20% to 40% in Switzerland even after defining large forest areas as protected nature reserves[429].

On the whole, imports of raw material for the paper industry could be substituted[430].

Quantitative criteria for sustainable forest management are generally met in Switzerland and the legal framework is considered appropriate. Nevertheless, qualitative criteria, for example the over-aged mountainous forests, the forest ecosystem health and vitality affected by air pollution, and a high degree of governmental subsidies of forestry activities are major concerns in the light of sustainable forest management in Switzerland[431]. On the other hand, native species cover more than 99% of the forest area.

Scant attention has been paid to another aspect: the closing of the cycle of nutrients that are withdrawn from forests mainly with bark and branches. First experiences have been gained in the processing of wood-based ashes for their use as fertiliser. In this respect, the forest industry activity is beginning to participate in the recycling of nutrients as base cations in the ashes are returned to the forests. In the future, this effort should be increased to complete the natural-industrial nutrient cycle[432].

An increased use of local wood in Switzerland would not only respect the environmental dimension of sustainable development but would also foster its social and economic dimensions by providing income in marginalised regions.

Discussing the use of wood from tropical forests, the world wide deforestation and the loss of biodiversity related to it are beyond the scope of this study.

16.2 Use of wood

Based on the concept of sustainability, a reduced use of virgin natural resources and a relative shift from non-renewable to renewable resources has to be fostered. Based on theoretical considerations, Messner (1999) states as management rules for the sustainable use of renewable resources:

> "*Fifth management rule*: In a *combined use of renewable and non-renewable resources* shocks in resource availability, quality losses in the resource use and forms of inefficient resource use are to be avoided in a long run. As, in effect, renewable resources will constitute the only

[429] after Mauch *et al.* 1995:65ff; Anonymous 2000c:12
[430] after Mauch *et al.* 1995:65ff; Hofer *et al.* 2001
[431] after Patosaari 2000:480ff; Brassel and Brändli 1999
[432] Korhonen *et al.* 2001:152; Winkler 1996:5

resource basis for future generations, the following guidelines have to be considered for the realisation of an inter-temporal continuous transformation path in resource use:

- The use of high-quality resources shall be reduced to a minimum level as soon as possible to ensure a transformation without major troubles towards an economy based on renewable resources. The principle of resource conservation is valid;
- The economic processes of an economy are to be organised and co-ordinated in a way that the lowest possible material and energy losses, a lowest possible affection of natural resources and natural areas and a lowest possible dispersion of material to reach the production targets occur. In this sense, the principle of thermodynamic efficiency in the production is valid" (Messner 1999:400ff; translated from German by WF).

An expanded use of renewable materials following from the replacement of non-renewable resources, however, leads, *ceteris paribus*, to an unacceptable pressure on land resources[433]. Therefore, the efficiency of resource utilisation needs to be increased to reduce or – even better – avoid this pressure[434].

However, the claim for dematerialisation – the fulfilment of a need using less material and energy input – does not make an exception for wood. Contrary to all other important materials like steel, aluminium or concrete, the resource efficiency of wood needs only to be increased by a factor of 2 – instead of a factor of 3 to 8 – to reach sustainable consumption levels in Europe[435]. Basically, this requires a more effective and efficient use of wood, including optimised process technology and products with longer service lives and an aptitude for repair, material recycling, and finally incineration with energy recovery[436].

Fraanje (1997) states that wood is often not used according to the 'appropriate application'-principle of resource cascading[437]:

"Appropriate application means that the resource is applied on the basis of its (typical) properties, at the highest quality level it is possible. In this view one should not make pulp out of tree directly, but first profit by specific qualities of massive wood (like strength). When making paper directly from vegetable resources, one should rather use flax or hemp,

[433] see calculations made by Müller 1996:881f
[434] Fraanje 1997:22; Schulz 1972:1416
[435] after Mauch *et al.* 1995:65ff
[436] see, *e.g.*, Werner 1999:74; Lafleur and Fraanje 1997:26f
[437] developed in Sirkin and ten Houten 1994; see also Chap. 9.5.2

instead of wood. Appropriate application has reference to the whole resource" (Fraanje 1997:24).

Use of primary wood in energy production is not economically feasible under current market conditions in Switzerland, owing to the high cost of labour and the infrastructure of fuel preparation[438]. Hence, only the use of wood as a material generates enough revenue to cover current forest exploitation costs.

16.3 Post-consumer wood

Handling post-consumer wood is situated between waste treatment/disposal and exploitation of a secondary material resource for material or energy purposes[439].

Reuse & recycling of post-consumer wood brings both advantages and disadvantages.

Among the advantages are[440]:

- Reintegration of waste into the socio-economic system;
- Careful treatment of resources and amplification of the resource basis;
- Less occupation of space in landfills;
- Lower thermal energy consumption and VOC-emissions from drying processes if used (dry) wood is recycled as material;
- Substitution of fossil fuels (CO_2-neutrality);
- Higher energy efficiency due to lower moisture content compared to 'fresh' wood if used for energy purposes;
- Destruction of harmful (organic) chemicals or their immobilisation and export from the ecosphere (in the case of inorganic chemicals) if used for energy purposes in cement kilns.

Among the disadvantages are:

- Dispersion of pollutants if recycled as material (see below);
- Generation of harmful emissions if used for energy purposes;
- Higher requirements of logistics and transport than primary material;
- Displacement of other wooden raw materials such as residues from thinning or sawmills leading to lower prices for products from primary processes with already low profits or even obtaining subsidies;
- Even less economically viable wood extraction and a correspondingly reduced potential substitution of fossil fuels.

[438] Hofer 2000a:19
[439] Anonymous 1995a; Anonymous 1996b; Voss 1998
[440] after Marutzky 1997:2204; Roeffael 1997

Environmentally relevant problems specifically related to the incineration of post-consumer wood are[441]:

- High ash content and low ash melting temperature;
- Fuel nitrogen leading to NO_x emissions;
- Sub-optimal combustion processes as a consequence of wood as solid fuel leading to emissions of NO_x, CO and particles;
- Chlorine in the fuel leading to HCl or even to PCDD/F emissions;
- Heavy metals (Pb, Zn) leading to respective emissions;
- Post-consumer wood can be contaminated with stones, sand, metals, etc.

For both material and energy uses of post-consumer wood, the potential content of harmful substances from chemical wood conservation is a major limitation. Most agents applied (biocides) are designed for a long life span. They remain in the wood even after the service life of the product is reached and can thus considerably limit the use and disposal options of post-consumer wood. Among the substances used for their biocidal character are mercury, zinc and arsenic compounds, creosote or pentachlorophenol (PCP). Some of them (mercury, arsenic, PCP, chloro-organic compounds) are prohibited today in most Western European Countries. However, they are present in current post-consumer wood fractions.

Processes have been developed to remove organic wood preservatives (creosote) from post-consumer wood to allow its use as secondary material in the particleboard industry. But such processes have not been put into industrial practice[442].

Other sources that limit the further utilisation of post-consumer wood, apart from chemical wood preservatives, are adhesives, toxic pigments of coatings, and mineral contaminants.

Up to now, no labelling scheme has been put into practice (although it is currently being discussed). Visual distinction of untreated and treated post-consumer wood, in the latter case differentiating between organic and inorganic active groups, is feasible only to a very limited extent. In practice, it is limited to certain homogenous fractions such as utility poles or railway sleepers[443]. Large-scale analytic techniques for the sorting of post-consumer wood according to its contamination are currently being investigated in pilot plants[444].

Limits to the recycling of post-consumer wood are not only set by the possible content of harmful substances but also by the unavoidable degrada-

[441] after Nussbaumer 1994
[442] see, *e.g.*, Kühne and Schwarz 1997
[443] after Peek 1998
[444] Peylo 1998a; Peylo 1998b; Weis *et al.* 1999

tion of material during reprocessing. Degradation during reprocessing includes, *e.g.*, a reduction in dimensions of wooden parts, an unavoidable amount of sawdust produced during reprocessing, more cubical dimensions of reprocessed secondary chips for particleboard production (if applying dry separation processes), or reduced fibre length of reprocessed fibreboards or paper, etc.

The above situation requires three basic strategies to extend the utilisation potential of post-consumer wood as a secondary resource:

- Labelling of wood products with declaration of used additives;
- Utilisation of homogenous materials concerning the type of wood or of the wood-based intermediate products used, and with regard to the treatment with substances that limit further utilisation;
- Simple structure of wood products to facilitate easy disassembly and the recuperation of homogenous fractions free of substances limiting further (thermal) utilisation.

16.4 Wood in the international discussion on climate change

Forestry and wood products hold a particular position in the current political discussion on climate change and on the implementation of the 'Kyoto-Protocol'[445,446].

On the one hand, the destruction of large forest areas taking place mainly in countries of the Southern Hemisphere is a considerable source of greenhouse gas emissions. On the other hand, aforestation, reforestation and forest conservation are considered valuable sinks for the storage of carbon[447].

Due to its carbon-neutrality[448], wood is given an important role for its carbon storing potential in applications with long service life[449] as well as for its substitution potential of non-renewable (fossil) fuels and of material whose extraction and processing heavily relies on non-renewable energy carriers[450].

[445] Anonymous 1997b

[446] see Fischlin 2000 for a summary of the current discussion

[447] Various 2000; Schärer and Jakobi 2000:484; Wegener *et al.* 1994

[448] Atmospheric carbon is fixed as cellulose and lignin in wood by photosynthesis and released during natural rotting or technical incineration processes (see also footnote 458 for literature on the subject).

[449] see, *e.g.*, Skog and Nicholson 1998

[450] see, *e.g.*, Gielen 1995; Sikkema and Nabuurs 1995; Matthews *et al.* 1996:294; Werner *et al.* 2004

16.5 Legal framework for a sustainable utilisation of post-consumer wood

Since the beginning of the debate on environmental issues in the 80's, the utilisation of post-consumer wood has been subject to increasing control and regulatory works in Switzerland. A number of ordinances[451] based on the Federal Law on Environmental Protection (USG) and some international conventions[452] regulate the utilisation of post-consumer wood[453] and corresponds to the threefold nature of post-consumer wood as waste[454], secondary material and fuel.

According to the Waste Concept of Switzerland[455], the following four strategies have to be followed for the use of wood in descending order of importance:

- Reducing waste at the source;
- Minimising pollutants in processes and in products;
- Reducing waste by improved recovery operations;
- Improving domestic treatment facilities for non-recoverable wastes.

In principle, material utilisation of waste has priority over thermal utilisation. Given the current conditions, this does not hold for post-consumer wood for two reasons:

1. Post-consumer wood fractions can be contaminated with heavy metals or organic substances. Instead of concentrating and disposing of these contaminants, they are more likely to be distributed and reintegrated into the anthroposphere. This leaves unsolved the problematic nature of contaminated post-consumer wood and can even accentuate problems in the future. The dilution of contaminated waste is forbidden under the Technical Ordinance on Waste (TVA, Art. 10);
2. For economic reasons, the potential of European forests as fuel suppliers is not exploited; currently, only the revenue arising from material uses justifies the cost of harvesting the wood. Recycling of post-consumer wood reduces harvesting and thus the amount of wood available for

[451] see *e.g.*, the Technical Ordinance on Waste (TVA), the Ordinance Relating to Environmentally Harmful Materials (StoV), the Ordinance on Air Pollution Control (LRV)

[452] The Basler Übereinkommen über die Kontrolle der grenzüberschreitenden Verbringung gefährlicher Abfälle und ihrer Entsorgung, SR 0.814.05 ('Basel Convention'), the OECD-resolution C(92)39/FINAL on the control of frontier passing traffic with wastes destined for recovery operations

[453] see, *e.g.*, Hofer 2000b for details

[454] in the sense of the Federal Law on Environmental Protection (USG) Art. 7, Abs. 6

[455] Anonymous 1992:5

incineration and substitution of fossil fuels. As a consequence, the substitution potential of wood concerning fossil fuels remains largely unutilised.

Hofer (2000) proposes four basic strategies for the management of post-consumer wood:

1. "Post-consumer wood generated in Switzerland will be supplied to thermal utilisation if feasible under sustainable conditions;
2. "Material utilisation within Switzerland and abroad will fulfil environmental requirements that are equivalent to the thermal utilisation of post-consumer wood in Switzerland;
3. "Disposal/treatment of post-consumer wood will comply with legal regulations;
4. "The quality of post-consumer wood will be improved on a mid- to long-term basis" (Hofer 2000a:19; translated from German by WF).

16.6 Rules for a sustainable utilisation of wood

Summarising the above chapters, the following principles for a sustainable use of wood and post-consumer wood in particular can be stated. They should be incorporated properly into LCA of wood products:

1. Wood production in forestry respecting qualitative and quantitative criteria for sustainable forest management;
2. Efficient and effective processing and use of wood as a material in wood industry;
3. Renunciation of chemical wood protection where possible (*e.g.*, by constructive means) and careful selection of additives like overlays, adhesives, coatings etc. (see also point 5);
4. Production of products that can easily be disassembled providing thus single-material fractions for easy recycling and incineration in appropriately equipped plants;
5. Maintenance of the incineration potential for the substitution of non-regenerative fossil fuels;
6. Recovery of the ashes stemming the incineration of post-consumer wood from power plants to return them to the forest ecosystem as fertiliser in an ecologically acceptable way;
7. Incineration of post-consumer wood and maximisation of the substitution of fossil fuels;
8. Maximisation of the amount of wood stored in long-term applications such as buildings.

Chapter 17

WOOD PROCESSING CHAIN AND ATTRIBUTION IN LCA

17.1 Attribution and allocation in LCA of wood and wood products

The specific conditions of wood as a naturally grown, renewable material and its twofold nature as a material and fuel make some specific, interconnected considerations necessary, which are related to attribution and allocation in LCA:

- The setting of system boundaries between the natural production function of woods and the technical system of forestry[456];
- The consideration of additional environmental and social functions of forestry such as protective functions in mountainous areas, regulative functions for watersheds, supply of non-wood products, recreation functions, habitat functions, etc. for the allocation of environmental burdens related to forestry[457];
- The consistent handling of the carbon uptake and of the embodied energy over the whole life cycle of a product[458];
- The combined combustion of post-consumer wood in municipal waste incinerators requiring the allocation of emissions to single fractions and the allocation of emissions from landfill[459];

[456] Lindfors *et al.* 1995b, Technical Report No 4; Fava *et al.* 1996:2-3ff; Tillman *et al.* 1994:22; Erlandson 1996:22
[457] Schweinle 2000:73ff
[458] De Feyter 1995; Jungmeier *et al.* 2001b; Fava *et al.* 1996:2-5ff/2-18ff
[459] Fava *et al.* 1996:2-9f/2-13ff

- The handling of the material and fuel aspect of wood in comparative studies with non-renewable materials or energy carriers[460];
- The reference use of forest area in comparative studies where non-regenerative materials are involved in the comparison[461];
- The allocation of coproducts, typically generated throughout the wood processing chain from thinning, saw mill and wood industry requiring coproduct allocation, allowing for material or energetic utilisation;
- The allocation related to reuse, recycling, and thermal energy recovery;
- The allocation of emissions and the recovered energy related to the incineration of wood, including the valuation of post-consumer wood as waste or fuel.

For the scope of this study, emphasis is put on the latter three points. Some hints on further literature related to the other points are given as footnotes.

17.2 Overview of LCA methodologies related to wood

Two guidelines for LCA of wood products have been published to date, one by the American Forest & Paper Association[462], and one as the result of the EU-research project Life-Sys Wood[463,464].

17.2.1 Life Cycle Inventory Analysis by the American Forest & Paper Association

The American Forest & Paper Association edited one of the earliest user guides on LCA and wood[465]. It was published before the approval of the series of standards ISO 14040.

Some of the key recommendations are summarised below:

- Allocation should be avoided by system expansion or subdivision of processes wherever possible (Fava *et al.* 1996:2-9; 3-2);
- Where allocation cannot be avoided, it must reflect the underlying physical relationships between the system inputs and outputs. Allocation procedures should preferably use measurable physical parameters such as mass, energy, or volume. (Fava *et al.* 1996:3-2f);

[460] Jungmeier *et al.* 2001c:2/5ff
[461] Fleischer and Schmidt 1996
[462] Fava *et al.* 1996
[463] Esser and Robson 1999a
[464] A third guide by Erlandson 1996 treats cradle-to-gate accounting; recycling issues are not covered.
[465] Fava *et al.* 1996

- Where physical causal relationships cannot be established or used as the basis for allocation, market prices or other market descriptors can be used (Fava *et al.* 1996:3-2);
- For closed-loop reuse & recycling, the life cycle burdens associated with manufacture and disposal must be allocated based on the average number of uses (*e.g.,* in the case of pallets) (Fava *et al.* 1996:3-4);
- The use of a model developed around the cellulose fibre as the discrete, physical module that is recycled or reused, is a useful concept for recycling allocations in forest product systems: partitioning equally all the 'virgin' inventory into the original plus additional uses (even in the case of not equivalent products), $(n+1)$, on a mass basis, allows for the allocation $1/(n+1)$ of the 'virgin' inventory to the virgin product system, and $n/(n+1)$ of the 'virgin' inventory to the remaining uses (Fava *et al.* 1996:3-5ff; the same example on paper recycling can be found in TR/ISO 14049, chap. 8.3.3). Hence, recycled material as input is always attributed part of the 'virgin' inventory, while with output material to be recycled, some of the 'virgin' inventory is exported;
- The basis for the allocation of woodland management activities is the volume of marketable wood in the woodlands (not a strict coproduct allocation for the thinning, but an ex-post allocation based on the total of wood harvested). Care should be taken to accurately allocate time-dependent burdens based on the proportion of the average volume of the trees harvested versus the average volume of the trees left in the woodlands (Fava *et al.* 1996:3-9).

The allocation procedure proposed for the use of fibre residues or wood residues remains unspecific – despite the above-stated proposition to apply the concept of 'the number of uses':

> "Collection of complete information to allocate the LCI burdens to wood products residues used in paper products may be difficult. Simplifications may be justified as these allocated burdens would likely be minor part of the overall paper product LCI burdens. In allocating wood product residues used in another wood product, a more detailed analysis may be required as the allocated burdens may make a significant contribution to the overall wood product LCI burdens" (Fava *et al.* 1996:3-10).

In addition, no clear guidance is given for the handling of solid waste management, *i.e.* the incineration of post-consumer wood with thermal energy recovery or its land filling:

> "Energy inputs and outputs and the environmental releases from solid waste disposal operations need to be allocated based on a causal relationship of the forest product being studied" (Fava *et al.* 1996:3-20).

17.2.2 Methodological Appendix of the Life-Sys Wood project

A second methodological guide for LCA of wood and wood products has been elaborated within the European project Life-Sys Wood[466]. This project was set up in 1995 within the EC FAIR programme combining the development of standard methodology with a co-ordinated LCA approach within wood industry (*e.g.,* common data basis, common data format, etc.).

The following quotations are taken from the Decision List elaborated for consistent life cycle inventory within the Life-Sys Wood project.

> "Allocation of processes to products will be done following the 3 options of the ISO standard:
> - Avoid allocation by subdividing processes (as far as possible),
> - Allocate on a physical basis if possible by mass,
> - Allocate on an economic basis if the allocation on physical basis makes no sense (Diamond example). The reason for this choice should be made clear (using causality principle)" (Esser and Robson 1999b:4).

These allocation principles correspond to the ones proposed in Fava *et al.* (1996).

Note that the proposed allocation on mass basis (point 2) as reflection of causal physical relationship cannot *a priori* be seen as an interpretation of ISO/EN 14041, where it says:

> "Where allocation cannot be avoided, the system inputs and outputs should be partitioned between its different products or functions in a way that reflects the underlying physical relationships between them; *i.e.* they shall reflect the way in which the inputs and outputs are changed by quantitative changes in the products or functions delivered by the system. *The resulting allocation will not necessarily be in proportion to any simple measurement such as the mass or molar flows of coproducts*" (ISO/EN 14041, chap. 6.5.3; italic by WF).

For a detailed discussion on the interpretation of ISO/EN 14041, chap. 6.5.3, see Chapter 5.6.5.

Furthermore, the decision list of the Life-Sys Wood specifies a general allocation procedure:

> "Using the rules above (and if allocation cannot be avoided by subdividing) allocation of inventory data of a process within the product system will be done on the following basis:

[466] Esser and Robson 1999a

1. Inventory data is allocated to coproducts only[467];
2. Inventory data is not allocated to by-products[468];
3. If waste is produced and it goes to a recycling process, then the inventory data is allocated to the product of recycling and not to the coproducts;
4. If waste is produced and it goes to final waste disposal (*e.g.*, incineration or landfill) then the inventory data of disposal is allocated to the coproducts" (Esser and Robson 1999b:4).

Within Life-Sys Wood, the second point was applied also to the CO_2 uptake during the growth of trees. Hence, coproducts were attributed also the CO_2 uptake of wood leaving the system under study, leading to negative overall CO_2-emissions of the coproduct[469]. This practice – basically a question of setting system boundaries between natural production and the technical system – was later rejected by several members of Life-Sys Wood[470,471].

The allocation principles in Esser and Robson (1999b) are further refined for recycling/reuse:

"In the case of recycling of wood products:

1. Disposal impacts are allocated to genuine subsequent products: (a genuine subsequent product example could be a wood transmission pole re-used as a gate post)
2. Benefits of recycling are allocated to secondary products; and
3. Consistent product system boundaries will be chosen:
   - after the demolition/collection process with the system boundary at the demolition site gate
   - transportation, sorting and reprocessing is allocated to the new product.
4. Waste disposal of the end product is allocated in principle only to this end product and not to former produced products in the chain" (Esser and Robson 1999b:12).

467 The coproduct is the intended output of the process with a defined market. The market price is (relatively) high (after Esser and Robson 1999b:4).

468 The by-product is not the intended output of the process, but it is a necessary output of the process and there is a market for the product. The market price is relatively low (after Esser and Robson 1999b:4).

469 see, *e.g.*, Esser and Robson 1999b:13; Richter 1999:43f

470 see, *e.g.*, Jarnehammer 2001; Jungmeier *et al.* 2001c:2/40f

471 Also in this book, carbon up-take and embodied energy are considered inherent material characteristics and thus are allocated according to physical causal relationship on a mass basis, independent of the allocation chosen for the allocation related to coproduction and reuse & recycling (see, *e.g.*, de Feyter 1995; Jungmeier *et al.* 2001b; Fava *et al.* 1996:2-5ff/2-18ff).

The proposed system boundaries and the allocation procedure correspond to the 'cut-off' procedure described in Chapter 11.1, but by-products with low market prices are introduced which are not assigned any up-stream environmental burdens.

Note further on that point 4 has to be substantiated considering ISO/EN 14041:

> "Reuse and recycling ...may imply that the inputs and outputs associated with unit processes for extraction and processing of raw materials and final disposal of products are to be shared by more than one product system" (ISO/EN 14041, chap. 6.5.4).

17.2.3 Conclusions

The following conclusions can be drawn from the above:

- No general consensus among the wood-specific guidelines including the standard ISO/EN 14041 exists on methodological questions related to reuse & recycling;
- In all methodological documents and guidelines (including the standard ISO/EN 14041), no systematic reasoning is proposed *why* methodological questions should be treated in the way proposed in guidelines and standards.

Chapter 18

MODELLING END-OF-LIFE OPTIONS OF BEECH WOOD RAILWAY SLEEPERS

18.1 Goal and scope definition

The second case study aims at evaluating different approaches to the modelling of two end-of-life options for beech wood railway sleepers. The two end-of-life options considered are:

- Incineration of the railway sleeper in an incineration plant with co-generation of electricity and thermal energy;
- Recycling of the railway sleeper as a constructive element in landscape architecture.

Modelling of end-of-life options can depend on decisions made in previous life cycle steps of a product, such as:

- Forestry processes;
- Transport from forestry;
- Sleeper production.

Hence, the same evaluation of different approaches to the modelling of these processes is necessary to allow for a consistent modelling of

- End-of-life options.

Finally, some

- Limited life-cycle models

are set up, combining modelling options of the above-mentioned processes to obtain 'consistent' product systems.

Evaluation is made based on the 'descriptiveness' of an approach in respect to:

- The material and market characteristics of wood;
- The logic of process-specific decision-making and planning;
- The wood-specific management rules for its sustainable use;

- The requirements of a product system and its LCI established in Chapter 3.4.

Conclusions for consistent modelling of the life cycle of a railway sleeper with special emphasis on end-of-life options are drawn from all the allocation situations evaluated, including the limited life-cycle models.

Environmental problems related to the impregnation of wooden sleepers with creosote are not within the goal of this study[472]. Possible recycling of used railway sleepers as railway sleepers in areas with lower strain is also disregarded for the sake of simplicity of the example[473].

18.2 Functional unit

The functional unit considered is the impregnated wooden part of the beech wood railway sleeper for Swiss standard gauge railway tracks, as assessed in Künniger and Richter (1998) (Figure 18-1). Its length is 2.7 m; details of the mass balance can be found in Table 18-1.

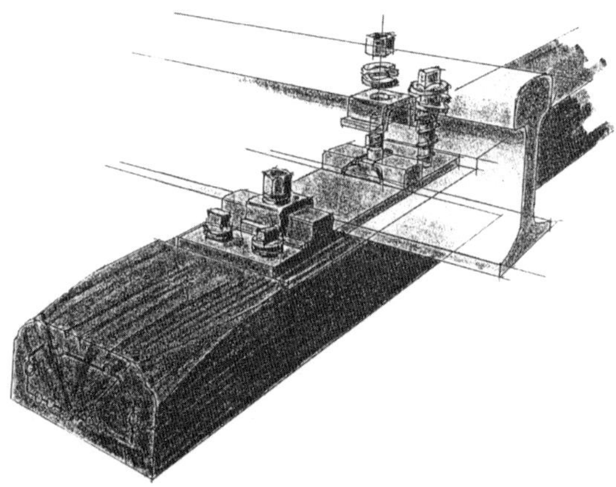

Figure 18-1. Swiss standard gauge beech wood railway sleeper (Künniger and Richter 1998: 15; reprinted with permission from the authors)

The track bed and its construction, maintenance and disposal as well as metallic auxiliary materials (accessories for securing the rails) are omitted for the purpose of this example.

Service life of the sleeper is not important for the modelling in this study.

472 for details, see Künniger and Richter 1998:27ff; Kohler *et al.* 2000

473 for details, see Künniger and Richter 1998:44ff

18.3 Model of the reference flow

The reference flow focuses on the processes possibly affected by allocation procedures used for the modelling of the two end-of-life options. All other mass and energy flows are omitted for the sake of simplicity of this example.

The life cycle steps considered are:

- Technical wood production;
- Transport from the forest to the production site;
- Sleeper production;
- (Impregnation);
- End-of life:
    - recycling as elements in landscape architecture;
    - incineration with co-generation of heat and electricity.

These life cycle stages and the respective allocation problems are described in separate chapters.

Figure 18-2 illustrates the main elements of the reference flow.

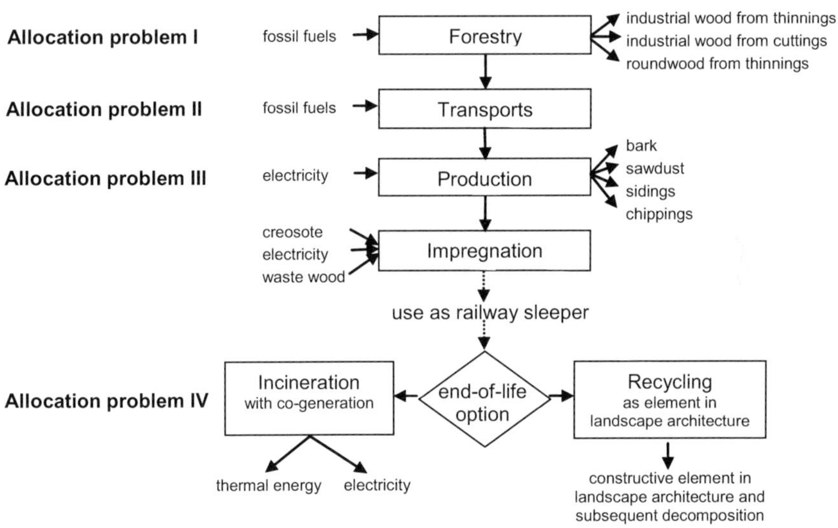

Figure 18-2. Model of the reference flow of the railway sleeper with the end-of-life options incineration with co-generation and recycling as element in landscape architecture

Finally, a (rudimentary) life cycle model is developed, covering all life cycle stages considered in this study. Impregnation is partly treated in allocation procedures for end-of-life options requiring it; impregnation is included in all models covering the whole life cycle of the sleeper.

Table 18-1 contains the wood flows throughout the life cycle model of one beech wood railway sleeper and the respective current market prices.

Table 18-1. Wood flows related to the production of one wooden railway sleeper (578 kg odm/m3; 751 kg/m3 with mc = 30%; basic densities) (own calculations based on Corbat 2001; Schweinle 2000; Schweinle 2001; Frühwald *et al.* 1996; Künniger and Richter 1998; Graf 2001 Keel 2001; Anonymous 2000b; Udry 2001)

Process	Input	mass kg odm	Output	mass kg odm	price CHF/kg
Nursery/Tending/ Plantation	8000 plants/ ha	-	No commercially used output	-	-
Thinning			Industrial wood (class I, crane size) (mc=80%)	91.5	0.090
			Log over bark (I) (mc=70%)	26.9	0.150
Cutting/Hauling			Industrial wood (mc=80%) (class I, crane size)	49.1	0.090
			Log over bark for sleepers (class I) (mc=70%)	119.0	0.150
Transport	Log over bark in forest	119.0	Log over bark, hauled (mc=70%)	119.0	*0.214*
Roundwood trimming	Round wood	119.0	Solid fuel wood (mc=30%)	3.63	0.110
			Log piece (mc=30%)	115.4	*0.233*
Debarking	Log piece	115.4	Bark (mc=30%)	8.05	0.046
			Log piece debarked (mc=30%)	107.4	*0.252*
Roundwood conversion	Log piece debarked	107.4	Sawdust (mc=30%)	3.38	0.018
			Squarings (mc=30%)	25.0	0.047
			Sidings (mc=30%)	20.0	1.112
			Raw sleeper (mc=30%)	59.0	*0.475*
Natural Drying	Log over bark hauled	119.0	Log over bark dried (mc=30%)	119.0	*0.552*
Sorting	Raw sleeper	59.0	Rejects (mc=30%)	1.72	0.110
			Sorted raw sleeper (mc=30%)	57.3	*0.565*
Mechanical processing	Sorted raw sleeper	57.3	Chippings, increment cores (mc=30%)	1.70	0.097
			Solid fuel wood (angle trimming; 3%) (mc=30%)	1.62	0.110
			Sleeper finished (mc=30%)	53.9	confid.
Impregnation (Creosote heating)	Residual wood [1)]	1.82	Sleeper finished, impregnated (mc=30%)	53.9	confi-dential
	Sleeper finished	53.9			
Resale for energy or construction purposes	Used impregnated sleeper	53.9	a) Used impregnated sleeper as element in landscape architecture	53.9	confi-dential
			b) Used impregnated sleeper as fuel	53.9	confid.

[1)] internal wood residues taken from the wood flows listed

italic: internal prices are rough estimates based on data on production costs by Gautschi (2001: 171/22) and on Corbat (2001)

18.4 Allocation problem I: technical wood production

Technical wood production covers the efforts from tree nurseries, tending, thinning, cutting, and hauling of the useable wood to the nearest forest road. Technical wood production has to be distinguished from natural wood production, *i.e.* the physiological processes of tree growth[474]. These physiological processes are normally not covered in LCA. An exception is usually the consideration of the CO_2 uptake during photosynthesis and the embodiment of solar energy[475].

Figure 18-3 gives an overview of the processes of technical wood production. It is based on an extensive analysis of forestry processes[476]. The definition of the process steps as well as their inputs and outputs are pre-set by the given structure of the data source.

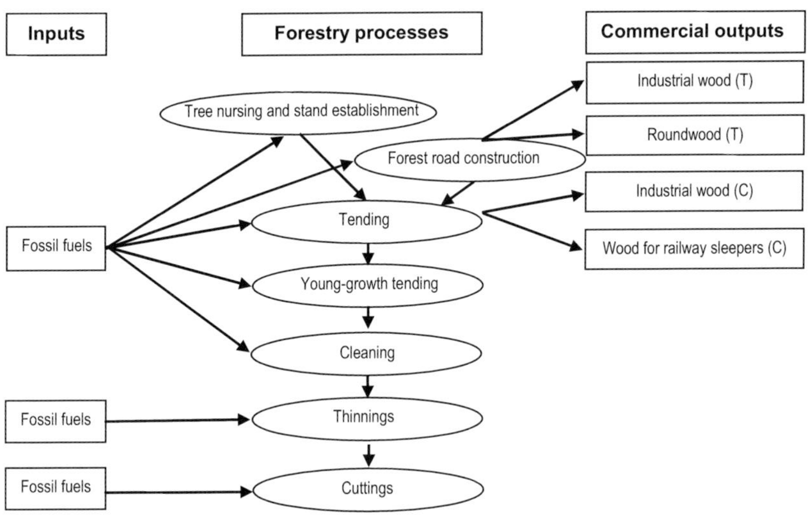

Figure 18-3. Forestry processes, their inputs and commercialised output related to the production of beech wood for railway sleepers

The following process steps are defined:
Tree nursing: the cultivation of tree seedlings in tree nurseries.

474 This study follows the distinction made by Fallscheer 1997; Frühwald *et al.* 1996; Jungmeier *et al.* 2001b; Schweinle 2000; Wegener *et al.* 1994.

475 see, *e.g.*, de Feyter 1995; Jungmeier *et al.* 2001c:2/11f; in this study, the natural CO_2-uptake is not considered. It is assumed that wood is CO_2-neutral over its entire life cycle.

476 Schweinle 2000; Schweinle 2001; these data sources are used, as they refer to the only extensive LCA-study on forestry processes covering all the four economically most important species for Central Europe: spruce, fir, beech and oak.

Stand establishment: the planting of the trees in the foreseen grounds and additional protective measures against damage caused by forest fauna. Planting is commonly made with planting machines mounted on forestry tractors.

Tending: Tending of forest plantations, *i.e.* of artificially established stands or crops, during the establishment stage. Tending mainly consists in grass cleaning with a hand-held motor grass mower.

Young-growth tending: The tending carried out in pre-thicket and thicket stands (both naturally and artificially regenerated) to encourage the even development of the trees and to improve the quality. Similar equipment is used as for tending.

Cleaning: Improvement felling during the thicket stage aiming at regulating the species composition, stocking level and stand stability, usually without utilisation of the felled tree. Selective cleaning is usually made with small chainsaws, whereas systematic cleaning is normally made with special equipment mounted on forestry tractors.

Thinnings: Several subsequent fellings made in an immature crop or stand in order primarily to accelerate diameter increment but also to salvage potential mortality and to improve the average form of the remaining trees. Thinnings vary in technique, in degree and in frequency of application. Motor-manual thinning is made with chainsaws; mechanised thinning is made with diesel-powered harvesters. Harvested trees are hauled with tractors to the nearest forest road.

For commercialisation, trees are sorted into several classes and subclasses such as industrial wood (for the use in paper or wood fibre industry), fuel wood, and roundwood, depending on diameter, branchiness, length, stature, and qualitative damages;

Cutting: Several subsequent cuts made in the mature stand as final harvesting. Again, industrial wood and roundwood are harvested, hauled and sorted into several classes. Cutting is usually made with big chainsaws for felling and with small chainsaws for debranching, whereas hauling is normally made with forestry tractors (or horses).

Several processes are not considered in this example, such as soil liming, the application of pesticides, debarking in the forest, and (wet timber) storage.

Several fractions of industrial wood and roundwood result from forestry. Simplifying assumptions are made for the following calculations because of the lack of more specific data:

- Forest management of beech aiming at yield class II (intense thinning) is adequate for the production of trees used in railway sleeper production;
- Only two fractions result from both thinning and cutting, one of industrial wood and another of roundwood;

- The two fractions of industrial wood from thinnings and cuttings are equivalent;
- Roundwood for sawn timber results from thinnings; roundwood for railway sleeper production results from cuttings; equal prices are assumed for both classes of roundwood.

Table 18-2 summarises the inputs of fossil fuels for each of the forestry processes, the yield and revenues achieved per hectare. Detailed data is presented in Appendix A3.

Table 18-2. Inputs of fossil fuels per hectare for the production of beech wood, volume, mass (odm) and market price (yield class II, intense thinning) (after Schweinle 2000; Schweinle 2001; Frühwald *et al.* 1996: 31; Anonymous 2000b; Udry 2001)

	Nursery, stand establ.	Tending	Young-growth tending	Clearing (selective)	Forestry roads	Expenditures without yield	Thinnings	Cuttings	Total
Inputs per hectare									
Diesel fuel (l/ha)	112				0.36	112	451	434	997
Gasoline for chainsaw (l/ha)		33.6	36	36		106	150	145	401
Oil for chainsaw (l/ha)		10.5	11.25	11.25		33	47	45	125
Yield per hectare (m^3 (u.b.)									
Industrial wood (m^3 (u.b.))							223	120	343
Roundwood (m^3 (u.b.))							66		66
Roundwood for sleepers (m^3 (u.b.))								291	291
Total m^3 (u.b.)							289	411	700
Yield per hectare (kg odm)									
Industrial wood (kg odm)							71'608	38'533	110'141
Roundwood (kg odm)							22'440		22'440
Roundwood for sleepers (kg odm)								98'940	98'940
Total kg odm							94'048	137'473	231'521
Revenues CHF per hectare									
Industrial wood (16.2 CHF/m^3 u.b.) 1)							3'612	1'941	5'553
Roundwood (52.1 CHF/m^3 u.b.) 2)							3'428		3'428
Roundwood for sleepers (52.1 CHF/m^3 u.b.) 3)								15'146	15'146
Total CHF/ha							7'040	17'087	24'127

1) Spanplattenholz; Laubholz, I. Klasse, kranlang (50.4 CHF/t odm; 16.18 CHF/m^3, mc = 80%)
2) Schwellen I. Klasse (52.1 CHF/m^3, mc = 70%)
3) Schwellen I. Klasse (52.1 CHF/m^3, mc = 70%)

Prices used for calculations are the average of trimestrial market prices taken from the Swiss Forestry Statistics, covering the years 1995 to August 2000[477,478].

477 Anonymous 2000b, Udry 2001

18.4.1 Allocation problems and scenarios considered

Attribution problems related to forestry processes basically are:

- The allocation of fossil fuel consumption caused before the first harvesting, *i.e.* tree nursing, tending, cleaning, and the construction of forestry roads;
- The allocation of fossil fuel consumption caused during thinnings and cuttings, where several products result.

For most applications of product LCA, this process structure does not allow one to cut off insignificant flows according to criteria for the setting of system boundaries (ISO/EN 14041, chap. 5.3.5).

Furthermore, subdivision of processes and system expansion as first step of the ISO allocation procedure with the aim to avoid allocation must be considered inadequate (ISO/EN 14041, chap. 6.5.3, point a). The second step of the allocation procedure (ISO/EN 14041, chap. 6.5.3, point b) is also not feasible; underlying physical relationship to be determined through marginal considerations cannot be established because processes are inherently joined.

Hence, the third step of the allocation procedure according to ISO/EN 14041 (chap. 6.5.3, point a) is applicable:

> "Where physical relationship alone cannot be established or used as the basis for allocation, the inputs should be allocated between the products and functions in a way which reflects other relationships between them. For example, input and output data might be allocated between coproducts in proportion to the economic value of the products" (ISO/EN 14041, chap. 6.5.3, point a).

This paragraph opens a variety of options for the allocation

a) related to the allocation factor; and
b) concerning the allocation of the processes with no direct product.

Ad a) Within forestry & LCA circles, several allocation factors have been discussed, such as mass (wet), mass oven-dried (mc = 0), volume, market price, and heat value. For the forestry processes, this study concentrates on volume as common measure of forestry and on market prices[479] of the products as allocation factors[480].

---

[478] As sorting criteria for raw wood changed in August 2000, the price-time series could not reasonably be followed any further.

[479] revenues; profits as driving force of economic processes would probably provide more consistent models; however, data on profits is hardly available over a whole production chain in which several companies are involved (see, *e.g.*, Huppes 1992; Huppes 1994)

[480] Data of the allocation based on mass is shown in Appendix A5, including a figure comparing the consequences of mass-, volume- and revenues-based allocation for the over-all

Ad b) Several lines of reasoning are possible for the allocation of the inputs of fossil fuels before thinnings. They correspond to a different mental model of the material and market characteristics of raw wood, and of the logic of forestry planning and management:

Over-all: This line of reasoning stresses the decision-processes during forestry planning. It is assumed that a forest stand is planned in its integrity over time and managed to achieve maximum wood volume or maximum revenues. Forest interventions are necessary over the whole lifetime of the forest stand to achieve this goal. As a consequence, all inputs are allocated equally over all products independently of the time of harvesting based on volume or relative share of revenues.

Step-by-step (*I*): This procedure follows a strict step-by-step procedure: each output is allocated the inputs that are directly preceding it. This means that the inputs for nursing, tending, clearing and the construction of forestry roads – processes with no direct wood output – are considered precursors of thinnings. Hence, all these inputs are summed up, added to the inputs for the thinnings and allocated to the products from thinnings based on volume or on the relative share of revenues. The products of the cuttings just carry the inputs necessary for the cuttings.

Step-by-step (*II*): This line of reasoning is a combination of the former two. It is argued that processes with outputs should get attributed the inputs directly linked with their production. So the inputs of the thinnings are allocated to the respective products on a volume basis or based on the relative share of revenues; the inputs of the cuttings are allocated to the products from cutting. The preceding processes are considered causally linked to both the products from thinnings and from cuttings. Hence, these inputs are allocated to all the products from the forest stand based on mass or revenues/m^3.

All-to-sleeper: A fourth procedure assumes the wood production for wooden sleepers to be the primary aim of forest management. Consequently, all inputs are attributed to the wood for railway sleepers.

---

procedure considered the most adequate for forestry processes. For beech, results of the volume- and mass-based allocation do not differ considerably. Therefore, only volume is used as allocation factor for further calculations.

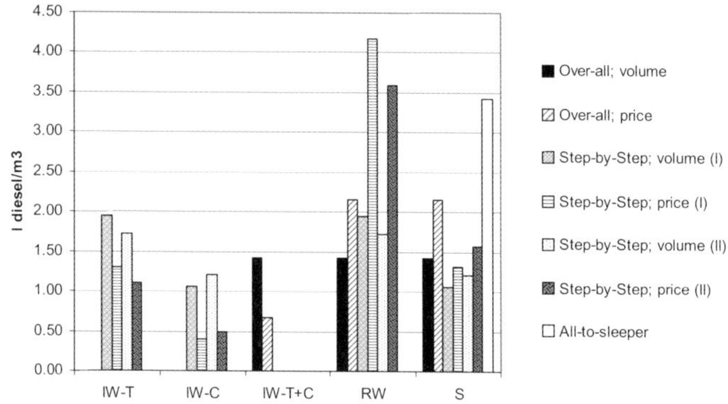

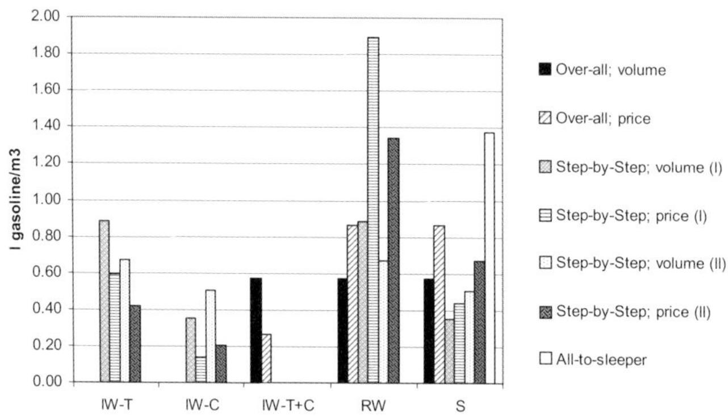

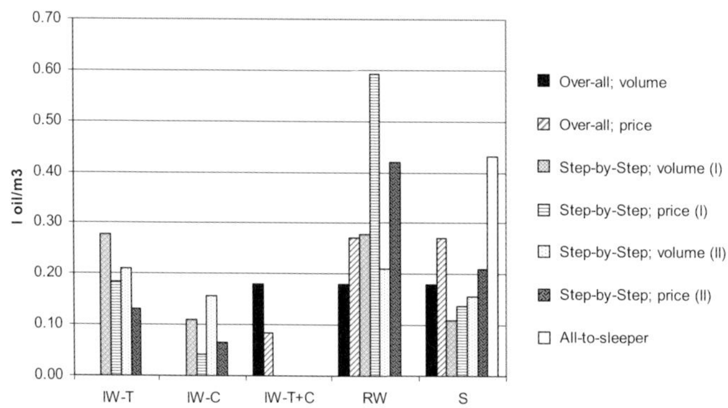

Figure 18-4 to 18-6. Collection of the inputs into the forestry processes and its allocation according to different models on forestry planning and management as well as the material and market characteristics of wood; based on Table 18-3 (mc = 80% for industrial wood; mc = 70% for roundwood)

IW-T: Industrial wood from thinnings; IW-C: Industrial wood from cuttings; IW-T+C: Industrial wood from thinnings and cuttings; RW: Roundwood from thinnings; S: Roundwood for sleepers from cuttings

None of these allocation procedures can *a priori* be excluded based neither on ISO/EN 14041 nor on the wood-specific LCA sustainable management rules outlined in Chapter 16.6.

Table 18-3 summarises the allocation of diesel fuel, gasoline, and engine oil used in forest management. Calculations are based on Table 18-2.

Figures 18-4 to 18-6 illustrate the consequences of the above-mentioned lines of reasoning on the allocation of inputs of forest management to the resulting productswood for railway sleepers from cuttings

Table 18-3. Collection of the inputs into the forestry processes and its allocation according to different mental model of forestry planning and management as well as the material and market characteristics of wood (mc = 80% for industrial wood; mc = 70% for roundwood)

	IW-T	IW-C	IW-T+C	RW	S
Diesel fuel into tractor (*l/m*^{3} (*u.b.*))					
Over-all; volume			1.43	1.43	1.43
Over-all; revenues			0.67	2.15	2.15
Step-by-step; volume (I)	1.95	1.06		1.95	1.06
Step-by-step; revenues (I)	1.30	0.41		4.17	1.32
Step-by-step; volume (II)	1.72	1.22		1.72	1.22
Step-by-step; revenues basis (II)	1.11	0.49		3.58	1.57
All-to-sleeper					3.43
Gasoline into chainsaw (*l/m*^{3} (*u.b.*))					
Over-all; volume			0.57	0.57	0.57
Over-all; revenues			0.27	0.87	0.87
Step-by-step; volume (I)	0.89	0.35		0.89	0.44
Step-by-step; revenues (I)	0.59	0.14		1.89	0.41
Step-by-step; volume (II)	0.67	0.50		0.67	0.50
Step-by-step; revenues (II)	0.42	0.21		1.34	0.67
All-to-sleeper					1.38
Engine oil into chain saw (*l/m*^{3} (*u.b.*))					
Over-all; volume			0.18	0.18	0.18
Over-all; revenues			0.08	0.27	0.27
Step-by-step; volume (I)	0.28	0.11		0.28	0.11
Step-by-step; revenues (I)	0.18	0.04		0.59	0.14
Step-by-step; volume (II)	0.21	0.16		0.21	0.16
Step-by-step; revenues (II)	0.13	0.06		0.42	0.21
All-to-sleeper					0.43

IW-T: Industrial wood from thinnings; IW-C: Industrial wood from cuttings; IW-T+C: Industrial wood from thinnings and cuttings; RW: Roundwood from thinnings; S: Roundwood for railway sleepers from cuttings

5% of the engine oil is assumed to drop onto the forest soil.

This data has been computed in an LCA-software. The complete inventory data as computed is shown in Appendix A4. The results of the impact assessment are presented in Chapter 18.4.3.

18.4.2 Inventory analysis

A detailed data inventory can be ordered from the author.

18.4.3 Impact assessment

Table 18-4 summarises the results of the life cycle impact assessment based on the 'effect-oriented classification' impact assessment method according to Heijungs *et al.* (1992, CML-method), updated according to Houghton *et al.* (1996).

Table 18-4. Impact assessment of different allocation procedures applied to forestry processes (mc = 80% for industrial wood; mc = 70% for roundwood)

	Green-house effect kg CO_2-e.	Ozone depletion kg R11-e.	Acidifi-cation kg SO_x-e.	Eutrophi-cation kg PO_4-e.	Photosmog kg ethy-lene-e.	Human toxicity kg	Eco-tox. soil kg	Eco-tox. water 10E6 m^3
Industrial wood (T/C/T+C)								
IW-T+C; over-all; vol.	6.09	8.22E-06	5.16E-02	8.01E-03	5.71E-02	2.87E-01	5.74E-06	204
IW-T+C; over-all; rev.	2.86	3.86E-06	2.42E-02	3.76E-03	2.68E-02	1.34E-01	2.69E-06	96
IW-C; step-by-step (I); vol.	4.37	5.75E-06	3.77E-02	5.88E-03	3.61E-02	1.87E-01	4.00E-06	144
IW-C; step-by-step (I); rev.	2.71	3.57E-06	2.34E-02	3.65E-03	2.25E-02	1.16E-01	2.49E-06	89
IW-C; step-by-step (II); vol.	5.22	7.08E-06	4.41E-02	6.84E-03	4.99E-02	2.49E-01	4.94E-06	175
IW-C; step-by-step (II); rev.	3.31	4.50E-06	2.79E-02	4.32E-03	3.20E-02	1.60E-01	3.14E-06	111
IW-T; step-by-step (I); vol.	8.52	1.17E-05	7.13E-02	1.10E-02	8.67E-02	4.27E-01	8.19E-06	289
IW-T; step-by-step (I); rev.	7.13	9.80E-06	5.96E-02	9.21E-03	7.25E-02	3.57E-01	6.85E-06	241
IW-T; step-by-step (II); vol.	7.31	9.83E-06	6.21E-02	9.66E-03	6.71E-02	3.39E-01	6.86E-06	244
IW-T; step-by-step (II); rev.	4.69	6.27E-06	4.00E-02	6.23E-03	4.19E-02	2.13E-01	4.37E-06	156
Roundwood (T)								
RW; over-all; vol.	6.09	8.22E-06	5.16E-02	8.01E-03	5.71E-02	2.87E-01	5.74E-06	204
RW; over-all; rev.	9.20	1.24E-05	7.79E-02	1.21E-02	8.62E-02	4.33E-01	8.67E-06	308
RW; step-by-step (I); vol.	8.52	1.17E-05	7.13E-02	1.10E-02	8.67E-02	4.27E-01	8.19E-06	289
RW; step-by-step (I); rev.	13.26	1.82E-05	1.11E-01	1.71E-02	1.35E-01	6.65E-01	1.27E-05	449
RW; step-by-step (II); vol.	7.31	9.83E-06	6.21E-02	9.66E-03	6.71E-02	3.39E-01	6.86E-06	244
RW; step-by-step (II); rev.	15.10	2.02E-05	1.29E-01	2.00E-02	1.35E-01	6.86E-01	1.41E-05	502
Roundwood for sleepers (C)								
S; over-all; vol.	6.09	8.22E-06	5.16E-02	8.01E-03	5.71E-02	2.87E-01	5.74E-06	204
S; over-all; rev.	9.20	1.24E-05	7.79E-02	1.21E-02	8.62E-02	4.33E-01	8.67E-06	308
S; step-by-step (I); vol.	4.37	5.75E-06	3.77E-02	5.88E-03	3.61E-02	1.87E-01	4.00E-06	144
S; step-by-step (I); rev.	5.48	7.21E-06	4.73E-02	7.38E-03	4.53E-02	2.35E-01	5.03E-06	180
S; step-by-step (II); vol.	5.22	7.08E-06	4.41E-02	6.84E-03	4.99E-02	2.49E-01	4.94E-06	175
S; step-by-step (II); rev.	6.76	9.22E-06	5.69E-02	8.82E-03	6.60E-02	3.28E-01	6.44E-06	228
S; all-to-sleeper	14.65	1.98E-05	1.24E-01	1.93E-02	1.37E-01	6.89E-01	1.38E-05	490

IW-T: Industrial wood from thinnings; IW-C: Industrial wood from cuttings; IW-T+C: Industrial wood from thinnings and cuttings; RW: Roundwood from thinnings; S: Roundwood for railway sleepers from cuttings

18.4.4 Interpretation focusing on methodological considerations

The interpretation of the results of the impact assessment emphasises on methodological considerations. Discussing the contributions of different processes is beyond the scope of the study. The different input processes are all combustion processes of fossil fuels, including their upstream processes. As allocation affects all input processes in the same way, interpretation would not reveal changes in the relative contributions of the different input processes when comparing different allocation procedures[481].

Figures 18-7 to 18-8 illustrate the allocation of the forestry processes based on the 'over-all' procedure[482]. The over-all procedure distributes all forestry processes over all products based on volume or revenues.

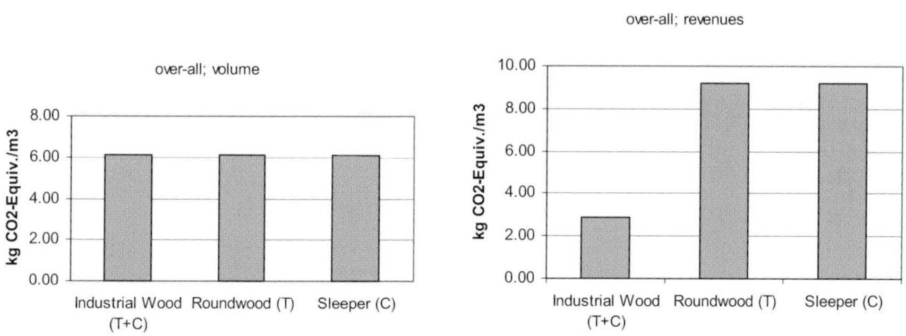

Figures 18-7 and 18-8. Effects of the over-all procedure applied based on volume or revenues, shown on the potential contributions to greenhouse effect

The 'over-all' procedure based on volume leads to equal shares for all products from forestry independently of their market prices or of the time of harvesting. Compared to the 'over-all' procedure based on revenues, the low-priced industrial wood carries higher environmental burdens whereas the higher-priced roundwood and the roundwood for sleepers are better off.

None of the two allocation variants can *a priori* be excluded for not respecting material- or market characteristics (see Chap. 15) nor for recommending 'improvement options' against wood-specific resource management rules (see Chap. 16).

The first step-by-step (I) procedure applies the step-by step allocation in a strict way. This means that the forestry processes preceding and including

[481] The same is of course valid for the 10% engine oil used as lubricant which is inventoried as biologically degradable oil lost to soil.

[482] Note that all figures in this chapter have the same scale.

thinnings are allocated to the products of thinnings based on volume or revenues, whereas the products from cuttings get attributed only the inputs to cuttings. The products from thinnings (T) carry much higher environmental loads than the products from cuttings (C) because of the nature of forestry processes. This is due to the planting, tending and clearing activities, which are allocated solely to the products of thinnings as its preceding processes (Figure 18-9 and Figure 18-10).

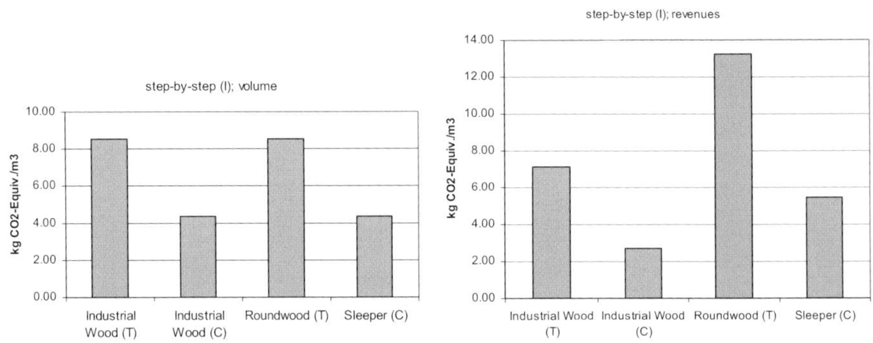

Figures 18-9 and 18-10. Effects of the step-by-step (I) procedure applied in a strict sense (I), based on volume or revenues, shown on the potential contributions to greenhouse effect

Allocation based on volume results in equal distributions of environmental burdens to the products of one forestry process. If allocation is based on revenues, roundwood (T) and roundwood for sleepers (C) carry more environmental burdens than industrial wood of the same processes. It result that roundwood from thinnings is attributed the highest environmental burdens, whereas roundwood for sleepers is attributed less environmental impacts than lower-value, earlier-harvested industrial wood from thinnings. Furthermore, industrial wood from thinnings carries more environmental burdens than the same industrial wood from cuttings. Although the same product this allocation procedure makes industrial wood from cuttings environmentally preferable to industrial wood from thinnings. This recommendation is not tenable from a market and material characteristic point of view.

All in all, this attribution procedure does not respect material and market characteristics of forestry products. Furthermore, it is not in line with the logic of decision-making in forestry.

The second step-by-step (II) procedure attributes the inputs used for thinnings and cuttings to the respective products. The preceding processes without yield are attributed to all the products resulting from forestry (Figure 18-11 and Figure 18-12).

The second step-by-step (II) procedure shows similar result like the strict step-by-step (I) procedure, although differences are generally smaller; products from cuttings carry less environmental burdens than the products from the preceding thinnings; industrial wood from cuttings results preferable to industrial wood from thinnings; roundwood from thinnings carries the highest environmental burdens if allocating based on the relative share of revenues; sleepers from cuttings carry about the same burdens as industrial wood from thinnings. The higher impacts per m^3 of the products from thinnings must be interpreted as a higher input of fossil fuels for a large amount of (thin) industrial wood and a small quantity of thick roundwood compared to large quantities of thick roundwood for sleepers and few (thin) industrial wood resulting from the cuttings.

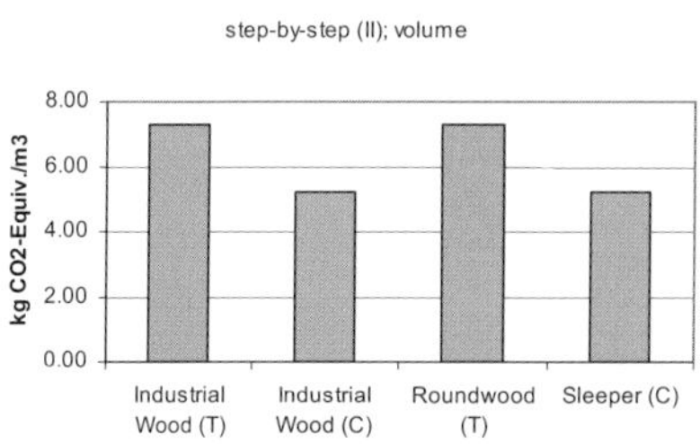

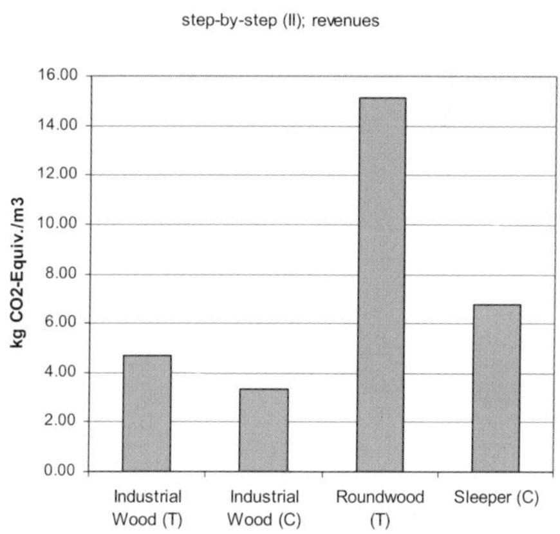

Figures 18-11 and 18-12. Effects of the modified step-by-step (II) procedure, based on volume or revenues, shown on the potential contributions to greenhouse effect

Like in the first step-by-step (I) procedure, the earlier-harvested products carry more environmental burdens than the later-cut products. This does not reflect the causal dependency of the cuttings from the thinning processes, which finally make the remaining trees to achieve their final dimensions. Furthermore, the 'preference' of industrial wood from cuttings to industrial wood from thinnings might result in recommendations not in line with the management rules for a sustainable use of wood (see Chap. 16.6).

The adequacy of the second step-by-step (II) procedure is still doubtful although it reflects the material and market characteristics as well as the logic of forestry management better than the first step-by-step procedure (I).

Figure 18-13 shows the results of the all-to-sleeper allocation procedure. The all-to-sleeper procedure allocates all the forest processes to the roundwood used for sleepers as main product.

This procedure results in by far the highest environmental impact attributed to roundwood for sleeper production; industrial wood and roundwood go without environmental burdens. This procedure is the other extreme of the equal distribution among all products (over-all procedure). The acceptance of this procedure is highly doubtful for its 'arbitrarity'. It neither respects technical nor economic causalities; also, it does not represent the logic of planning and decision-making in forestry.

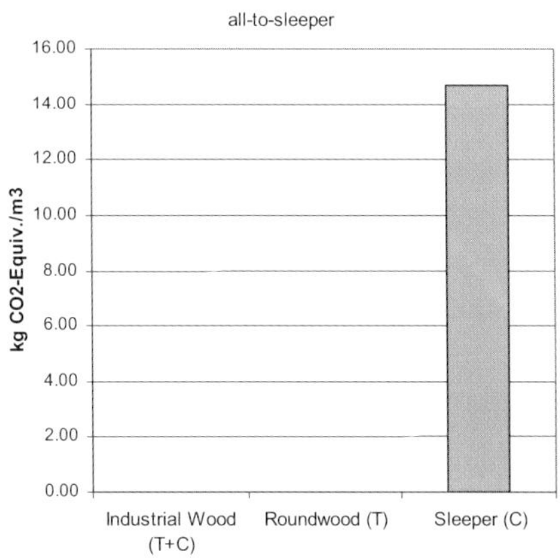

Figure 18-13. Effects of the all-to-sleeper procedure, shown on the potential contributions to greenhouse effect

Furthermore, industrial wood and roundwood for sawn timber would enter the wood processing chain without any environmental burden. This would certainly raise the suspicions of representatives of other industries if wood products are compared to products from other materials in comparative LCAs.

18.4.5 Conclusions

From the allocation procedures investigated, only the over-all procedure describes material and market characteristics of forest products in a reasonable way, respecting the logic of decision-making in forest planning and management. Furthermore, the resulting model of forestry is in line with the requirements established in Chapter 3.4: the model is simple, transparent, actor-based (respecting the logic of forestry planning and management), it can be site-specific (if data is available), and risk related to temporal structural uncertainty is no predominant subject.

The second step-by-step (II) procedure can be considered second-best option. This procedure has some limitations in its 'descriptiveness' and requires careful interpretation of the results if misleading recommendations should be avoided.

The first step-by-step (I) procedure and the all-to-sleeper procedure must be considered not 'descriptive' of the decision situation[483]. Thus, they are inadequate for the modelling of forestry processes in LCA.

Not only the choice of an allocation procedure but also the choice of the allocation factor is of interest. It has been an almost eternal methodological discussion among LCA-experts if allocation of joint coproduction processes should be based on arbitrarily chosen physical/chemical parameters or on economic parameters.

Volume on the one hand represents a simple-to-calculate parameter that requires no additional data for calculations in LCA. However, the parameter volume is not as definite as it might look like at first glance because of the inhomogeneous, anisotropic character of wood.

Revenues on the other hand are a more-complicated-to-calculate parameter that requires often hard-to-obtain, additional data. Because of limited knowledge on the specific origin of a wood log, it is often difficult to adequately model forestry processes and their products. As a consequence, the wooden forestry products can only by assigned different wood classes on a best guess basis. This implies uncertainties when allocating forestry processes based on revenues.

Mass (wet) and mass (mc = 0%) would also be possible allocation factors. Apart from some low-market-value fractions (*e.g.,* fuelwood and industrial wood for particleboard production) volume is the common measure for forestry processes. Therefore, mass has not been considered in this study. However, the selection of mass (kg odm) leads to significantly different allocation factors than volume or revenues. In the case of beech wood, the allocation based on mass (mc = 0%) provides results similar to the allocation based on volume (see also Appendix A5)[484].

The limited knowledge on the specific forestry processes related to a log (or lot) is not a specific problem for allocation procedures that are based on economic parameters. This problem is related to the complex and long-term

[483] see Chap. 3.2.1

[484] The moisture content of industrial wood and roundwood of beech is quite similar (Frühwald *et al.* 1996:31).

forestry planning and site-specific[485] forest management, resulting in various products with different use options. Furthermore, reliable unallocated LCI-data for various tree species in Central Europe from single sources is not publicly available at all. Solving this point must be an issue for the establishment of public databases.

The best guess assignment of the wooden forestry products to different wood classes as well as the fluctuations of price ratios of different wood classes within one species add further uncertainties to the uncertainty related to the volume of wood (varying densities).

The choice of volume or price/revenues per cubic meter is a subjective, arbitrary choice. It is related to the decision-maker's mental model how the socio-economic system works. Given the fact that the price range for products from forestry is quite large, choosing revenues as an allocation factor might seem adequate. This choice would be consistent with the definition of coproducts, by-products and waste, where market prices are used to determine the outputs that get allocated environmental interventions. This point becomes crucial further down in the wood processing chain where bark, sawmill, wood chips, sidings, and other products with low market prices but established buyers and processing chains result besides high-value products like sawn timber. The choice of an allocation factor for the forestry processes might also influence the conceptualisation of the allocation problems related to technical wood production and thus also the choice of an allocation factor for these processes.

18.4.6 Data quality and sensitivity

The data sources used in the forestry part are summarised in Table 18-5.

Table 18-5. Data sources used in the forestry part

Processes	Data sources
Forestry	Schweinle 2000; Schweinle 2001; Frühwald *et al.* 1996: 31; Wegener *et al.* 1994
Energy processes	Frischknecht *et al.* 1996
Chainsaw	EDIB, based on emission measurements at EMPA
Transport	Maibach *et al.* 1995, revised
Market prices	Corbat 2001, Anonymous 2000b, Udry 2001

[485] Production conditions of forestry are controllable only in a limited way and are not standardised in a technical sense. This asks for non-standardised, adaptable and flexible forest management and commercialisation of the resulting products depending on the wood classes (and their prices) resulting from the interventions. Growth tables and yield tables can just give some guidance during planning. Correspondingly, assumptions made for the modelling of forestry processes in LCA must be considered highly simplifying.

Wood is an inhomogeneous material. Data on moisture-content dependent densities is readily available but vary considerably between different data sources[486]. Results can vary depending on the density and shrinkage assumed for oven-dried wood, as calculations are usually based on oven-dried wood (mc = 0%). Due to the inhomogeneity of wood caused through local site conditions and variations within the same log, wood technological data can vary in a range ± 5%, affecting results respectively[487].

In this example, time series exist for the determination of relative changes of the prices used for allocation. An analysis of relative wood prices based on Anonymous (2000b) and Udry (2001) shows that standard deviation of price ratios of different wood products of one tree species from forestry lies below 10% (trimestral prices from 1995-2000). This range is mainly due to the market turbulence after 1999 as a consequence of the storm 'Lothar', which caused windfalls three times the quantity of annual harvesting.

Data quality can be considered best available and adequate for the purpose of the study.

18.5 Allocation problem II: transport from forest to production site

As a second allocation problem, transport from forestry to the production site needs to be assessed. Usually, a truck with a total weight of 28 tons is used; distance from the forest to the sawmill is assumed to be 50 km.

Again, various lines of reasoning are possible for the allocation of transport, among them:

All-to-sleeper: transport is considered necessary to bring a log to the sawmill where the railway sleeper is produced. All other products are (unwanted) by-products; hence, all environmental interventions related to transport are allocated to the sleeper.

Over-all: transportation brings the log to the production site where several products are produced. Therefore, the transport is allocated to all resulting products on a mass basis or based on the relative share of revenues of the different products.

Step-by-step: transport brings the log to the production site where it first undergoes roundwood trimming. The resulting products are a log piece for sleeper production and solid fuel wood. These two products get allocated the

[486] data on the density of oven-dried beech varies: 680 kg/m³ (Remmler and Fischer 1996:65), 660 kg/m³ (Frühwald *et al.* 1996:31), 640-720 kg/m³ (Sell 1997:29)

[487] Richter 2001

environmental interventions related to transport on a mass basis or based on the relative share of revenues.

Table 18-6 shows the allocation factors for the allocation of the transport to the different coproducts according to the above-stated lines of reasoning. A transport distance of 50 km and a truck with 28 tons total weight are assumed. Note again that these lines of reasoning reflect different mental models of causal relations. Figure 18-14 illustrates the results of Table 18-6.

Table 18-6. Allocation factors for transport from forestry to the production site (10.118 tkm for a distance 50 km, total weight of the truck 28 t) according to several allocation principles

	step-by-step (mass)	step-by-step (revenues)	over-all (mass)	over-all (revenues)	all to sleeper
Log piece	0.970	0.994	0.468 [1)]	0.699 [1)]	1
Solid fuel wood	0.030	0.006			
Further products			0.532	0.301	

[1)] includes wood for the heating of the creosote

The interpretation is made based on this inventory data; any assessment method would provide the same relative results.

All allocation principles cause the main environmental interventions to be attributed to the log, except the over-all allocation based on mass. In this case, the further products resulting from the production processes carry even more environmental interventions than does the sleeper.

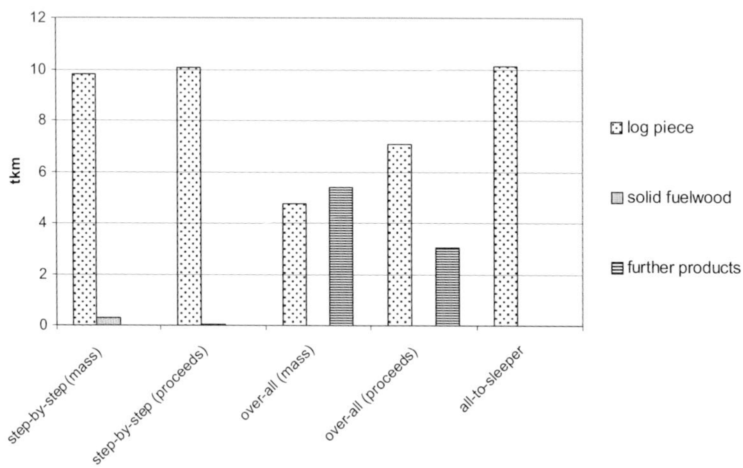

Figure 18-14. Attribution of environmental interventions related to transport from forestry to the production site for one railway sleeper according to several allocation principles

The step-by-step procedures are the least in line with the causalities of considered transportation activities. There is no plausible reason why one of the coproducts of the sawmill processes besides the railway sleeper should be attributed environmental interventions related to transport and the others go free of environmental interventions. Hence, the step-by-step procedure cannot be recommended for the attribution of transportation.

None of the other two allocation procedures can *a priori* be excluded for not being in line with the market and material characteristics of wood (Chap. 15) or for not being in line with the wood-specific management rules for a sustainable development (Chap. 16.6).

18.6 Allocation problem III: Production processes for the sleeper production

The production processes for the sleeper production start with the log at the gate of the sawmill and include roundwood trimming, debarking, roundwood conversion, natural drying, sorting, and mechanical processing. Then, the raw sleeper is impregnated with creosote heated with production residues. Finally, metallic fittings are mounted[488].

The following considerations do not include the metallic parts or the impregnation. Impregnation is integrated into the models of the whole life cycle (Chap. 18.8) and partly into the modelling related to recycling/incineration (Chap. 18.7).

Mass flows of the production processes and respective market prices are summarised in Table 18-1.

18.6.1 Allocation problems and scenarios considered

Wood processing consists of a chain of subsequent joint coproduction processes. Again, various lines of reasoning are possible for allocating the environmental interventions related to railway sleeper production, among them:

Step-by-step I: a step-by-step allocation procedure is applied, allocating environmental interventions related to a process to the products directly resulting from it on a mass basis. Previous upstream processes are not reallocated.

The second option of this procedure uses the relative share of revenues resulting from the direct outputs of a process as an allocation factor. Internal

[488] see Künniger and Richter 1998 for details

prices of the part finally resulting in the sleeper are estimated based on internal data from a production facility and on average statistical data on the cost structure of the wood processing industry[489]. Again, previous (upstream) forestry processes are not reallocated.

Step-by-step II: a step-by-step allocation procedure is applied, allocating environmental interventions related to one process including upstream processes to the products directly resulting from it on a mass basis. So, previous processes are reallocated.

Table 18-7. Allocation factors for the allocation of railway sleeper production processes according to different lines of reasoning

Processes (factors)	step-by-step I (mass)	step-by-step I (revenues)	step-by-step II (mass)	Step-by-step II (revenues)	over-all (mass)	over-all (revenues)	all-to-sleeper
Allocation factors for previous upstream forestry processes [1]							
Previous processes allocated to the roundwood for sleepers (forestry)							
Sleeper	1	1	0.467	0.603	0.478	0.699	1
Further products	0	0	0.533	0.397	0.532	0.301	0
Allocation factors for over-all procedures							
Sleeper production (all processes) (11.27 MJ)							
Sleeper					0.478	0.699	1
Further products					0.532	0.301	0
Allocation factors for step-by-step procedures							
Roundwood trimming (9.00 MJ)							
Sleeper	0.970	0.994	0.467	0.603			
Further product	0.030	0.006	0.533	0.397			
Debarking (0.90 MJ) [2]							
Sleeper	0.930	0.986	0.482	0.606			
Further product	0.070	0.014	0.518	0.394			
Roundwood conversion (0.269 MJ)							
Sleeper	0.566	0.623	0.518	0.615			
Further product	0.434	0.377	0.482	0.385			
Sorting (0 MJ)							
Sleeper	0.971	0.994	0.915	0.986			
Further product	0.029	0.006	0.085	0.014			
Mechanical processing (1.10 MJ)							
Sleeper	0.942	0.992	0.942	0.992			
Further product	0.058	0.008	0.058	0.008			

[1] transport from forest to the production site are already allocated to the products of the production processes (see Chap. 8.5)

[2] 10% of total roundwood trimming are assumed (Künniger and Richter 1998: 171)

The second option of this procedure uses the relative share of revenues resulting from the direct outputs of a process as an allocation factor. Internal

[489] Corbat 2001; Gautschi 2001; see *italic* in Table 18-1

prices of the part finally resulting as the sleeper are estimated based on internal data from a production facility and on average statistical data on the cost structure of the wood processing industry[490]. Again, previous (upstream) forestry processes are reallocated.

Over-all: it is assumed that the production processes provide several products; the temporal sequence of their production is disregarded. Allocation is made based on the mass ratios of the products or based on the relative share of revenues.

All-to-sleeper: all processes are considered to take place with the single goal of producing a railway sleeper.

Table 18-7 summarises the allocation factors for the lines of reasoning outlined above. The inputs of these processes are electricity (medium voltage; UCPTE)[491].

Internal transport and infrastructure are disregarded in this calculation for the goal and scope of the study.

18.6.2 Inventory analysis

A detailed data inventory can be ordered from the author.

18.6.3 Impact assessment

Table 18-8 shows the results of the life cycle impact assessment based on the 'effect-oriented classification' impact assessment method according to Heijungs *et al.* (1992, CML method), updated according to Houghton *et al.* (1996).

Figure 18-15 illustrates the effects of different allocation procedures for the railway sleeper production processes. For this illustration, the potential contributions to the global greenhouse effect are chosen in an arbitrary way. Contributions to other impact categories are equally distributed, as the allocation procedures affect all inputs and (environmental) outputs in the same way.

[490] Corbat 2001; Gautschi 2001; see *italic* in Table 18-1

[491] Data is taken from Künniger and Richter 1998; electricity for the mounting of bands and anti-splits are not considered.

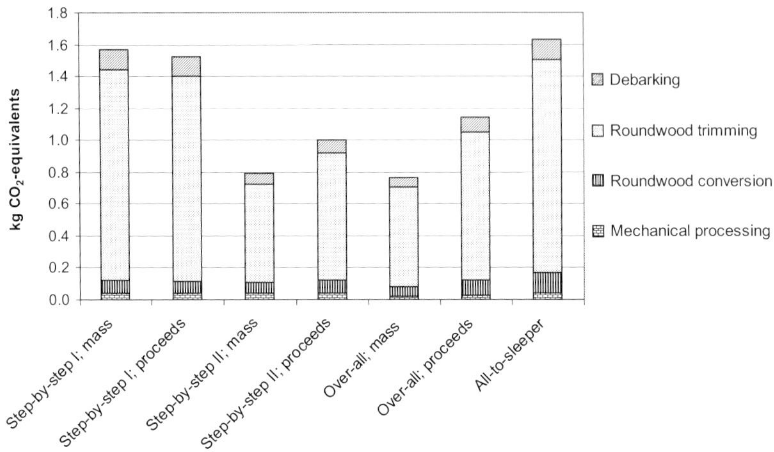

Figure 18-15. Impact assessment of different allocation procedures applied to production processes of a beech wood railway sleeper; potential contributions to global greenhouse effect; (wooden part excluding impregnation and metallic auxiliary materials) (per sleeper)

18.6.4 Interpretation focusing on methodological considerations

The similarity of the results of the all-to-sleeper allocation procedure and the step-by-step (I) allocation procedures based on mass or on revenues is striking. The over-all allocation procedures and the second step-by-step (II) allocation procedures lead to much less environmental impacts attributed to the railway sleeper. This is due to the fact that these allocation procedures allocate the comparatively high energy input of roundwood trimming to all products finally resulting from the production processes of the railway sleeper (see Table 18-7). In contrast to those, both step-by-step (I) allocation procedures attribute the majority of the environmental interventions to the sleeper, as only insignificant other coproducts with low market prices (solid fuel wood) result directly from roundwood trimming.

The selection of revenues or mass as allocation factor can lead to significant differences in the results. Whereas the step-by-step (I) allocation procedures leads to a difference -3% of the mass-based allocation for the revenues-based allocation, this difference is + 27% for the step-by-step (II) allocation procedures and + 49% for the over-all procedure.

Table 18-8. Impact assessment of different allocation procedures applied to production processes of a beech wood railway sleeper (wooden part, excluding impregnation, metallic auxiliary materials) (per sleeper)

	Green-house effect kgCO_2-e.	Ozone depletion kg R11-e.	Acidification kg SO_x-e.	Eutrophication kg PO_4-e.	Photo-smog kg ethylene-e.	Human toxicity kg	Eco-tox. soil kg	Eco-tox. Water 10E6 m^3
Step-by-step I; mass								
Mechanical processing	0.039	1.57E-08	2.79E-04	9.52E-06	8.81E-06	3.97E-04	7.06E-09	1.0
Roundwood conversion	0.083	3.30E-08	5.86E-04	2.00E-05	1.85E-05	8.35E-04	1.48E-08	2.0
Roundwood trimming	1.320	5.27E-07	9.34E-03	3.19E-04	2.95E-04	1.33E-02	2.37E-07	32.2
Debarking	0.131	5.22E-08	9.27E-04	3.17E-05	2.93E-05	1.32E-03	2.35E-08	3.2
Total	1.573	6.28E-07	1.11E-02	3.80E-04	3.52E-04	1.59E-02	2.82E-07	38.4
Step-by-step I; revenues								
Mechanical processing	0.037	1.49E-08	2.65E-04	9.04E-06	8.36E-06	3.77E-04	6.71E-09	0.9
Roundwood conversion	0.075	3.00E-08	5.32E-04	1.82E-05	1.68E-05	7.58E-04	1.35E-08	1.8
Roundwood trimming	1.288	5.14E-07	9.12E-03	3.12E-04	2.88E-04	1.30E-02	2.31E-07	31.4
Debarking	0.124	4.93E-08	8.74E-04	2.99E-05	2.76E-05	1.25E-03	2.22E-08	3.0
Total	1.524	6.08E-07	1.08E-02	3.69E-04	3.41E-04	1.54E-02	2.73E-07	37.2
Step-by-step II; mass								
Mechanical processing	0.037	1.49E-08	2.65E-04	9.04E-06	8.36E-06	3.77E-04	6.71E-09	0.9
Roundwood conversion	0.069	2.74E-08	4.87E-04	1.66E-05	1.54E-05	6.94E-04	1.23E-08	1.7
Roundwood trimming	0.620	2.47E-07	4.39E-03	1.50E-04	1.39E-04	6.26E-03	1.11E-07	15.1
Debarking	0.064	2.55E-08	4.53E-04	1.55E-05	1.43E-05	6.46E-04	1.15E-08	1.6
Total	0.790	3.15E-07	5.59E-03	1.91E-04	1.77E-04	7.97E-03	1.42E-07	19.3
Step-by-step; II revenues								
Mechanical processing	0.039	1.57E-08	2.79E-04	9.52E-06	8.81E-06	3.97E-04	7.06E-09	1.0
Roundwood conversion	0.082	3.26E-08	5.78E-04	1.98E-05	1.83E-05	8.24E-04	1.47E-08	2.0
Roundwood trimming	0.801	3.20E-07	5.67E-03	1.94E-04	1.79E-04	8.08E-03	1.44E-07	19.5
Debarking	0.080	3.21E-08	5.70E-04	1.95E-05	1.80E-05	8.12E-04	1.44E-08	2.0
Total	1.002	4.00E-07	7.09E-03	2.42E-04	2.24E-04	1.01E-02	1.80E-07	24.5
Over-all; mass								
Mechanical processing	0.019	7.41E-09	1.31E-04	4.49E-06	4.15E-06	1.87E-04	3.33E-09	0.5
Roundwood conversion	0.062	2.48E-08	4.40E-04	1.50E-05	1.39E-05	6.27E-04	1.12E-08	1.5
Roundwood trimming	0.622	2.48E-07	4.40E-03	1.50E-04	1.39E-04	6.27E-03	1.12E-07	15.2
Debarking	0.062	2.48E-08	4.40E-04	1.50E-05	1.39E-05	6.27E-04	1.12E-08	1.5
Total	0.764	3.05E-07	5.41E-03	1.85E-04	1.71E-04	7.71E-03	1.37E-07	18.6
Over-all; revenues								
Mechanical processing	0.028	1.11E-08	1.96E-04	6.71E-06	6.21E-06	2.80E-04	4.98E-09	0.7
Roundwood conversion	0.093	3.70E-08	6.57E-04	2.24E-05	2.08E-05	9.36E-04	1.67E-08	2.3
Roundwood trimming	0.928	3.70E-07	6.57E-03	2.24E-04	2.08E-04	9.36E-03	1.67E-07	22.6
Debarking	0.093	3.70E-08	6.57E-04	2.24E-05	2.08E-05	9.36E-04	1.67E-08	2.3
Total	1.142	4.56E-07	8.08E-03	2.76E-04	2.55E-04	1.15E-02	2.05E-07	27.9
All-to-sleeper								
Mechanical processing	0.040	1.58E-08	2.81E-04	9.60E-06	8.88E-06	4.00E-04	7.12E-09	1.0
Roundwood conversion	0.133	5.30E-08	9.40E-04	3.21E-05	2.97E-05	1.34E-03	2.38E-08	3.2
Roundwood trimming	1.328	5.30E-07	9.40E-03	3.21E-04	2.97E-04	1.34E-02	2.38E-07	32.4
Debarking	0.133	5.30E-08	9.40E-04	3.21E-05	2.97E-05	1.34E-03	2.38E-08	3.2
Total	1.633	6.52E-07	1.16E-02	3.95E-04	3.65E-04	1.65E-02	2.93E-07	39.8

18.6.5 Conclusions

The all-to-sleeper allocation procedure does not seem to be justifiable in view of the yield of sleeper production processes of 46% and in view of the coproduction of a respectable amount of high-value coproducts (*e.g.*, 20 kg of sidings). The importance of the high-value coproducts can be underlined from a statement of the owner of the production plant. He commented that it was sometimes decided during processing that apt (high-quality) logs are not cut to railway sleepers but further cut to high-quality sidings[492].

Similar reasoning can be made for the step-by-step (I) allocation procedures. Their results are very similar to the all-to-sleeper allocation procedure. The railway sleeper carries the large majority of the environmental interventions because the environmentally relevant processes provide very little low-value coproducts besides the material from which the sleeper is made. The step-by-step allocation (I) procedures must be rejected for the same reasons as the all-to-sleeper allocation procedure.

The step-by-step (II) allocation procedure allocates also the upstream processes based on mass or on the relative share of revenues. Correspondingly, a process is allocated to all the downstream products. This reflects that none of the upstream processes can be omitted for the efficient and effective production of the (high-value) products; they must be considered causally linked to the production of all the products. Thus, these procedures depict the causalities within the sleeper production and the added value generated.

Also the over-all allocation procedures distribute the environmental interventions to all the products resulting from the production processes. In distinction to the step-by-step (II) allocation procedure, the over-all allocation procedure also attributes downstream processes to products occurring further up the processing chain. Indeed, this does not reflect the causalities within the processing chain.

However, the results of the over-all procedure are similar to the step-by-step (II) allocation procedure for the particularities of the specific production processes. The most influencing particularity is that it is the first process step, the roundwood trimming, which contributes the most to the environmental interventions. Consequently, the over-all procedure and the step-by-step (II) provide more or less the same results.

Under these considerations, the step-by-step (II) allocation procedure (first choice) and the over-all allocation procedures (second choice) provide the most consistent models, being in line with the characteristics of the production processes.

[492] Corbat 2001

Given the large range of market prices to be achieved for the products from sleeper production, the allocation based on revenues seems to be more appropriate. However, allocation based on mass cannot *a priori* be excluded.

18.6.6 Data quality and sensitivity

The data sources used in the production-process part are summarised in Table 18-9.

Table 18-9. Data sources used in the production process part

Processes	Data sources
Production processes	Künniger and Richter 1998
Energy processes	Frischknecht *et al.* 1996
Market prices	Corbat 2001, Keel 2001
Internal prices	Corbat 2001, Gautschi 2001

Process-related data relies on site-specific data from 1998. Its quality can be considered best available for the actual conditions in Switzerland, and adequate for the purpose of this study.

Market prices used for calculations are actual recommended prices from the Schweizerische Vereinigung für Holzenergie (VHe). Internal prices are taken from a statistical economic analysis of the wood processing chain in Switzerland (dated 2000). The transposition of this average data to the production of wooden railway sleepers must be considered a rough estimate. For confidentiality reasons, a part of the internal price is not published. The estimation of cost of the transport from forestry to the sawmill, which is used for the estimation of the internal added-value chain (10% of round wood cost according to Gautschi 2001:22), is not sensitive for the results[493].

18.7 Allocation problem IV: Recycling or incineration with energy recovery

After service life, the railway sleepers are removed from the railway track. In this chapter, the modelling of the two end-of-life scenarios recycling or incineration are looked at more closely.

It is assumed that the modelling of multifunctional systems achieved by system expansion is not allowed for the goal and scope of the assessment.

[493] affects the step-by-step (II) allocation procedure based on revenues; the net effect of multiplying the estimated transport costs from forestry to the production site by factor 10 changes the over-all result less than 1.5%.

18.7.1 Allocation problems and scenarios considered

The recycling option and the incineration option pose different modelling problems. Whereas an owner of used railway sleepers can sell them for about € 10.- per piece for uses in landscape architecture, incinerating them in a waste incinerator costs him/her an average of € 10.- per piece[494]. Whereas a railway sleeper used in landscape architecture is assumed to rot completely without using the embodied energy of wood and creosote, this potential can be used in waste incinerators if thermal energy and/or electricity are co-generated.

Among the allocation problems that arise concerning the modelling of recycling or incineration of railway sleepers are:

- Should the railway sleeper used in landscape architecture carry environmental burdens from sleeper production?
- Should substitution effects be considered if the railway sleeper is used in landscape architecture instead of concrete elements?
- Should substitution effects be considered for the production of heat and/or electricity if the sleeper is incinerated, *e.g.*, the substitution of thermal energy and/ or electricity of other plants?
- Should the environmental opportunity 'costs' be considered if wood is used in a way which inhibits thermal energy production and the substitution of fossil fuels as a consequence?
- Should environmental burdens from the use of the sleeper in landscape architecture partly be attributed to the railway part, as the secondary use fulfils waste disposal function?
- Should incineration be considered a bi-functional process providing waste treatment and generation of thermal energy and/or electricity, attributing part of the incineration process to the sleeper, part to the generation of thermal energy and/or electricity?

Several lines of reasoning for the allocations related to recycling or incineration are discussed for each end-of-life option, making reference to the above questions:

Recycling of railway sleepers as constructive material and landscape architecture

Cut-off (*LSA*): the primary aim of the railway sleeper production is the production of railway sleepers. Hence, all environmental interventions

[494] Graf 2001; Hofer 2000a:17; in Switzerland, the use of used railway sleepers in landscape architecture has been restricted to specified companies, based on a revision of the Ordinance on Substances/Verordnung über umweltgefährdende Stoffe (StoV).

related to production processes are attributed to the sleeper. If anybody takes the sleeper for further uses, the use as railway sleeper does not carry any environmental interventions related to its disposal. The secondary user gets the used sleeper as an element in landscape architecture (LSA) free of environmental burdens[495].

50/50: the uses of the railway sleeper in the railway track and as element of landscape architecture are part of the same life cycle of the impregnated beech wood. Correspondingly, all environmental interventions related to the raw material acquisition and production of the sleeper (as shaping and impregnation are also a precondition for the use of the sleeper as element in landscape architecture) are equally distributed to the two uses. Rotting of the sleeper is assumed as the end-of-life scenario for the second use of the sleeper; as wood carbon is considered a material-inherent property, no additional emissions are allocated to the first life cycle as biogenic carbon is disregarded in these calculations (this reasoning also applies to other substitution based procedures)[496,497,498].

Value-corrected substitution (*VCS*): The production processes provide two products in a cascade. Allocation is made based on the relative price difference of the input materials. The environmental interventions (EI_1) allocated to the sleeper as product 1 are calculated from the total environmental interventions (EI_T) of the railway sleeper production as follows:

$$EI_1 = EI_T * \frac{p_1 - p_2}{p_1} \qquad (18.1)$$

with:
p_1 : market price of product 1 (impregnated sleeper)
p_2 : market price of product 2 (used sleeper)

The second use, the railway sleeper as an element in landscape architecture, carries the remaining environmental interventions[499].

Environmental opportunity 'costs' of inhibiting thermal energy recovery (*Op-cost*): The recycling of the railway sleeper as an element in landscape architecture inhibits thermal energy recovery. The potential of (creosote-

[495] see Ekvall and Tillman 1997:160; Fava *et al.* 1991:79ff
[496] see Fava *et al.* 1991:79ff
[497] Further processes solely related to the sleeper such as its metallic parts, or the mass flows related to the track bed would be allocated fully to the sleeper.
[498] Boguski *et al.* 1994:152f outline another approach, assuming equal functionality. They propose to base allocation on relative shares of mass x service life. This approach is not considered any further.
[499] see Chap. 10; also Werner and Richter 2000a; Werner and Richter 2000b

treated) wood as substitute for fossil energy carriers is not used. Hence, the avoided incineration and the production of an equivalent amount of thermal energy (and electricity) are accounted for[500].

Substitution of concrete element production (*Sub-con*): The railway sleeper provides two functions, a) the fixation of the railways, and b) the retention of soil in landscape architecture. It is assumed that a concrete element would otherwise have fulfilled function b). Hence, the production and disposal of an equi-functional concrete element is subtracted from the inventory of the railway sleeper production processes to obtain a monofunctional model.[501].

Environmental opportunity 'costs' and substitution effects considering the material and energy aspects of wood (*Op-cost MEA*): This procedure is a combination of the opportunity-cost considerations on the energy-side (Op-cost) and the avoided-burden procedure on the material side (Sub-con)[502].

Disposal in waste incinerator with co-generation of heat and electricity

Cut-off (*WI*): waste incineration (WI) is considered a waste treatment process. Hence, the wooden sleeper gets all environmental interventions related to waste incineration[503]. One could of course also argue that the thermal energy and/or electricity should also be taken into account. This is done by the Strict coproduct allocation described further down.

Closed-loop: The incineration processes are considered a part of the life cycle of the railway sleeper. The environmental interventions are allocated to the sleeper, whereas the electricity generated is subtracted from the life cycle inventory of the sleeper. The thermal energy generated is not considered any further, as it is assumed that the sleeper production facility is not connected to the district heating system ('open-loop')[504].

Strict coproduct allocation (*SC-PA*): waste incineration is considered a bi-functional process, providing waste treatment and generation of thermal energy and/or electricity. Incineration is considered waste treatment as long as the market price of the impregnated wood in treatment is negative. The process is allocated to the thermal energy and/or electricity produced as soon as the market price of the material treated becomes positive. Hence, the incineration and co-generation processes are allocated partly to the sleeper,

[500] see, *e.g.*, Azapagic and Clift 1999a:360ff; Tillman *et al.* 1994:23ff; Heintz and Baisnée 1991:43f; Jungmeier *et al.* 2001b
[501] *ibid.*
[502] combining arguments of, *e.g.*, Azapagic and Clift 1999a:360ff; Tillman *et al.* 1994:23ff; Heintz and Baisnée 1991:43f; Jungmeier *et al.* 2001b
[503] see Ekvall and Tillman 1997:160; Fava *et al.* 1991:79ff
[504] see ISO/EN 14'041, chap. 6.5.4

Table 18-10. Operationalisation of different lines of reasoning for the modelling of end-of-life options of a wooden railway sleeper (per sleeper); case-specific descriptions, see assumptions

Recycling of railway sleepers as constructive material in landscape architecture	
Cut-off (LSA)	No further environmental interventions are attributed to the railway sleeper.
50/50	Half of all processes related to wood and sleeper production including transport from forest to production site are allocated to the sleeper (mathematically: production sleeper – ½ production wooden landscape element); no impact is considered for rotting (biogenic CO_2 is disregarded in model). Two product systems are chosen from the previous options: m) Forestry processes: over-all (revenues); transport: over-all (revenues); production processes: step-by-step II (mass) + impregnation p) Forestry processes: over-all (volume); transport: over-all (mass); production processes: step-by-step II (mass) + impregnation Further sleeper-specific processes, *e.g.*, related to auxiliary materials, transport to the site where the sleepers are laid, maintenance are disredarded in this example.
VCS Value-corrected substitution	Allocation is made based on the relative share of the market price of the new sleeper and the price of the used sleeper as future element in landscape architecture. The environmental interventions allocated to the sleeper are calculated based on Eq. (17.1). The resulting allocation factor is 0.786 of all the up-stream processes related to the wooden sleeper production including impregnation. Again, two partial product systems are chosen from the previous options: m) Forestry processes: over-all (revenues); transport: over-all (revenues); production processes: step-by-step II (mass) + impregnation p) Forestry processes: over-all (volume); transport: over-all (mass); production processes: step-by-step II (mass) + impregnation Further processes, *e.g.*, related to auxiliary materials, transport to the site where sleepers are laid, maintenance, etc. would fully be attributed to the sleeper (not considered in this example).
Op-cost Environmental opportunity 'costs' of inhibiting thermal energy recovery	The incineration of the railway sleeper is avoided. On the other hand, thermal energy recovery is inhibited. Thus, the incineration processes are subtracted; the average amount of middle-voltage electricity (111.25 MJ; UCPTE-mix) and thermal energy from an industrial light oil combustion plant (1 MW) (289.25 MJ) are added to the LCI of the railway sleeper. Energy could be recovered if the sleeper was incinerated in a waste incineration plant ($\eta_{electricity} = 0.1$; $\eta_{thermal\ energy} = 0.26$)[505,506]; see also: substitution of energy production (Sub-en).
Sub-con Substitution of concrete element production	Production and disposal of an equi-functional concrete element (138.6 kg; 1500 kg/m^3) and its disposal in a landfill for inert material are subtracted from the inventory of the railway sleeper production processes; transport are assumed to be equal for the installation of the railway sleeper and for the concrete element (although weights are different).
Op-cost MEA Environmental opportunity 'costs' and substitution effects (material and energy)	The production of an equi-functional concrete element and the incineration of the sleeper are avoided. Instead, the average amount of middle voltage electricity (111.25 MJ; UCPTE-mix) and thermal energy from an industrial light oil combustion plant (1 MW) (289.25 MJ) are added to the LCI of the railway sleeper.

continued

505 based on Zimmermann *et al.* 1996:96; process data from Künniger and Richter 1998:180, complemented based on Frischknecht *et al.* 1996

506 It is further on assumed that part of the creosote has evaporated; the remaining amount of creosote in the sleeper is assumed to be 8.3 kg per sleeper (Künniger and Richter 1998:180).

Table 18.10 continued

Disposal in waste incinerator with co-generation of heat and electricity	
Cut-off (WI)	Waste incineration is fully attributed to the product system of the railway sleeper see also: cut-off (LSA)
Closed-loop	Incineration processes are accounted; the electricity generated in the waste incineration plant is considered closed-loop and is subtracted from the LCI of the railway sleeper (111.25 MJ; UCPTE-mix). Heat generated is considered open-loop as the sleeper production is assumed not to be connected to a district heating system; no credit is given for thermal energy production; see also: cut-off (WI).
SC-PA Strict coproduct allocation	The incineration of a wooden railway sleeper costs about 10 € and generates electricity and thermal energy with a market price of about 6 €. Thus, the incineration processes are divided. The sleeper is attributed 10 / (10 + 6) = 0.63 of the total environmental interventions related to incineration[507].
Sub-en Substitution of energy production	The energy generated by incineration substitutes other energy production processes. Hence, the average amount of middle-voltage electricity (111.25 MJ; UCPTE-mix) and thermal energy from an industrial light oil combustion plant (1 MW) (289.25 MJ) are subtracted from the LCI of the railway sleeper; see also: Environmental opportunity 'costs' of inhibiting thermal energy recovery (Op-cost).
Op-cost (MEA) 'Environmental opportunity costs' and substitution effects (material and energy aspects of wood)	The production of an equi-functional concrete element and the incineration of the sleeper are caused. On the other hand, the average amount of middle voltage electricity (111.25 MJ; UCPTE-mix) and thermal energy from an industrial light oil combustion plant (1 MW) (289.25 MJ) are avoided and subtracted from the LCI of the railway sleeper; see also: Op-cost (MEA) above.

partly to the thermal energy and/or electricity generated. The allocation factor is calculated as the portion of the disposal cost compared to all revenues of the process, *i.e.* the sum of the market prices of thermal energy, the electricity produced and the disposal cost[508].

Substitution of energy production (*Sub-en*): The railway sleeper production provides two functions, a) the fixation of the railways, and b) the generation of a fuel. A substitution for function b) is assumed to obtain a monofunctional model. It is assumed that the waste treatment process provides thermal energy and/or electricity, which substitutes respective products from the most expensive alternative plant. This assumption is true for Switzerland where the landfilling of burnable waste is prohibited; it is not necessarily true for countries with a shortage of incineration capacities. In the latter case, the incineration of other waste fraction would be substituted for – with no effect on alternative plants. To depict Swiss conditions, the production of thermal energy and/or energy in an alternative plant is subtracted from the inventory of the railway sleeper processes[509].

[507] price assumed for electricity: 0.11 CHF/kWh; for thermal energy: 0.07 CHF/kWh

[508] see Chap. 6.2; Huppes 1994:84f; Lindeijer and Huppes 2001:309ff; Guinée *et al.* 2004

[509] see, *e.g.*, Azapagic and Clift 1999a:360ff; Tillman *et al.* 1994:23ff; Heintz and Baisnée 1991:43f; Jungmeier *et al.* 2001b

Environmental opportunity 'costs' and substitution effects considering the material and energy aspects of wood (*Op-cost MEA*): The reciprocal reasoning on substitution effects as made in the above Op-cost MEA is also feasible if the sleeper is burnt in an incineration plant: In this case, the production of an equi-functional concrete element and the incineration of the sleeper are accounted for. On the other hand, the average amounts of electricity and thermal energy generated in an alternative plant are subtracted from the LCI of the railway sleeper (see also Op-cost (MEA) above).

The lines of reasoning outlined above are inventoried in Table 18-10.

Table 18-11 summarises the scenarios considered. Note that these descriptions are case-specific adaptations of the generic allocation procedures (see above for explanations).

Table 18-11. Allocation procedures for the modelling of the end-of-life options 'incineration with co-generation' and 'reuse as element in landscape architecture' (per sleeper)

Element in landscape architecture	
Cut-off (LSA)	± 0; 2nd life cycle carries burdens from recycling
50/50	– ½ * (wood processes incl. impregnation); 2 functions
VCS	$(p_1 - p_2)/p_1$ * (wood processes incl. impregnation)
Op-cost	+ fossil energy + electricity – incineration process
Sub-con	– concrete element and its disposal
Op-cost MEA	+ fossil energy + electricity – incineration process – concrete element and its disposal
Incineration with co-generation	
Cut-off (WI)	+ incineration process (no credits for energy)
Closed-loop	+ incineration process – electricity generated; (heat = open-loop)
SC-PA	+ incineration process * $(p_{waste\ treatment}/(p_{waste\ treatment} + p_{electricity} + p_{heat})$
Sub-en	– fossil energy – electricity + incineration process
Op-cost MEA	– fossil energy – electricity + incineration process + concrete element and its disposal

18.7.2 Inventory analysis

A detailed data inventory can be ordered from the author.

18.7.3 Impact assessment

Table 18-12 summarises the results of the life cycle impact assessment based on the 'effect-oriented classification' impact assessment method according to Heijungs *et al.* (1992, CML-method), updated according to Houghton *et al.* (1996).

Table 18-12. Impact assessment of different allocation procedures related to recycling or incineration of a creosote-treated beech wood railway sleeper (per sleeper)

	Green-house eff. kg CO_2-e.	Ozone depletion kg R11-e.	Acidifi-cation kg SO_x-e.	Eutrophi-cation kg PO_4-e.	Photo-smog kg ethy-lene-e.	Human toxicity kg	Eco-tox. soil kg	Eco-tox. water 10E6 m^3
Recycling of railway sleepers as constructive material in landscape architecture								
Cut-off (LSA)								
Total	0.0	0.00E+00	0.000	0.00E+00	0.00E+00	0.000	0.00E+00	0
50/50 (p)								
oa(p)/oa(p)/sbs II (p) + Imp	-14.6	-8.72E-06	-0.120	-7.54E-03	-9.40E-02	-0.503	-5.48E-06	-1465
Total	-14.6	-8.72E-06	-0.120	-7.54E-03	-9.40E-02	-0.503	-5.48E-06	-1465
50/50 (m)								
oa(v)/oa(m)/sbs II (m) + Imp	-13.7	-7.66E-06	-0.113	-6.56E-03	-8.89E-02	-0.474	-4.81E-06	-1411
Total	-13.7	-7.66E-06	-0.113	-6.56E-03	-8.89E-02	-0.474	-4.81E-06	-1411
VCS (p)								
oa(p)/oa(p)/sbs II (p) + Imp	-6.2	-3.73E-06	-0.051	-3.23E-03	-4.02E-02	-0.215	-2.35E-06	-627
Total	-6.2	-3.73E-06	-0.051	-3.23E-03	-4.02E-02	-0.215	-2.35E-06	-627
VCS (m)								
oa(v)/oa(m)/sbs II (m) + Imp	-5.9	-3.28E-06	-0.048	-2.81E-03	-3.81E-02	-0.203	-2.06E-06	-604
Total	-5.9	-3.28E-06	-0.048	-2.81E-03	-3.81E-02	-0.203	-2.06E-06	-604
Op-cost (Inhibiting thermal energy recovery)								
Heat oil into industrial heating	25.5	2.92E-05	0.059	5.11E-03	2.53E-02	0.167	2.10E-05	611
Electricity medium voltage (UCPTE)	16.4	6.55E-06	0.116	3.97E-03	3.67E-03	0.166	2.95E-06	401
Incineration of creosote-treated sleeper	-28.9	-5.39E-08	-0.083	-1.17E-02	-1.30E-03	-0.104	-2.12E-08	-3
Total	13.0	3.57E-05	0.092	-2.60E-03	2.77E-02	0.229	2.40E-05	1009
Sub-con (concrete element)								
Concrete into landfill	-0.1	0.00E-00	-0.001	-1.17E-04	-1.28E-04	-0.001	-0.00E-01	0
Concrete (without reinforcement)	-9.5	-2.58E-06	-0.032	-3.81E-03	-3.11E-03	-0.050	-1.62E-06	-193
Total	-9.5	-2.58E-06	-0.033	-3.92E-03	-3.24E-03	-0.052	-1.62E-06	-193
Op-cost MEA (Environmental opportunity 'costs' considering material and energy aspects of wood)								
Heat oil into industrial heating	25.5	2.92E-05	0.059	5.11E-03	2.53E-02	0.167	2.10E-05	611
Electricity medium voltage (UCPTE)	16.4	6.55E-06	0.116	3.97E-03	3.67E-03	0.166	2.95E-06	401
Incineration of creosote-treated sleeper	-28.9	-5.39E-08	-0.083	-1.17E-02	-1.30E-03	-0.104	-2.12E-08	-3
Concrete into landfill	-0.1	-0.00E-01	-0.001	-1.17E-04	-1.28E-04	-0.001	-0.00E-01	0
Concrete (without reinforcement)	-9.5	-2.58E-06	-0.032	-3.81E-03	-3.11E-03	-0.050	-1.62E-06	-193
Total	3.5	3.31E-05	0.060	-6.52E-03	2.45E-02	0.177	2.23E-05	816

continued

Table 18-12 continued

	Green-house eff. kg CO_2-e.	Ozone depletion kg R11-e.	Acidifi-cation kg SO_x-e.	Eutrophi-cation kg PO_4-e.	Photo-smog kg ethy-lene-e.	Human toxicity kg	Eco-tox. soil kg	Eco-tox. water 10E6 m^3
	Disposal in waste incinerator with co-generation of heat and electricity							
Cut-off (WI)								
Incineration of creosote-treated sleeper	28.9	5.39E-08	0.083	1.17E-02	1.30E-03	0.104	2.12E-08	3
Total	28.9	5.39E-08	0.083	1.17E-02	1.30E-03	0.104	2.12E-08	3
Closed-loop								
Incineration of creosote-treated sleeper	28.9	5.39E-08	0.083	1.17E-02	1.30E-03	0.104	2.12E-08	3
Electricity medium voltage (UCPTE)	-16.4	-6.55E-06	-0.116	-3.97E-03	-3.67E-03	-0.166	-2.95E-06	-401
Total	12.4	-6.50E-06	-0.033	7.70E-03	-2.37E-03	-0.061	-2.92E-06	-398
SC-PA								
Incineration of creosote-treated sleeper	18.2	3.40E-08	0.052	7.35E-03	8.21E-04	0.066	1.33E-08	2
Total	18.2	3.40E-08	0.052	7.35E-03	8.21E-04	0.066	1.33E-08	2
Sub-en (Substitution of energy production)								
Heat oil into industrial heating	-25.5	-2.92E-05	-0.059	-5.11E-03	-2.53E-02	-0.167	-2.10E-05	-611
Incineration of creosote-treated sleeper	28.9	5.39E-08	0.083	1.17E-02	1.30E-03	0.104	2.12E-08	3
Electricity medium voltage (UCPTE)	-16.4	-6.55E-06	-0.116	-3.97E-03	-3.67E-03	-0.166	-2.95E-06	-401
Total	-13.0	-3.57E-05	-0.092	2.60E-03	-2.77E-02	-0.229	-2.40E-05	-1009
Op-cost MEA (Environmental opportunity 'costs' considering material and energy aspects of wood)								
Heat oil into industrial heating	-25.5	-2.92E-05	-0.059	-5.11E-03	-2.53E-02	-0.167	-2.10E-05	-611
Electricity medium voltage (UCPTE)	-16.4	-6.55E-06	-0.116	-3.97E-03	-3.67E-03	-0.166	-2.95E-06	-401
Incineration of creosote-treated sleeper	28.9	5.39E-08	0.083	1.17E-02	1.30E-03	0.104	2.12E-08	3
Concrete into landfill	0.1	0.00E-01	0.001	1.17E-04	1.28E-04	0.001	0.00E-01	0
Concrete (without reinforcement)	9.5	2.58E-06	0.032	3.81E-03	3.11E-03	0.050	1.62E-06	193
Total	-3.5	-3.31E-05	-0.060	6.52E-03	-2.45E-02	-0.177	-2.23E-05	-816

18.7.4 Interpretation focusing on methodological considerations

Recycling of railway sleepers as constructive material in landscape architecture

The results of the modelling approaches differ for each of the environmental impact categories as a consequence of the different types of processes involved. Two rankings of the different modelling approaches can be distinguished. The difference of the ranking is caused by the Sub-con procedure accounting for the substitution of a concrete element.

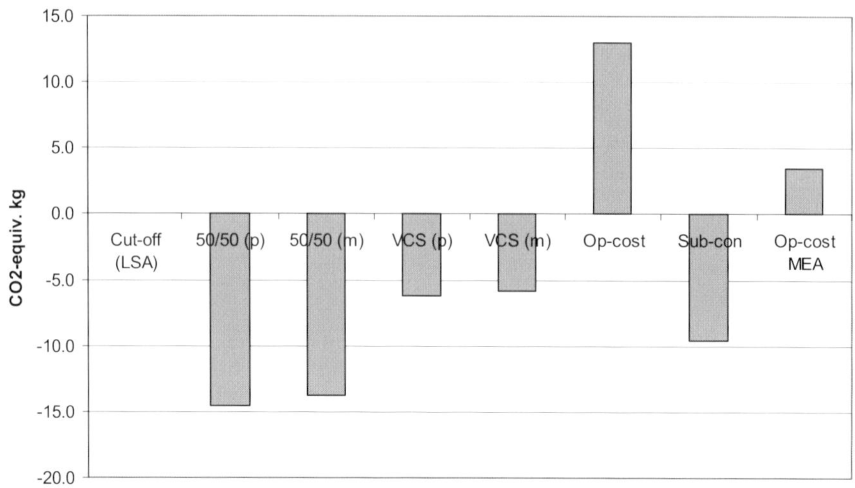

Figure 18-16. Potential contributions to the greenhouse effect of differing modelling approaches related to the recycling of railway sleepers as constructive material in landscape architecture (per sleeper)

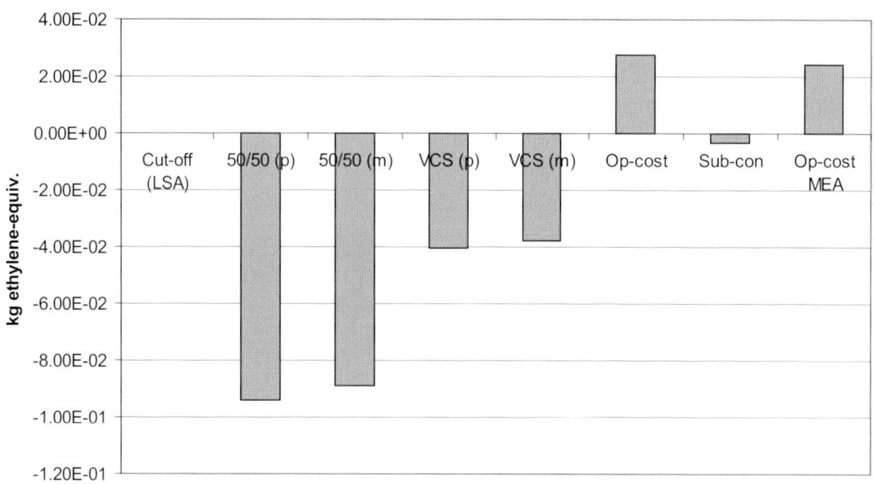

Figure 18-17. Potential contributions to photosmog of different modelling approaches related to the recycling of railway sleepers as constructive material in landscape architecture (per sleeper)
(p) allocation based on revenues; (m) allocation based on mass

The potential contributions to the greenhouse effect and to eutrophication are represented in Figure 18-16. The rest of the impact categories are represented by the potential contributions to photosmog (Figure 18-17).

Figure 18-16 and Figure 18-17 show that three of the six allocation procedures lead to negative environmental impacts.

The 50/50 and VCS allocation procedures redistribute environmental interventions partly to the secondary use. The choice of the allocation factor (revenues or mass) does not have a significant impact on the results of these alternatives. Anyhow, the 50/50 procedure possibly leads to a more consistent model if allocation of processes of previous life-cycle stages is also made based on mass (50/50(m)), whereas the VCS using economic data is more consistent if previous processes are attributed based on revenues (VCS (p)).

The Sub-con procedure, accounting for the substitution of a concrete element, leads to negative environmental interventions. But the assumed substitution of a concrete element is only part of the avoided or additionally caused environmental interventions related to the recycling of the railway. A proper 'avoided burden' procedure would also consider the avoided incineration processes and account for the thermal energy and electricity processes that would result from the incineration of the sleeper (see Op-cost MEA).

The Op-cost procedure, accounting for the environmental opportunity 'costs' related to the energetic aspect of wood, leads to further environmental impacts attributed to the railway sleeper; the avoided incineration of the sleeper does not compensate for the additional impacts for thermal energy and energy production. But as the Sub-con procedure focuses on the material aspect of wood, the Op-cost procedure only considers part of the substitution effects.

The Op-cost MEA procedure accounts for both the environmental opportunity 'costs' of inhibiting thermal energy recovery (Op-cost) and the substitution of a concrete element (Sub-con). The result of the Op-cost MEA procedure is the sum of the Op-cost and the Sub-con procedure. It results in additional environmental opportunity 'costs' to be attributed to the sleeper if using it as element in landscape architecture (!).

The cut-off procedure does not add or subtract any environmental interventions to the product system of the railway sleeper.

Disposal in Waste Incinerator with co-generation of heat and electricity

Rankings also differ for the modelling approaches applied to the disposal of the sleeper in a waste incinerator with co-generation. Equal rankings of the modelling options are achieved for the potential contributions to:

- Greenhouse effect, acidification, photosmog, human toxicity, and ecotoxicity water;

- Ozone depletion and the use of abiotic resources;
- Eutrophication;
- Eco-toxicity soil.

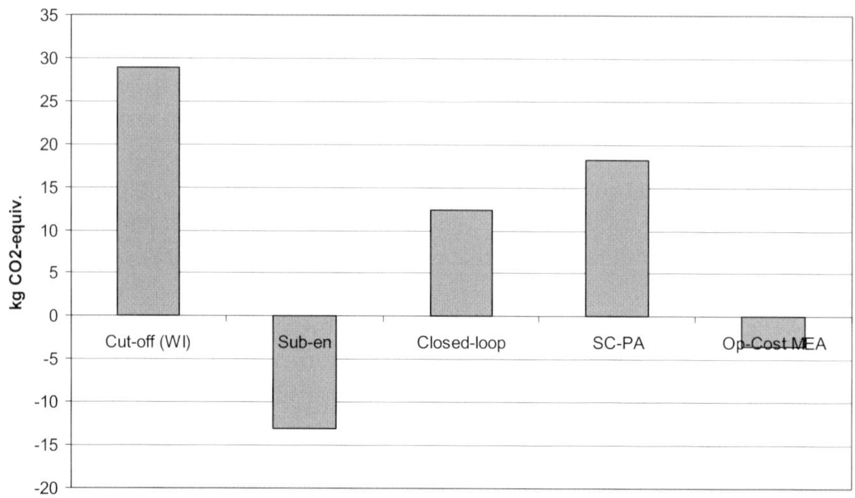

Figure 18-18. Potential contributions to the greenhouse effect of differing modelling approaches related to the disposal of the wooden railway sleeper in an incineration plant with co-generation of thermal energy and electricity (per sleeper)

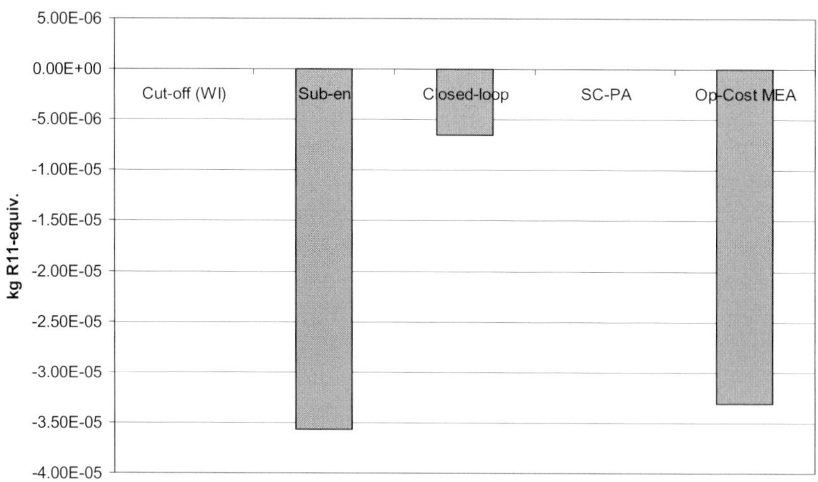

Figure 18-19. Potential contributions to ozone depletion of different modelling approaches related to the disposal of the wooden railway sleeper in an incineration plant with co-generation of thermal energy and electricity (per sleeper)

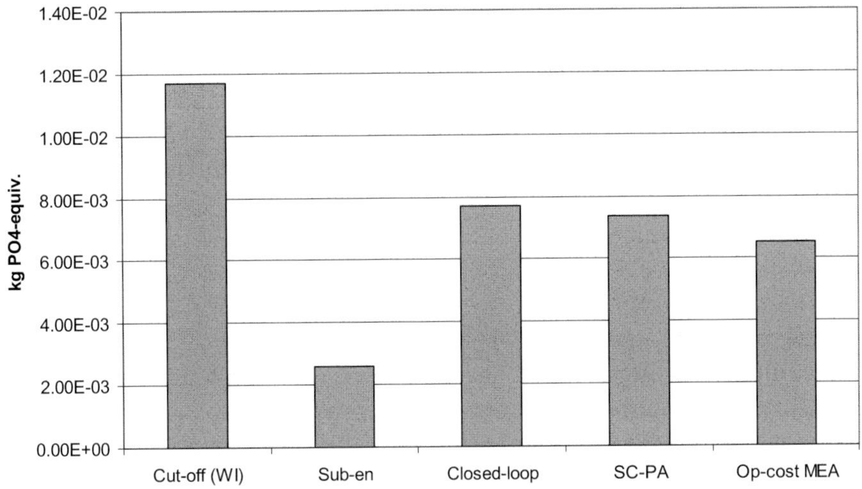

Figure 18-20. Potential contributions to eutrophication of different modelling approaches related to the disposal of the wooden railway sleeper in an incineration plant with co-generation of thermal energy and electricity (per sleeper)

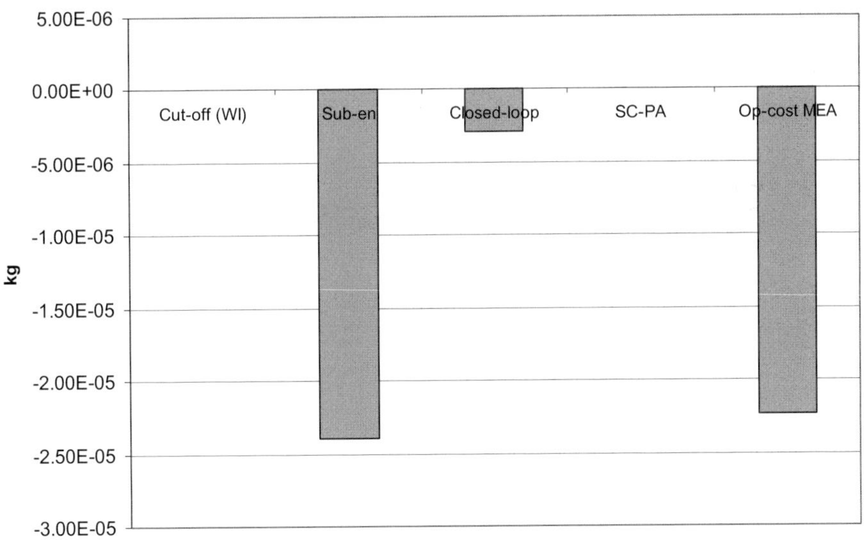

Figure 18-21. Potential contributions to eco-toxicity soil of different modelling approaches related to the disposal of the wooden railway sleeper in an incineration plant with co-generation of thermal energy and electricity (per sleeper)

Again, the representatives of the four rankings are chosen arbitrarily.

The results of the impact assessment of the incineration option are not very definite. Except for the cut-off (WI) procedure, all allocation procedures lead to both positive and negative environmental interventions to be attributed to the railway sleeper. Only the cut-off (WI) procedure accounts solely for positive (additional) environmental interventions. This result is not very surprising, given the differing types of processes covered and different ways of accounting of the different allocation procedures. However, some comments can be made with regard to the assumptions underlying each allocation procedure.

The cut-off (WI) procedure considers the incineration process as pure waste treatment. The co-generation of heat and electricity in modern Swiss municipal waste incineration plants is disregarded. This assumption is highly questionable, given the state-of-the-art character of these technologies. Additionally, no incentive is given to a decision-maker to deliver the sleepers to an incineration facility with co-generation. This would be against the management rule for the sustainable use of wood claiming thermal energy recovery for the substitution of non-renewable energy carriers.

The Sub-en procedure focuses on the potential of thermal energy recovery and the substitution of non-renewable energy carriers. But the material aspect of wood and the related substitution processes are disregarded. This omission can be an advantage if the determination of hypothetical substitutes for the burnt wooden sleeper is considered too difficult, *e.g.*, for the risk-aversion of the decision-maker. But the 'picture' is incomplete if the material aspect of wood is disregarded; no recommendations can be made with regard to the adequacy of recycling or incineration of wooden sleepers.

The closed-loop procedure, giving credits for the generation of electricity ('closed-loop') and disregarding thermal energy production ('open-loop'), provides an inconsistent model. There exists no causal economic or technical reason justifying this arbitrary different handling of electricity and thermal energy production (see Chap. 18.7.5).

The strict coproduct allocation procedure (SC-PA) provides a consistent attribution of the incineration processes to the sleeper as well as to the generated electricity and thermal energy. Inherent to the SC-PA, environmental opportunity 'costs' or any substitution effects are disregarded.

The Op-cost MEA procedure, accounting for the substitution effects and the environmental opportunity 'costs' related to the material and energy aspect of wood, provides results reciprocal to the recycling of used railway sleepers as elements in landscape architecture. The Op-cost MEA is the only allocation procedure that allows statements on the usefulness of recycling versus incineration with co-generation of electricity and thermal energy. Interestingly enough, incineration with co-generation is favourable to the greenhouse effect (supporting the management rule No 5 for the sustainable use of wood (Chap. 16.6) and to various other impact categories. However, it leads to additional potential impacts to eutrophication.

18.7.5 Conclusions

The cut-off procedure gives preference to the recycling of the sleeper although environmental opportunity 'costs' including substitution effects are positive for this option (Op-cost MEA). This is not in line with the management rule No 5 for the sustainable use of wood of Chapter 16.6.

Additionally, the market price of a railway sleeper is not determined freely, as there exist, in fact, two market prices for it, one for resale and one for disposal, more or less independent of its condition. This brings additional uncertainty into modelling if the secondary use is not pre-set. The cut-off procedure can also lead to discussions in comparative studies if, *e.g.*, elements for landscape architecture are assessed. According to cut-off philosophy, the recycled railway sleeper would be available with no environmental burden attributed, although the cutting of wood and especially the impregnation are preconditions for this application. Representatives of other material industries would possibly argue strongly against this line of reasoning.

To summarise, the cut-off procedure does not address the management rules for a sustainable use of wood as well as the material and current market characteristics of wood in an adequate way.

The 50/50 procedure assumes equal functionality ('value') of the function of a railway sleeper as railway sleeper itself and as element of landscape architecture (although there exists another rationale for the 50/50 procedure in marginal LCA (Chap. 6.5.4). This underlying mental model is most probably not shared by many people.

The application of the value-corrected substitution can only partly be recommended. As preconditions for the applicability of the value-corrected substitution, the following points must be given:

- Primary material production processes (and final disposal processes) can be assumed equal for each type of products produced from the resource;
- The assumption that recycled material substitutes primary material production processes must be plausible.

The difficulty consists in the fact that primary material production processes of equi-functional wooden elements for landscape architecture – though they exist – would certainly not correspond to the production and impregnation processes of railway sleepers. It is much more likely that salt-treated wooden palisade elements would be used instead. This contradicts the first of the above-mentioned points. So, processes are attributed to the secondary use, which are definitely caused by the use of the sleeper in its primary purpose as sleeper.

Furthermore, substitution effects most probably will not occur throughout the wood processing chain but affect other materials, most probably concrete elements (contradicting the second point, above). Hence, the basic assumptions on which the VCS are based are inconsistent with the material and market characteristics of used wooden sleepers.

The closed-loop procedure applied to the incineration process treats thermal energy and electricity co-generation inconsistently. This could only be justified if the aspect that electricity is a more valuable form of energy as it can be easily be stored and transported, is considered should be stressed in an LCI. This does not necessarily make sense in the case of densely populated areas with thermal heat demand, such as Switzerland.

The strict coproduct allocation procedure (SC-PA) can only be applied if environmental opportunity 'costs' and substitution effects shall not be considered for any reason. A limited 'descriptiveness' of the LCA in view of the management rules for the sustainable use of wood has to be accepted.

If environmental opportunity 'costs' shall be considered – management rule No 5 for a sustainable use of wood requires this – substitution effects should be considered for both the energy and material aspects of wood. This is done by the Op-cost MEA procedure. All other procedures relying on substitution effects or environmental opportunity 'costs' only cover part of the real and hypothetical substitution effects and of the environmental opportunity 'costs'. They can lead to misleading recommendations.

A difficulty consists in determining the substitution effects related to the material aspect of wood. Consensus should be achieved among the stakeholders of an LCA. The Op-cost MEA procedure is the only procedure that allows for statements on the usefulness of recycling compared to incineration of wood in a product LCA (if one does not build up two separate models which depict one of the possible substitutions each – which is of course contradiction the criterion that a model has to be complete).

18.7.6 Data quality and sensitivity

The data sources used in the recycling and incineration part are summarised in the following Table 18-13.

Data quality can be considered best available and appropriate for the goal and methodological scope of this study, including recycling and incineration.

For the interpretation of the results, some restrictions have to be made related to the modelling of the concrete element. The assumptions made for the modelling of the concrete element are rough approximations. The assumption of a concrete element as substitute for a railway sleeper in landscape architecture is also a best guess. These limitations seem justifiable in view of the goal and scope of the study. Conclusions drawn from numerical results from allocation procedures involving the concrete element should be handled with caution.

Table 18-13. Data sources used in the recycling and incineration part

Processes	Data sources
Incineration process	Künniger and Richter 1998, based on LRV; Frischknecht *et al.* 1996
Energy processes	Frischknecht *et al.* 1996
Concrete element, production and disposal	Frischknecht *et al.* 1996
Further production processes	see Chap. 18.4 to Chap. 18.6; Künniger and Richter 1998: 174 for impregnation
Economic data	Graf 2001; Hofer 2000a: 17

Prices for incinerating railway sleepers vary considerably for each (municipal) waste incineration plant[510]. Conclusions from models relying on allocation procedures relying on economic data should be handled with caution.

No time series has been available for the prices of used railway sleepers and for the costs of incineration, including prices for the generated thermal energy and electricity. Correspondingly, no statement can be made about the correlation of these prices compared to the other ones used in this case study.

Calculations assume a CO_2-neutral life cycle of wood[511]. This means that no methane with a considerably higher greenhouse effect potential than CO_2 is generated during the rotting of the recycled wooden sleeper. Less optimistic modelling would lead to higher greenhouse effect potentials for the material recycling options.

No sensitivity analysis is made for the methodological scope of the study.

18.8 Modelling whole life cycles

In this chapter, the significance of an allocation procedure for end-of-life options is investigated. Whole life cycles are modelled 'from cradle to grave' for both the recycling and incineration options of the railway sleeper.

18.8.1 Scenarios selected

For the following considerations, life cycle models are set up for some of the modeling approaches discussed in the previous chapter. Allocation

[510] see, *e.g.,* Hofer 2000a:17
[511] see footnote 472

problems encountered in previous (up-stream) processes are solved aiming at over-all consistency of the whole life cycle model considering the underlying assumptions of each of the allocation:

- Forestry processes are allocated to industrial wood and roundwood independently of its time of harvest ('over-all') based on volume or relative share of proceeds
- Transports of the roundwood for sleeper production from forest to the production site are allocated to all the products of the log based on mass or on the relative share of proceeds ('over-all')
- Production processes are subdivided and allocated on a step-by-step basis based on mass (mc = 0) or relative share of proceeds
- End-of-life scenarios are modeled according to allocation procedures outlined in Section 3.2. The approaches treated further are Cut-off, VCS, the Op-Cost MEA, and the SC-PA

In addition to the above procedures, cut-off procedures are included as a reference because they represent the way most of the LCA studies are currently conducted.

Table 18-14 summarises the life-cycle models.

Table 18-14. Life cycle models combined for the recycling option and the incineration option with the co-generation of thermal energy and electricity

	Material use				Energetic use			
	Cut-off (LSA)	VCS (p)	Op-cost (MEA) (p)	Op-cost (MEA) (m)	Cut-off (WI)	Op-cost (MEA) (p)	Op-cost (MEA) (m)	SC-PA
Forest [1]	over-all volume	over-all revenues[2]	over-all revenues[2]	over-all volume[3]	over-all volume	over-all revenues[2]	over-all volume[3]	over-all revenues[2]
Transport	all-to-sleeper	over-all revenues	over-all revenues	over-all mass	all-to-sleeper	over-all revenues	over-all mass	over-all revenues
Production	all-to-sleeper	step-by-step II revenues	step-by-step II revenues	step-by-step II mass	all-to-sleeper	step-by-step II revenues	step-by-step II mass	step-by-step II revenues
Impregnation	all	all	all	all	all	all	all	all
Recycling	Cut-off (LSA)	VCS	Op-cost (MEA)	Op-cost (MEA)				
Incineration					Cut-off (WI)	Op-cost (MEA)	Op-cost (MEA)	SC-PA

[1] using input quantities of wood to the production processes (119 kg odm) which are reallocated according to the allocation rules applied to production processes and/or final disposal
[2] redistributed according to the step-by-step II allocation procedure based on revenues for the production processes; (re-) allocation factor = 0.603 (see Table 18-7)
[3] redistributed according to the step-by-step II allocation procedure based on mass for the production processes; (re-) allocation factor = 0.467 (see Table 18-7)

Note that in the product systems based on the cut-off procedure the forestry processes (119 kg wood odm) are fully accounted to the sleeper. This is assumed to be the logical consequence of the all-to-sleeper production used for the allocation of the production processes in these models. In the product systems based on the step-by-step II allocation procedure for the production process, the forestry processes are re-allocated (see also Table 18-7).

Note further on that this life cycle model does not cover several aspects. For instance, transport processes related to recycling and incineration activities would also be subject to allocation procedures, similar to the transport from forestry to sleeper production. Metallic auxiliary materials for securing the ties to the sleepers and the material and energy flows related to the track bed are also not considered in these calculations.

18.8.2 Inventory analysis

A detailed data inventory (LCI) can be ordered from the author.

18.8.3 Impact assessment

Table 18-15 summarises the results of the life cycle impact assessment based on the 'effect-oriented classification' according to Heijungs *et al.* (1992), updated according to Houghton *et al.* (1996).

Table 18-15. Impact assessment of different life cycle models for the recycling or incineration of a creosote-treated beech wood railway sleeper, resulting from different allocation procedures (per sleeper)

	Greenh. effect kg CO_2-e.	Ozone depletion kg R11-e.	Acidification kg SO_x-e.	Eutrophication kg PO_4-e.	Photosmog kg ethylene	Human toxicity kg	Eco-tox. soil kg	Eco-tox. water 10E6 m^3
Life cycle of the railway sleeper and recycling as constructive element in landscape architecture								
Cut-off (LSA)								
Forestry	2.1	2.88E-06	0.018	2.80E-03	2.00E-02	0.100	2.01E-06	7.13E+01
Transport	2.2	3.04E-06	0.019	2.74E-03	4.09E-03	0.041	1.66E-06	3.05E+02
Sleeper Production	1.6	6.52E-07	0.012	3.95E-04	3.65E-04	0.016	2.93E-07	3.98E+01
Impregnation	24.7	1.23E-05	0.203	1.04E-02	1.67E-01	0.876	7.79E-06	2.63E+03
Recycling	0.0	0.00E+00	0.000	0.00E+00	0.00E+00	0.000	0.00E+00	0.00E+00
Total	30.7	1.89E-05	0.252	1.63E-02	1.91E-01	1.034	1.18E-05	3.04E+03
VCS (p)								
Forestry	1.9	2.62E-06	0.016	2.55E-03	1.82E-02	0.091	1.83E-06	6.49E+01
Transport	1.6	2.12E-06	0.014	1.91E-03	2.86E-03	0.029	1.16E-06	2.13E+02
Sleeper Production	1.0	4.00E-07	0.007	2.42E-04	2.24E-04	0.010	1.80E-07	2.45E+01
Impregnation	24.7	1.23E-05	0.203	1.04E-02	1.67E-01	0.876	7.79E-06	2.63E+03
Recycling	-6.2	-3.73E-06	-0.051	-3.23E-03	-4.02E-02	-0.215	-2.35E-06	-6.27E+02
Total	22.9	1.37E-05	0.189	1.19E-02	1.48E-01	0.791	8.62E-06	2.30E+03
Op-cost MEA (p)								
Forestry	1.9	2.62E-06	0.016	2.55E-03	1.82E-02	0.091	1.83E-06	6.49E+01
Transport	1.6	2.12E-06	0.014	1.91E-03	2.86E-03	0.029	1.16E-06	2.13E+02
Sleeper Production	1.0	4.00E-07	0.007	2.42E-04	2.24E-04	0.010	1.80E-07	2.45E+01
Impregnation	24.7	1.23E-05	0.203	1.04E-02	1.67E-01	0.876	7.79E-06	2.63E+03

continued

Table 18-15 continued

	Greenh. effect kg CO_2-e.	Ozone depletion kg R11-e.	Acidification kg SO_x-e.	Eutrophication kg PO_4-e.	Photosmog kg ethylene	Human toxicity kg	Eco-tox. soil kg	Eco-tox. water 10E6 m^3
Recycling	3.5	3.31E-05	0.060	-6.52E-03	2.45E-02	0.177	2.23E-05	8.16E+02
Total	32.7	5.06E-05	0.300	8.56E-03	2.12E-01	1.183	3.33E-05	3.75E+03
Op-cost MEA (*m*)								
Forestry	1.0	1.34E-06	0.008	1.31E-03	9.33E-03	0.047	9.37E-07	3.33E+01
Transport	1.0	1.42E-06	0.009	1.28E-03	1.91E-03	0.019	7.78E-07	1.43E+02
Sleeper Production	0.8	3.15E-07	0.006	1.91E-04	1.77E-04	0.008	1.42E-07	1.93E+01
Impregnation	24.7	1.23E-05	0.203	1.04E-02	1.67E-01	0.876	7.79E-06	2.63E+03
Recycling	3.5	3.31E-05	0.060	-6.52E-03	2.45E-02	0.177	2.23E-05	8.16E+02
Total	31.0	4.85E-05	0.286	6.63E-03	2.03E-01	1.127	3.20E-05	3.64E+03
Life cycle of the railway sleeper and waste incineration with co-generation of thermal energy and electricity								
Cut-off (*WI*)								
Forestry	2.1	2.88E-06	0.018	2.80E-03	2.00E-02	0.100	2.01E-06	7.13E+01
Transport	2.2	3.04E-06	0.019	2.74E-03	4.09E-03	0.041	1.66E-06	3.05E+02
Sleeper production	1.6	6.52E-07	0.012	3.95E-04	3.65E-04	0.016	2.93E-07	3.98E+01
Impregnation	24.7	1.23E-05	0.203	1.04E-02	1.67E-01	0.876	7.79E-06	2.63E+03
Incineration/co-gener.	28.9	5.39E-08	0.083	1.17E-02	1.30E-03	0.104	2.12E-08	2.81E+00
Total	59.5	1.89E-05	0.335	2.80E-02	1.92E-01	1.138	1.18E-05	3.05E+03
Op-cost (*MEA*) (*p*)								
Forestry	1.9	2.62E-06	0.016	2.55E-03	1.82E-02	0.091	1.83E-06	6.49E+01
Transport	1.6	2.12E-06	0.014	1.91E-03	2.86E-03	0.029	1.16E-06	2.13E+02
Sleeper production	1.0	4.00E-07	0.007	2.42E-04	2.24E-04	0.010	1.80E-07	2.45E+01
Impregnation	24.7	1.23E-05	0.203	1.04E-02	1.67E-01	0.876	7.79E-06	2.63E+03
Incineration/co-gener.	-3.5	-3.31E-05	-0.060	6.52E-03	-2.45E-02	-0.177	-2.23E-05	-8.16E+02
Total	25.7	-1.57E-05	0.180	2.16E-02	1.63E-01	0.829	-1.14E-05	2.11E+03
Op-cost (*MEA*) (*m*)								
Forestry	1.3	1.73E-06	0.011	1.69E-03	1.20E-02	0.060	1.21E-06	4.30E+01
Transport	1.0	1.42E-06	0.009	1.28E-03	1.91E-03	0.019	7.78E-07	1.43E+02
Sleeper production	0.8	3.15E-07	0.006	1.91E-04	1.77E-04	0.008	1.42E-07	1.93E+01
Impregnation	24.7	1.23E-05	0.203	1.04E-02	1.67E-01	0.876	7.79E-06	2.63E+03
Incineration/co-gener.	-3.5	-3.31E-05	-0.060	6.52E-03	-2.45E-02	-0.177	-2.23E-05	-8.16E+02
Total	24.3	-1.74E-05	0.169	2.01E-02	1.56E-01	0.787	-1.24E-05	2.02E+03
SC-PA								
Forestry	1.9	2.62E-06	0.016	2.55E-03	1.82E-02	0.091	1.83E-06	6.49E+01
Transport	1.6	2.12E-06	0.014	1.91E-03	2.86E-03	0.029	1.16E-06	2.13E+02
Sleeper production	1.0	4.00E-07	0.007	2.42E-04	2.24E-04	0.010	1.80E-07	2.45E+01
Impregnation	24.7	1.23E-05	0.203	1.04E-02	1.67E-01	0.876	7.79E-06	2.63E+03
Incineration/co-gener.	18.2	3.40E-08	0.052	7.35E-03	8.21E-04	0.066	1.33E-08	1.77E+00
Total	47.3	1.75E-05	0.292	2.24E-02	1.89E-01	1.072	1.10E-05	2.93E+03

18.8.4 Interpretation focusing on methodological considerations

Figures 18-22 to 18-33 illustrate the outcome of the impact assessment, selecting several impact categories on an arbitrary basis[512]. On the left side,

[512] having in mind severe scientific reservations against some of the impact categories published in Heijungs *et al.* 1992 but not presented

the contributions are listed without recalculating net effects; the net effects are shown in the right figures.

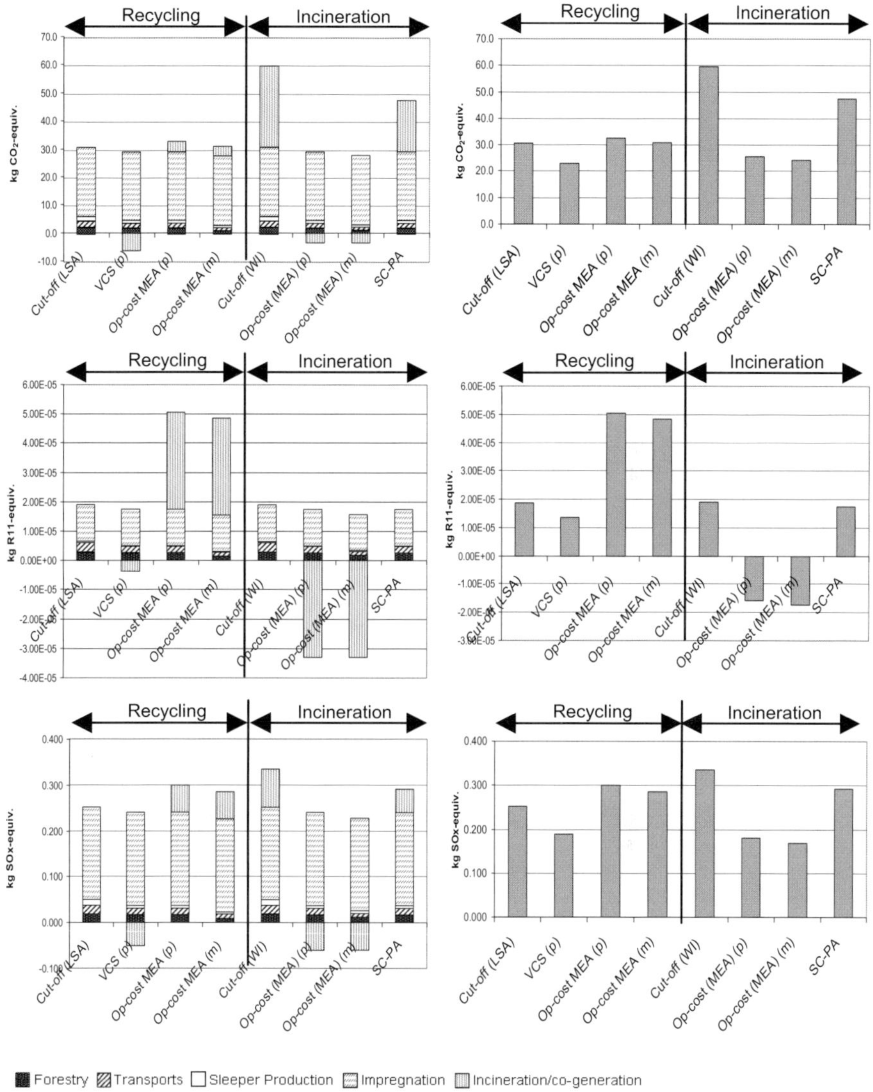

Figure 18-21 to 18-27. Impact assessment of different life cycle models for recycling or incineration of a creosote-treated wooden railway sleeper, resulting from different allocation procedures applied: greenhouse effect, ozone depletion, acidification (per sleeper)

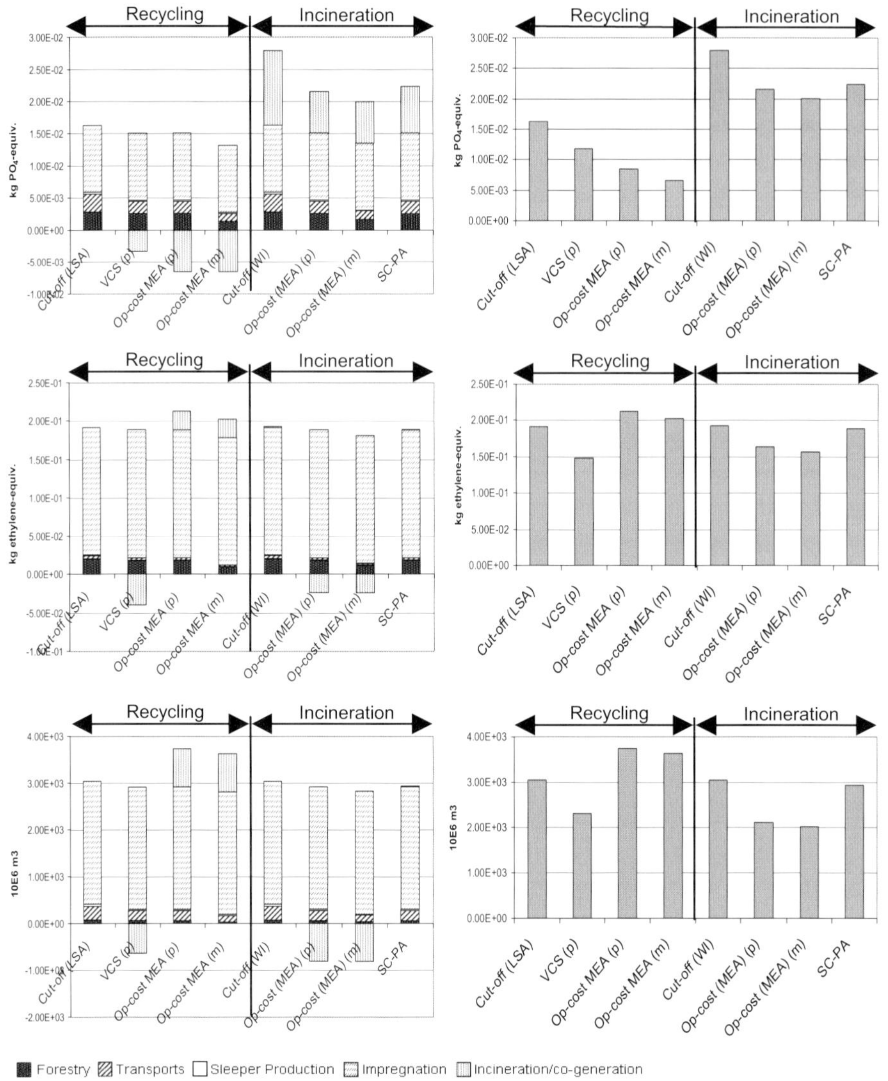

Figures 18-28 to 18-33. Impact assessment of different life cycle models for the recycling or incineration of a creosote-treated beech wood railway sleeper, resulting from different allocation procedures applied: eutrophication, photosmog, eco-toxicity in water (per sleeper)

For all impact categories, the effects are mainly caused by the impregnation processes (the production of creosote) and by the modelling of end-of-life processes. Transport from forestry to the production site of the sleeper are of secondary importance, whereas electricity production for the sleeper production processes as well as forestry processes are insignificant for all the impact categories.

The effects of different allocation procedures can more or less be ranked clearly. For the recycling alternative, the life cycle with the VCS applied scores generally lower than the widely used cut-off (LSA) procedure. Considering environmental opportunity 'costs' and substitution effects (Op-cost MEA) of the recycling of the sleeper leads to higher environmental impacts for all impact categories except eutrophication[513] (see also Chap. 18.4.6).

For the incineration alternative, the life cycle with the widely used cut-off (WI) procedure scores highest for all the environmental impact categories considered. The procedures considering environmental opportunity 'costs' and substitution effects (Op-cost MEA) lead to the lowest environmental impacts attributed to the railway sleeper. The strict coproduct allocation (SC-PA) leads to environmental interventions in between the cut-off (WI) procedure and the opportunity-cost procedures.

Note that the cut-off procedure favours recycling of the sleepers as landscape element, whereas the calculations based on the opportunity-cost procedures recommend the incineration and co-generation of thermal energy and energy. The cut-off procedure leads to wrong recommendations, as this procedure does not adequately depict causalities within the wood processing chain and the management rules for the sustainable use of wood (see Chap. 16.6).

The choice of mass or revenues as allocation factors does not have a significant impact on the result for the opportunity-cost procedures.

The choice of the strict coproduct allocation, affecting mainly the end-of-life phase, has significant impact on the results.

Data quality and sensitivity have already been discussed in each of the chapters on the different life cycle stages. The model is not complete and only set up for methodological considerations; no practical recommendations should be drawn from the above data[514].

[513] caused by the NO_x-emissions from incineration

[514] for this purpose, see Künniger and Richter 1998

Chapter 19

CONCLUSIONS

19.1 Wood-specific methodological conclusions

The following methodological conclusions and recommendations can be drawn for the modelling of end-of-life options of wood products in LCA:

- Considering (direct) substitution effects and (indirect) opportunity 'cost' are necessary components for the consistent modelling of end-of-life options in descriptive product LCA (if one aggress to the management rules for a sustainable use of wood outlined in Chapter 16.6). In product LCA, statements on the effects and usefulness of recycling can only be made if the material and the energy aspects of wood and the respective substitution effects and environmental opportunity 'costs' are taken into account;
- The wood-specific criteria established in Chapter 16.6 combined with the general requirements of a product system and its LCI (Chap. 3.4) provide a consistent framework for the evaluation and pre-selection of allocation procedures in LCAs of wood products. However, they do not allow for an unambiguous determination of an allocation procedure as the underlying mental models and values are always disputable ;
- The choice of an allocation procedure for modelling end-of-life options is more relevant for the result than the choice of an allocation factor for the coproduct allocation throughout the wood processing chain. This is because not only the allocation of the end-of-life process itself is at stake but also processes related to substitutes or environmental opportunity 'costs' (environmental opportunity effects);
- The following questions are closely linked: which (co-) products get attributed environmental interventions of a process? How many environmental interventions should these coproducts get attributed? Allocation

based on prices seems preferable to mass-based allocation procedures, given the wide price range of the coproducts occurring throughout the wood processing chain and the established demand structure for them;
- It has turned out that unallocated process data for forestry is very hard to obtain. Taking into account the importance of choosing allocation procedures to get consistent life cycle models, the elaboration and publication of unallocated process data for forestry (and other processes) is of predominant importance;
- Treating inputs like bark, sawdust, squarings, chippings and coarse chips as 'by-products' with no environmental interventions from up-stream processes attributed leads to tenable results in streamlined LCA (if one agrees to the preferability of market prices-based allocation for production processes).

19.2 General methodological conclusions

The following methodological conclusions and recommendations can be drawn from the example for the modelling of end-of-life options in product LCA:
- There is no simple answer to a simple question in a complex context. Modelling end-of-life options (including recycling material as inputs) takes recourse to methodological decisions made all over the product system. It is a real challenge to establish a model respecting material and market characteristics of the materials involved as well as their corresponding management rules of their sustainable use;
- Consistent life-cycle models are more than the sum of their inventories. Environmental opportunity 'costs' and substitution effects might have to be considered to depict the environmental relevance of product life cycles (also in descriptive LCA);
- Given the wood-specific considerations used to choose an adequate allocation procedure, it can be assumed that no generic allocation procedure is definable that would adequately depict the material and market characteristics of all materials available, the specific decision logic for each of their life cycle steps, and the management rules for the sustainable use of a material;
- Several allocation procedures are applicable for different life cycle steps within the same LCA. The allocation procedure chosen for each step widely depends on the decision-maker's mental model of the material and market characteristics, on the specific planning and decision structure assumed for the processes to be allocated, and on the management rules for the sustainable use of the materials;

- Revenues are a feasible option for an allocation factor, although data on revenues is sometimes hard to obtain, especially for intermediate products that are further processed internally. Assumptions based on published prices or estimations can solve this problem. Profits cannot be used for allocation in LCA, as data on profits is not obtainable over the whole life cycle of a product if various companies are involved;
- Mental models and values are always subjective and thus disputable. For the acceptance of an LCA, stakeholder agreements on allocation factors, allocation procedures and further elements of an LCI as a model are essential to fulfil its role as decision-support tools (see Chapter 23).

Taking these conclusions into account will allow one to depict the environmental advantages of wood as well as its weaknesses in a much more adequate way. This allows one to derive recommendations from LCA that a) take the material and market characteristics of wood into account and that b) are in line with national post-consumer wood policies.

VI.

CONCLUSIONS AND OUTLOOK

Chapter 20

REVIEW OF THE THESES

In this chapter, the theses set up in Chapter 1.5 are revisited.

Thesis I: The definition of a product system as an attribution of material and energy flows to a product out of a complex reality cannot be made in an unambiguous way. Modelling the product system contains subjective elements, i.e. the use of mental models and value choices. An objectively right, generic way of modelling a product system in the life cycle inventory analysis is not definable.

It has been demonstrated in Chapter 5 and Chapter 6 that various mental models and values guide the attribution of material and energy flows to a product under study in the life cycle inventory analysis of LCA. Among the most important mental models influencing modelling in the life cycle inventory analysis are (Chapter 7) the:

- Life cycle of a product itself;
- Demarcation of the technosphere and ecosphere;
- Internal structure of the technosphere;
- Material and market characteristics of the materials and products involved;
- Organisational principle of the socio-economic system;
- Range of the decision-maker's responsibility in view of the management rules for sustainable development;
- Role of environmental information for sustainable development.

The decision-maker's values expressed as preferences, which are involved in the definition of the product system and its LCI, are stated in the following decisions:

- Attribution of material and energy flows to the product system in the sense of 'less is better';
- Level of relevance for cutting off 'irrelevant' material and energy flows;

- Valuation of outputs (or functions) of a multifunctional process as coproducts and waste;
- Valuation of the functionality of input and output materials when deciding on modelling closed-loop or open-loop recycling;
- Material-specific management rules for a sustainable use of the materials involved;
- Temporal preferences expressed in the decision-maker's attitude towards risk if changes within the socio-economic system have to be modelled.

All these mental models and value guide the setting up of the product system as attribution problem in LCA. These mental models and values have to be addressed and depicted in a product system in a proper way if the results of an LCA should be relevant for a decision-maker (Chapter 3.2.3).

Material and market characteristics of different materials can differ considerably; also their role and management strategies for a sustainable development of the world can differ, *e.g.*, for renewable and non-renewable materials (thesis II). Also based on the two case studies, it must be concluded that no generic methodology for the life cycle inventory analysis can reasonably be defined. Modelling options within the life cycle inventory analysis must be used to its full extend to address the decision-maker's mental models and values in an adequate way. The requirements of product system and its LCI summarised in Chapter 3.4 provide a useful basis for the re-evaluation of methodological decisions.

Thesis II: Modelling characteristics of LCA only partially address the environmental dimension of sustainable development. Recommendations derived from the LCA-model could be perceived sub-optimal from a 'real world' perspective if attribution of material and energy flows to the product(s) under study is not made considering the specific characteristics of the decision situation, i.e. the material and market characteristics of the materials involved, and the 'descriptive' power of LCA in view of the material-specific management rules for their sustainable use.

'Sustainable development' is a very broad concept for the future development of our society (Chapter 3.3.2). This is also true for its environmental dimension. All the current life cycle impact assessment methods address this environmental dimension only partially. Among the shortcomings, which are not properly covered, are (Chapter 3.3.5):

- Demand-related issues;
- Resource consumption;
- Mechanical impacts;
- Risk-related considerations;
- Effects on genetic biodiversity;

- Sustainable management of material and energy flows within the technosphere.

From the perspective of defining a product system, the last point is of particular interest. Sustainable management rules of material and energy flows within the technosphere cover, *e.g.*:

- Thermodynamic efficiency of the circular flows for non-renewable resources;
- The limiting of changes of entropy of a material over subsequent uses;
- Current and future substitution rates of materials, especially the rate of (the intended) substitution of non-renewable with renewable materials.

The 'descriptiveness' of the whole LCA-model depends very much on how these aspects are integrated in the life cycle inventory analysis. Having the a-temporal, spatially non-differentiated input-output model of LCA in mind, this means that:

- Entropic aspects have to be taken into account when modelling recycling material (*e.g.*, by value-corrected substitution for aluminium, Chap. 13);
- Environmental opportunity 'costs' and substitution effects have to be taken into account when modelling life cycles – and particularly end-of-life options – of renewable materials (Chap. 19 for wood).

Hereby, it has been demonstrated that the decision-maker's values also influence the life cycle inventory analysis of LCA. These values cannot solely be restricted to the impact assessment step of LCA.

Thesis III: Decisions related to modelling the product system in the life cycle inventory analysis such as module definition, setting system boundaries, allocation and the implementation of the guiding principle 'causal relationship' for the attribution of material and energy flows to the product, and also the definition of the functional unit are interdependent. They cannot be treated separately and in subsequent order. They have to be addressed reflecting the decision-maker's mental models and values in the decision situation in a consistent way.

Most methodological decisions encountered during the setting up of the product system in the life cycle inventory analysis are related to the only accounting principle: the a-temporal, spatially undifferentiated attribution of material and energy flows, resulting in an a-temporal, spatially undifferentiated input-output model (Chap. 2.3.2). This attribution is not unambiguous (Thesis I) .

It has been shown that decisions encountered during the definition of the product system – especially decision situations related to multifunctional processes – can be solved by different solutions (Chapters 5 and 6). For instance, they can be solved by:

- Redefining modules;

- Extending the system and adding additional functions to the functional unit (thus redefining the functional unit);
- Fully accounting the multifunctional process and subtracting surplus functions as equi-functional single-functional processes (the avoided-burden approach);
- Cutting processes off or by including them entirely (*e.g.*, the cut-off procedure for the modelling of reuse & recycling, thus by the setting of system boundaries);
- Splitting processes and 'allocating' them partially according to the decision-maker's mental model of causality.

The modelling option applied should lead to a product system that finally respects the decision-maker's mental models and values stated under thesis I. As such, all the above-mentioned elements of modelling the life cycle of a product during the life cycle inventory analysis interdepend and cannot be treated separately or in subsequent order.

Thesis IV: Structural uncertainty ('ignorance') related to changes within the socio-economic system over time are a predominant aspect of modelling and allocation related to reuse & recycling in the life cycle inventory analysis. Attitude towards risk is an essential determinant of modelling and allocation related to reuse & recycling within LCA.

Reuse and recycling are imperatively related to the course of time. In the case of reuse & recycling, temporal aspects affect:

- The recycling processes that will be used for the treatment of the post-consumer material;
- The technology or process that will be substituted in the future by the material to be recycled;
- The percentage of the material collected for recycling and the yield of the recycling processes;
- The stability over time of the allocation factor chosen.

It has been shown in that different attitudes towards risk can lead to different methodological choices for all the points mentioned above (Chap. 5.8.5).

The influence of the decision-maker's attitude towards risk, *e.g.,* when choosing an allocation procedure for reuse & recycling, implicitly brings temporal preferences into modelling. Nonetheless, the a-temporal, spatially undifferentiated character of LCA must be reminded when deciding on methodological issues related to reuse & recycling:

- Ignoring past environmental interventions in the a-temporal model of descriptive LCA leads to systematic errors;
- Taking into account past environmental interventions leads to systematic errors in change-oriented, marginal LCA.

Chapter 21

CONSEQUENCES FOR LCA AS A DECISION SUPPORT TOOL

LCA is caught in the trap of having to address mental models and values of the decision-maker on the one hand, and the claim of compiling the environmental relevance of products in a reliable and objective way on the other hand. The claim of objectivity cannot be redeemed by LCA, not only because of value-based impact assessment methods but also because of the influence of mental models and values throughout the modelling process of the life cycle inventory analysis. From this perspective, the goal and scope dependency of an LCA has highest actuality.

This has several consequences for LCA as a decision support tool:

- In comparative LCA, the application of equal allocation procedures does not lead to LCA results on which products made from different materials can be compared in an adequate way. Each product and material must be modelled according to its specific material and market characteristics as well as according to its particular management rules for its sustainable use. A generic LCA methodology, including preferences on methodological options, is not definable if material and market characteristics of different products and materials should adequately be depicted.
  This point is also of highest importance for establishing of environmental labelling-schemes based on quantitative LCA-results (Type III-labels; ISO/TR 14025);
- The evaluation of changes provides insight into the suitability of the models developed. A partial validation of the product system as model of a product life cycle can be made if improvement options deducible from an LCA contradict material and market characteristics of the materials involved, or if 'recommended' improvement options lead to sub-optimal solutions if reassessed from a meta-perspective. This is particularly true for descriptive LCA;

- Modelling consistent life cycles of a product is a complex problem. This is particularly true for end-of-life options. Unaggregated published LCI data is essential for the development of consistent product systems. Practice has shown that such data is hard to obtain. This matter of fact must be considered for any establishment of public databases for LCI-data.
  But the quality of information generated by a model does not only depend on the modelling procedures itself but also on the data quality. Schaltegger and Burritt (2000:240ff) and Schaltegger (1997a) have stressed problems with LCI data quality. These problems relate to the lack of representativeness and site-specificity of average LCI data, the perverse incentives of data providers if data is collected by public institutes or producer associations, to public cost of data collection by public institutes being against the 'polluter pays' principle, and to the inefficient way of central data collection compared to decentralised data collection, resulting in high cost and poor information quality (see also von Hayek 1945);
- With regard to product design-related decision-making, the current way of conducting an LCA has to be considered ineffective. This is caused on the one hand by the fact that decisions affecting the environmental function efficiency of a product are strategic decisions, made by the company management much before the design stage of a product. During this stage, insufficient data is known that could be integrated in quantitative LCA (Chap. 3.1.2).
  On the other hand, the largely applied descriptive LCA serves to identify environmental hot spots of a product in a general way. Such results can be integrated as requirements or desirable features in checklists and guidelines for product design. Still, it must be kept in mind that descriptive LCA (also streamlined LCA) can lead to wrong recommendations (or hot spots), as descriptive LCA also include past environmental interventions. For instance, recycling material can carry additional environmental burdens from former primary material production[515]. This could lead to the identification of a hot spot that would not be considered a hot spot in marginal future-bound considerations.
  Descriptive considerations are not very adequate for assessing consequences of design alternatives. For such a purpose, the decisions to be made during product design must be understood as investment options and thus as investment decisions. From this perspective, not functional

[515] It could be argued that in the case of recycling no primary material production should be attributed to recycling material. But given the a-temporal model of LCI, this would lead to systematic errors for descriptive LCIs (see Chap. 5.8.3)

units as defined in ISO/EN 14040, chap. 3.5, but *decision alternatives* should be considered objects to be assessed. The environmental consequences of decision alternatives can then be assessed in a marginal LCA, covering the whole life cycle of the materials and energy carriers involved.

LCA as a decision support tool could be most powerful if it followed an actor-based, site-specific marginal approach applied in a specific decision situation. With the increasing amount of site-specific data available, stemming from Environmental Management Systems, even data quality will raise the general information quality of such an LCA approach[516]. This position has to be consolidated if current disillusionment on the usefulness and capabilities of LCA is to be overcome.

[516] for additional incentives to provide site-specific LCI-data, see Schaltegger and Burritt 2000:255ff, Schaltegger 1997a

Chapter 22

CONSEQUENCES FOR A REVISION OF ISO/EN 14041

22.1 Critical points in ISO/EN 14041

As a conclusive summary, the following recommendations for allocation related to open-loop recycling have to be made for the next revision of the standard ISO/EN 14041. They concern:

a) The option to allocate primary material production processes and final production to several subsequent product systems (ISO/EN 14041, chap. 6.5.4);
b) The recommendation to base allocation on scrap value in relation to primary material (ISO/EN 14041, chap. 6.5.4);
c) The missing methodological distinction between descriptive and change-oriented (marginal) types of LCA in the series of standards ISO 14040;
d) The missing recommendation in ISO/EN 14041 to explicitly consider material and market characteristics for the modelling, and for the choice of an open-loop recycling allocation procedures in particular;
e) The confusing distinction of setting system boundaries and allocation of multifunctional processes.

Ad a) The distribution of environmental interventions of primary material production and final disposal to several product systems (ISO/EN 14041, chap. 6.5.4) is a valuable modelling option for certain materials to incorporate the management rules for their sustainable use in LCA in an adequate way. This option must not be eliminated during the next revision of the standard ISO/EN 14041. The case study on aluminium windows underlines this point.

Ad b) As a second option to handle attribution problems related to reuse & recycling, ISO/EN 14041 mentions:

> "economic value *e.g.*, scrap value in relation to primary material" (ISO/EN 14041, chap. 6.5.4).

as possible allocation factors. This proposal has also been made by Brouwer and Lindeijer (1993). As can be followed in Table 6-3, this procedure is not additive and is thus not in line with the 100%-rule of ISO/EN 14041, chap. 6.5.2, point 2.

Furthermore, the elementary metallic composition of the material, responsible for the future inherent properties of the secondary material (and thus for the value of the material), are only one factor among others for the determination of scrap prices. They vary considerably due to the homogeneity (large vs. small pieces), impurities (dirt, plastics etc.), humidity, amount available at one site, transport distance, scrap availability etc. Furthermore, they can vary due to fluctuations of the exchange rate of the local currency compared to the currency of the reference market. This can lead to 'absurd' recommendations.

Allocation procedures that distribute primary material production and final disposal to several product systems based on devaluation should be based on the determination of the material devaluation considering equal (re-) processing stages, *e.g.*, primary material ingot vs. secondary material ingot.

Ad c) Descriptive (attributive) and change-oriented (marginal) LCA have not been distinguished so far in ISO/EN 14041. However, for both types attribution rules and allocation procedures for coproduction and recycling differ considerably.

Ad d) The product system as model of the life cycle of a product, and particularly its respective LCI, are very reductionistic in view of the complex character of its modelling object and objective[517]. Considering material and market characteristics of the materials involved as well as the management rules for their sustainable use are of predominant importance for the attribution of material and energy flows to the functional unit under study. Only in this way, the 'causality' of the material and energy flows attributed to a product will be met, allowing for the analogy assumption between the LCA model and the 'real world' (see Figure 1-1). This is especially important for the choice of allocation procedures related to recycling to avoid suboptimal or 'absurd' recommendations. This matter of fact should be prominently stated in the corresponding chapter of a revised standard ISO/EN 14041.

Ad e) Each decision on the attribution of material and energy flows to an input-output model is finally a decision on setting system boundaries. The

[517] see Chap. 2.3.2

current way is misleading, in which ISO/EN 14040 distinguishes between setting system boundaries, allocation and the allocation related to reuse & recycling. For a future revision of the standards, it is recommended to describe the definition of a product system as an attribution problem during which different decision situations have to be passed following certain attribution rules to be stated. A conceptual proposition is made in the following chapter.

22.2 Outline of attribution rules in inventory analysis for a revision of ISO 14041

Methodological distinctions and considerations should be made during the revision of the standard ISO/EN 14041:

- In dependency of the goal and scope of the study, a preliminary set of life-cycle stages or processes possibly to be included in the product system should be selected out of the list in ISO/EN 14041, chap. 5.3.3;
- Attribution of material and energy flows entering or leaving the preliminary set of life-cycle stages should preferably be made effect-related based on cut-off criteria (Figure 22-1). Effects of cut-off criteria include relative contribution to total mass, total energy, or total environmental impact (see ISO/EN 14041, chap. 5.3.5). In descriptive (average) LCA, average data is used; marginal LCA are modelled considering the relevance of additional or avoided (marginal) processes.

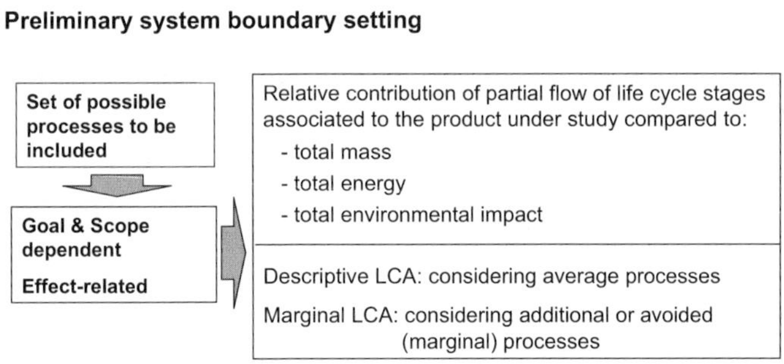

Figure 22-1. The first effect-based attribution rule: relevance of partial flows

A second set of attribution rules is required if effect-related attribution is not feasible: in the case of simultaneous multifunctional processes such as coproduction or combined waste treatment (Figure 22-2).

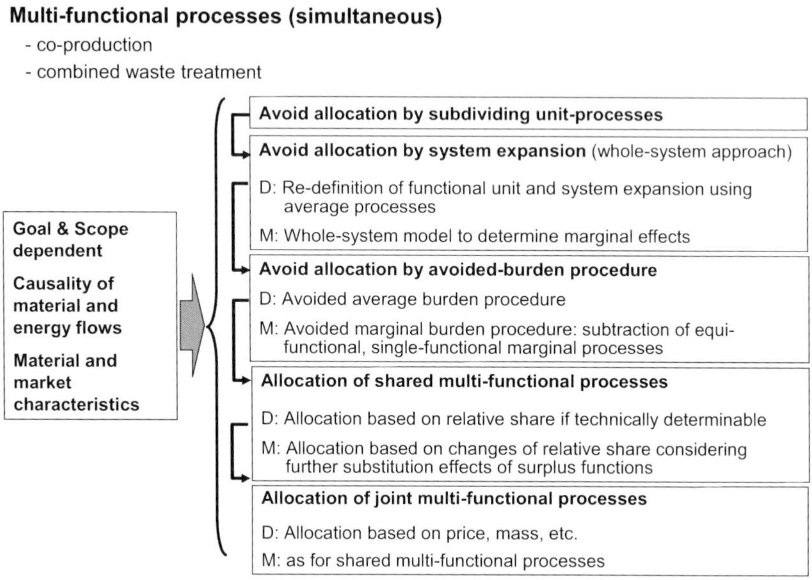

Figure 22-1. The second, cause-related attribution rule: causal relationship of partial flows

Following ISO/EN 14041, chap. 6.5.3, allocation should be avoided by subdividing unit processes or by system expansion wherever possible.

In practice, allocation of multifunctional processes cannot be solved by subdividing unit processes although the process to be allocated can be further refined, reducing the significance of the choice of the allocation procedure.

In the descriptive type of LCA, system expansion solves the allocation problem by re-defining the functional unit. The multifunctional process is fully accounted whereas additional functions are added to the preliminary functional unit.

In the change-oriented (marginal) approach, the whole-system of several functions is modelled to determine effects caused by the change to be assessed within the whole-system. In this case, the setting-up of a whole-system model requires the application of a mental model of the cause-related organisational principle of the socio-economic system, as the dynamics of the system have to be modelled.

A second option to avoid allocation is the avoided-burden procedure. In descriptive LCA, the multifunctional process is fully accounted, but substituted mono-functional (average) processes of the surplus-functions are subtracted[518]. Also in marginal LCA, the avoided-burden procedure is applicable. Again the multifunctional process is fully accounted for but substituted mono-functional (marginal) processes of the surplus-functions are subtracted.

If system expansion and whole-system models as well as the avoided-burden procedure are not feasible for the goal and scope of the study, the attribution problem related to simultaneous multifunctional problems has to be solved on a process level by allocation. If processes are shared – thus providing several functions in an unfixed ratio – the relative share of a function can be determined by modifying the relative share of the products.

In marginal LCA, the effects induced by the modification of the relative share are attributed to this change, induced by the (additional) functional unit. Substitution effects induced by changes of the surplus-functions have to be taken into account by the avoided-burden (or additionally-caused-burden) procedure. For descriptive LCA, the relative amount of environmental interventions to be attributed to each function can mathematically be determined under certain (rare!) circumstances (see Chap. 5.6.5).

The above-mentioned procedure is not feasible in the case of joint coproduction where the ratio of products or functions is fixed and cannot be modified by the decision-maker.

In descriptive (average) LCA, allocation cannot be made other than applying 'arbitrary' criteria such as revenues, profits, volume, mass, fat content, etc. The decision-maker's mental model of the organisational principle will decide on the allocation factor chosen.

For marginal LCA, this third option according to ISO/EN 14041, chap. 6.5.3 is not feasible as it contradicts the underlying accounting principle of marginal LCA. Instead, the same procedure as for combined processes outlined above applies: the effects induced by the modification of the relative share are attributed to this change that is induced by the (additional) functional unit. Substitution effects caused by changes of the surplus-functions have to be taken into account by the avoided-burden (or additionally-caused-burden) procedure.

[518] Note that the avoided-burden approach is subsumed under system expansion in the current version of ISO/EN 14'041, annex B3. This is considered to be not very consistent, as whole-system approaches based on system expansion, in the proper sense, lead to different models for both descriptive and marginal LCA than the avoided-burden approaches. In most cases, the avoided-burden approach is probably more practical and feasible than system expansion. Especially for marginal LCA but also for descriptive LCA, system expansion, in the strict sense, is possibly more a theoretical option. Consequently, the terminology and hierarchy of the attribution rules related to the multifunctionality problem set by the standard ISO/EN 14'041 should be reconsidered.

A third set of attribution rules has to be applied to handle multifunctionality related to recycling (Figure 22-3).

For the discussion of multifunctionality related to recycling, two aspects have to be considered:

- The multifunctionality of primary material production and final disposal enabling secondary uses of a material;
- The bi-functional character of the recycling processes serving as waste treatment and secondary material extraction.

Following the general allocation procedure according to ISO/EN 14041, chap. 6.5.3, allocation shall be avoided by subdividing processes. In the case of recycling, this procedure is not feasible if primary material production and final waste disposal are considered joint coproduction processes to be shared by several subsequent product systems. However, if primary material production and final waste disposal are treated according to the sunk-effects approach (see Chap. 5.8.3), then recycling processes can be subdivided[519].

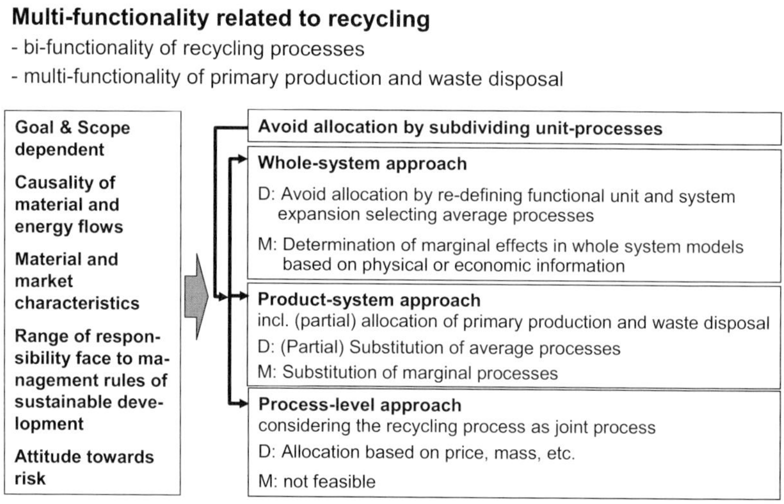

Figure 22-2. A third set of attribution rules for the handling of multifunctionality related to recycling

If subdividing processes does not lead to an adequate and consistent product system[520], recycling can be treated on the three levels outlined in Chapter 6.

[519] as proposed, *e.g.*, by Anonymous 1994b; see Chap. 6.5.2 for more detailed description of the procedure

[520] *e.g.*, if the asset approach described in Chap. 5.8.3 seems more adequate, or if a zero-point of the market price is not determinable, see Chap. 6.2

Recycling can be handled enlarging the system to a whole cascade of product systems.

For descriptive LCA, adding additional functions to the functional unit while expanding the system solves the allocation problem. Partial allocation of all the processes involved in the cascade can be achieved by applying allocation factors derived *e.g.,* from the price for the different functions. Although environmental impacts can be attributed to single product systems, the whole cascade has to be determined. For marginal LCA, modelling a whole-system based on physical or economic causalities and determining marginal effects throughout the model solves the allocation problem.

Recycling can also be handled on product system level, considering upstream primary material production processes and downstream waste treatment processes (following ISO/EN 14041, chap. 6.5.4).

In descriptive LCA, (partial) substitution can be applied for allocating (average) primary material production and final waste processes to a specific product system. As a precondition, the allocation factor chosen has to represent the devaluation of a material over the product system under study. Recycling processes have to be allocated consistently with the whole allocation procedure[521]. The cut-off procedure (or modifications of if) can be applied if the application of the sunk-effects approach seems adequate[522].

In marginal LCA, substitution of marginal processes can be applied, leading possibly to quite long chains of substituting processes[523]. Substitutions do not necessarily occur within a specific material cascade but can also affect completely different materials[524].

As third option, recycling processes can be handled on process level.

For descriptive LCA, multifunctionality is thus reduced to allocating the recycling process, disregarding the multifunctionality of primary material production of final disposal. In this case (apart from subdividing recycling processes; see above), recycling processes can be allocated on a price basis, using the relative share of the revenues from the waste treatment function and from the sale of the recovered material[525].

For marginal LCA, the process-level approach is not feasible as changes have to be assessed, and the two functions of the recycling process are inherently joint. However, subtraction based on substitution considerations can be

[521] see, *e.g.*, the value-corrected substitution, proposed by Werner and Richter 2000a; Werner and Richter 2000b; see also Chap. 10

[522] for a description of the procedures, see Chap. 11.1

[523] see, *e.g.*, Weidema 1998b; Weidema 1999

[524] Assumptions on substitutive effects largely depend on the decision-maker's mental model of the organisational principle of the socio-economic system.

[525] Huppes 1994

applied, finally solving the allocation problem as substitution on product system level (see above).

The level of handling allocation related to recycling, the choice of an allocation procedure with its underlying mental model of the organisational principle of the socio-economic system and on the range of the decision-maker's range of responsibility should be selected according to the criteria mentioned in Chapter 3.4 (see also Chap. 6.6). Additionally, they should be consistent with the handling of simultaneous multifunctionality problems.

Temporal changes of the socio-economic system should be dealt with according to the decision-maker's attitude towards risk. Nonetheless, the a-temporal, spatially undifferentiated character of LCA must be reminded when deciding on methodological issues related to reuse & recycling:

- Ignoring past environmental interventions in the a-temporal model of descriptive LCA leads to systematic errors;
- Taking into account past environmental interventions leads to systematic errors in change-oriented, marginal LCA.

Chapter 23

TOWARD A GROUP-MODEL BUILDING PROCESS IN LCA

The previous Chapter 22 summarises the methodological consequences of this book. In this chapter, the procedural consequences for the conduct of an LCA are outlined from a meta-perspective.

So far, a unitary-actor perspective for the decision-maker and a 'neutral' modeller were assumed for the methodological considerations in this book. Allocation procedures were evaluated according to the most realistic representational ('descriptive') content of the model in the eyes of the decision-maker. For various reasons, this is in fact a very simplifying perception of LCA in the decision-making process:

- Decision-making in companies mostly includes more than one actor;
- Modellers are not 'neutral' but bring in their own knowledge (mental models among other types of knowledge, see below) and values;
- The decision problem does probably not only affect the organisation itself, represented by the decision-maker(s), but also other stakeholders.

This chapter follows several goals. The modelling process of an LCA is looked at from the meta-perspective to:

- Position LCA in a product-related stakeholder model;
- Position the topics of mental models and value choices in the life cycle inventory (LCI) during the conduct of an LCA;
- Outline a general group model-building procedure, which has to be designed according to a specific LCA-project;
- Outline a possible validation framework for the evaluation of group model-building processes.

This chapter contains some theoretic reasoning. Its application and evaluation in practice is a future research need (see Chap. 24).

23.1 Stakeholders of an LCA

Strong client involvement has often been considered a useful way to create model ownership and to ensure implementation[526]. Depending on the goal of the LCA-study, an even broader frame can be advisable requiring a systematic selection and involvement of different stakeholders.

In the stakeholder model of management sciences, the company is at the centre of interest. A stakeholder is usually a representative of a group of persons who has a general interest (concerned or affected) in the activities of a company[527].

Product-oriented environmental management focuses on products, which live through an entire life cycle, and hence, the concept connects stakeholder interests throughout this cycle. For the application of the stakeholder model in the context of LCA, the company-related perspective has to be abandoned for a product-related perspective. Based on the above-mentioned stakeholder-definition and the scope and structure of LCA, the definition of a stakeholder is:

> "individuals or groups with own goals and needs who can actively influence individual stages of the product life cycle in environmental terms, or who are influenced by these stages" (Lundie 1999:114; translated from German by WF).

If stakeholders demonstrate a growing environmental concern, the addressing of such demands might lead to an improved company performance and to a competitive advantage[528]. Case studies show that gaining some form of competitive advantage is an important motive for managers to engage in product stewardship. Such an advantage could be the improvement of the company's image, strengthening bonds with customers, or getting better knowledge of one's products and processes, hence improving transparency[529]. Such advantages can influence a company's efficiency and effectiveness as well as social legitimacy and autonomy in the political/legal environment in a positive way[530].

To acknowledge stakeholder interests, legitimisation will form an important driving force. In environmental problem framing, not only mana-

[526] see, *e.g.*, Lane 1994

[527] for the term 'stakeholder', no unambiguous definition can be found in literature; see, *e.g.*, Freeman 1984:1967; Pfeffer and G.R. Salancik 1978; Staehle and Nork 1992; Thompson 1967

[528] according to Greenley and Foxall 1997; Hart 1995

[529] see de Bakker *et al.* 2002

[530] see Hill 1990; Schaltegger and Sturm 1992:12ff; also Chapter 5.7.5

gerial decision-makers try to conceptualise their environment, but also the involved stakeholders develop their own points of view. The way a company can legitimise its (planned) activities is likely to influence its stakeholders' replies.

Environmentally proactive companies will probably view a range of stakeholders as important and will actively address those stakeholders. Stakeholder selection in the context of LCA is discussed in Chapter 23.3.

Stakeholders can be classified according to different criteria such as intensity of influence[531], internal or external stakeholders[532], origin[533], or interest and culture[534]. As Lundie (1998) has worked out, the distinction of internal and external stakeholders based on Mintzberg (1983), Meffert and Kirchgeorg (1998), Freeman (1984) and Schierge (1996) is the most suitable one in the context of LCA.

Table 23-1 summarises stakeholders with direct and indirect relations to the life cycle of a product.

Table 23-1. Stakeholders with an (in-) direct relation to the stages of a product life cycle (after Lundie 1998:120; modified and added)

Stakeholders with a direct relation to the stages of a product life cycle	Stakeholders with an indirect relation to the stages of a product life cycle
Suppliers of	*Science*
Raw and process materials, fuels	Experts for LCA
Semi-finished and final products	Environmental sciences
Companies	*State*
Owner(s)	Ministry of Environment
Management (purchase, product development, marketing, production)	Other governmental institutions in the environmental field
Employees	*Businesses*
(Re)Distribution	Consultants/modeller(s)
Carriers	Banks and insurances
Clients	Investors
Traders	*Citizen initiatives*
Final consumers (represented by consumer organisations)	Mainly local and regional interest groups
	Environmental protection organisations
Disposal companies	Local, regional and global organisations
Recycling companies	*Public*
Owners of municipal waste incinerators	Representatives of prevailing, social opinions
Owners of landfills	

531 Achtleitner 1985:76

532 Mintzberg 1983:26-30; Meffert and Kirchgeorg 1998:95; Freeman 1984:11-22; Schierge 1996:48-101

533 Portisch 1996:19-20

534 Jörgensen 1996

It is beyond the scope of this book to discuss the possible roles of different stakeholders of an LCA in detail. The interested reader is referred to Lundie (1998:121ff).

Nonetheless, it seems noteworthy that depending of the goal of an LCA, different stakeholders are involved in different ways. Brief evaluation of the environmental consequences during material selection in product design will require a smaller stakeholder involvement than strategic decisions related to a product portfolio of a company. Products comparison (and lobbying) by industrial associations will emphasise on other stakeholders than the environmental assessment of public infrastructure projects.

Not all the selected stakeholders are involved in the same way. Involvement can consist in:

- Telephone interviews;
- Questionnaires (written, by e-mail);
- Workshop participation;
- Delphi-technique[535];
- Workbooks[536];
- Plant visits;
- Project group memberships;
- Internal or external peer reviews.

In all cases, stakeholder involvement in LCA has various purposes[537]:

- Combining knowledge from different sources with different viewpoints (*e.g.*, factual knowledge, methodological knowledge, data; see below) to enhance mutual learning;
- Sharing values related to environmental preferences related to the material and energy flows within the technosphere as well as with regard to the ecosphere to foster consensus;
- Raise the understanding and acceptance of the outcomes of an LCA, and thus commitment ('ownership');
- Foster implementation.

Although desirable from a general point of view, a more intense stakeholder involvement considerably complicates the model building process, requires commitment of the commissioner(s) and/or decision-maker, and has its costs in terms of money and time. Besides that, the outcome is not straightforward; goals and motivation of different stakeholders can differ leading not necessarily to consensus and acceptance of an LCA. Social skills

[535] Dalkey 1969

[536] explaining methodology, results of questionnaires and preliminary models for feedback purposes of the participants; see, *e.g.*, Vennix and Gubbels 1992:93

[537] see also Morecroft 1994

of the modeller – even the engagement of a neutral facilitator – are essential for a fruitful group modelling-process (see below). Anyhow, the choice of and the way of the stakeholders' involvement must be carefully considered, given the goal and scope of an LCA.

23.2 Knowledge involved in LCA

Combining knowledge from different sources and considering different viewpoints has been claimed to be a primary purpose of stakeholder involvement in LCA. Three main kinds of knowledge are involved in LCA[538]:

Structural knowledge: this kind of knowledge comes from the available theoretical knowledge that is expressed with the help of scientific concepts. Usually, the only sources of structural knowledge are the mental models with some subjects – the decision-maker(s) and/or other stakeholders – have about the system to be modelled. In this case, the knowledge is expressed in intuitive terms and in ordinary language.[539]

Quantitative knowledge: stated in reference modes, temporal series or, in general, empirical behaviours and in the knowledge concerning the initial conditions in which the real system is placed. In other words, empirical knowledge that is available with regard to the variations of the relevant magnitudes of the socio-economic system over time and the particular values of these magnitudes in a given situation.

Operational knowledge: the specific LCA skills and practical knowledge that the modeller uses when he/she integrates the other two kinds of knowledge, obtaining a special formal representation, the LCA model. The LCA model represents the life cycle of a product as product system with its related environmental impacts. The intention is that the LCA-model will be able to state 'real' environmental improvement options.

It is essential to have these three kinds of knowledge coherently included in an LCA. While empirical behaviours, *e.g.*, recorded by measurements of processes anchor in reality, mental models give information which is not so much quantitative but structural. Mental models are sometimes supplemented with particular applications of available theories but, if none is available, they constitute the only source of structural knowledge concerning the systems being modelled. From this point of view, mental models are a basic and indispensable source of knowledge in the process of building a product system with its related environmental impacts.

[538] after Vázquez *et al.* 1996:24
[539] Vázquez *et al.* 1996:24 referring to Forrester 1986

At this point, the four important attributes of mental models should be pointed out (after Vázquez *et al.* 1996:25; Ulli-Beer 2002):

- Mental models on complex aspects of our world can differ; they are the result of selective perception caused by various aspects such as a individual's personality, training, education, professional background, and social position within an organisation;
- Mental models are not fixed; they change with experience, action and discussion, and also during the conduct of an LCA;
- Mental models are not simple; they contain rich and relevant information about the basic components and structural relationships of the systems in which subjects are involved;
- The structural information that mental models provide about certain systems that – like the socio-economic one – are the result of human actions guided by these same mental models, is usually reliable.

In order to get an LCA-model accepted, it is necessary that the hypotheses used to build the model be compatible with available scientific or heuristic knowledge. Secondly, these hypotheses must be captured adequately with the representational tool LCA; all this information must be processed properly to obtain conclusions that will fit the empirical behaviour of the socio-economic system and its interrelation with the ecosphere.

Seen from this perspective, the LCA-model is the final result of the progressive refinement and formalisation of a (shared) mental model. This refinement and formalisation is not merely quantitative but also structural. Unfortunately (also for group model-building purposes), the explicit structural representation of a product system in LCA is very poor, as the model is a simple a-temporal, spatially undifferentiated input-output model (Chap. 2.3.3). Herein lies the reasoning:

- Taking structural aspects on material and energy flows as well as the decision-maker's values into account for the choice of attribution and allocation procedures is crucial for the representiveness of the model;
- In-depth discussion of mental models between the stakeholders on aspects related to the product system as well as 'real world' consequences of methodological decisions becomes crucial within the modelling process.

Grasping the mental models (and values) of the relevant stakeholders of an LCA are thus a *conditio sine qua non* for the representiveness of the model of a specific decision situation, for the ownership and acceptance of the model and its results as prerequisite for implementation. This requires:

1. an 'instrumental conception' of LCA instead of an LCA as 'true representation'[540]; and
2. the understanding of the conduct of an LCA as group model-building process (instead of an external consultant's job).

In such an 'instrumental conception' of LCA, stakeholder participation as in-depth dialogue and mutual learning become key elements of the modelling process. Only this can lead to the 'ownership' of the model by the stakeholders and improve the possibility that the outcomes of an LCA lead to improvement actions[541].

23.3 LCA as group model-building process

Group model-building is a method suggested by Vennix (1996) for stakeholder involvement in system dynamics[542,543]. It refers to a model building process in which a group of stakeholders are deeply involved in the process of model construction. The group members describe their mental models of a situation and analyse the causes and consequences of some selected measures. In the first part of the modelling process, the primary focus is descriptive and diagnostic and aims to construct maps of shared reality of interactions. In a second part, the mappings of the mental models can be analysed further and specified to develop a quantitative system dynamics model.

The following adaptation of the group modelling process for LCA is inspired by the group model-building procedure using system dynamics by Vennix (1996) and Vennix and Gubbels (1992) taking into account Lundie (1998). The modelling process is divided into nine steps (Figure 23-1).

In principle, these steps follow the structure according to ISO 14040ff. Nonetheless, emphasis lies on stakeholder participation in a group model-building approach. In the following, the steps are described in more detail.

[540] see Morecroft and Senge 1994, particularly Lane 1994; Senge and Sterman 1994; Diehl 1994; also Heiskanen 2002:428; in fact, the conduct of an LCA for 'the sake of knowledge generation' is a doubtful venture, as it stipulates a 'degree of objectivity' that cannot be granted for the epistemological reasons outlined in this book.

[541] Morecroft 1994; Lane 1994

[542] for further literature on group model-building, modelling in for learning organisations, etc, see Morecroft and Sternman 1994

[543] other group process techniques include Delphi (Dalkey 1969; Delbecq *et al.* 1975; Rohrbaugh 1979), Nominal Group Technique (Delbecq *et al.* 1975; Bouchard 1969; Diehl and Stroebe 1987; Rohrbaugh 1981) and Social Judgement Analysis (Rohrbaugh 1979; Rohrbaugh 1981; Reagan-Cirincione 1991); for a discussion of these techniques, see Vennix and Gubbels 1992; the group model-building process proposed by Vennix 1996 combines elements of the three techniques.

Step 1: Problem identification and overall-goal definition
↓
Step 2: Project design/stakeholder selection
↓
Step 3: Goal and Scope Definition of the LCA
↓
Step 4: Data collection and inventory analysis
↓
Step 5: Impact assessment
↓
Step 6: Interpretation
↓
Step 7: External Review
↓
Step 8: Approval of the Study
↓
Step 9: Implementation

Figure 23-1. Nine steps of an LCA as group model building process

Step 1: Problem Identification and Overall-Goal Definition

At the very beginning of an LCA, a couple of important questions need to be addressed. Staring from an initial contact with a potential client organisation the question needs to be answered whether LCA is suitable to the problem in question. The decision situation itself, the client's motivation and expectancies, the resources available as well as the possibilities of LCA as a tool have to be evaluated. One of the most important decisions is whether a descriptive or a marginal LCA is suitable for the project. If LCA is not suitable or cannot reasonably be applied, the project should better be abandoned for alternative methods.

Step 2: Project Design/Stakeholder Selection

If LCA is held suitable, a number of questions with regard to the design of the project need to be addressed. The first relates to the distinction between qualitative and quantitative approaches to LCA: which of the different qualitative and quantitative approaches to LCA will be most appropriate (or in which combination)? The second question relates to the participant selection based on the stakeholder considerations made above: how many people should be involved in the group model-building process, who, and how? This question gets important because it may significantly affect the effectiveness of the project. Moreover, the answer to the second question is important because it is one of the aspects determining the design of the group model-building process. Further determinants for the design of the group model-building process are, *e.g.*, limited time of managers or the political sensitivity of the topic.

In principle, a variety of 'group' model-building processes are possible, depending on the degree of stakeholder involvement, *e.g.*:

- Personal union of the decision-maker and the modeller using average data from software packages, *e.g.*, for brief environmental evaluation during material selection;
- Decision-maker(s) and modeller(s) in a commissioner – client relationship; internal and external production managers and technicians are involved in data collection and internal peer reviews (for non-participatory internal strategic decision-making);
- Decision-maker(s) and modeller(s) in a commissioner – client relationship with a project steering group, constituted of internal stakeholders such as representatives from product development, marketing, production, etc. (for participatory internal strategic decision-making);
- Decision-maker(s) and modeller(s) in a commissioner – client relationship with a project steering group, constituted of external stakeholders such as representatives from different materials and products in comparative LCAs;
- Decision-maker(s) and modeller(s) in a commissioner – client relationship with a project steering group and/or workshops, consisting of internal and external stakeholders directly and/or indirectly involved in the product life cycle (for marketing purposes and environmental positioning of products[544], far-reaching strategic decision-making, lobbying, participatory decision-making in projects of local or national authorities).

Table 23-1 contains a list of possible direct and indirect stakeholders of an LCA. Lundie (1998) proposes the two determinants 'motivation' and 'qualification' as selection criteria for stakeholder participation[545]. For an efficient and effective group model-building process, both high motivation and good qualification of the participants are essential.

Stakeholders and other – also non-academic – experts are valuable sources of structural and quantitative knowledge. In group model building, it is far too short-sighted to claim that structural and quantitative knowledge is restricted to scientists, whereas the societal stakeholders (including scientists) only bring in their values[546].

A primary aspect of group model-building processes is the combination of different types of knowledge (see Chap. 23.2) and different knowledge

[544] *e.g.*, Plinke *et al.* 1999
[545] Lundie 1998:130ff
[546] as assumed, *e.g.* by Rotmans and von Asselt 1996:332

Table 23-2. Stakeholders of an LCA, their goals, focuses, available knowledge and constraints

	Decision-maker(s)	Modeller(s)	Other Stakeholders
Goals	Problem-specific goals from the decision-makers' perspective Effective and efficient conduct of the LCA Sound background for decision-making	Effective and efficient conduct of the LCA Descriptiveness of the models Methodological soundness	Problem-specific goals from the stakeholders' perspective Effective and efficient conduct of the LCA
Focus/relevant elements/values	Concept of sustainability (related to environmental impacts, sustainable management of material and energy flows) Concept of personal responsibility Attitude towards risk Further economic, social (and environmental) values related to the decision situation	Concept of sustainability (related to environmental impacts, sustainable management of material and energy flows) Concept of the decision-makers' responsibility Attitude towards risk Internal logical consistency of the LCA	Concept of sustainability (related to environmental impacts, sustainable management of material and energy flows) Concept of the decision-makers' responsibility Attitude towards risk
Knowledge available	Economic and social constraints of the decision situation Structural and quantitative knowledge on specific internal and external technical processes ("know-how"), causalities	Modelling in LCA (cut-off, attribution and allocation rules and their consequences, among others) Procedural knowledge and knowledge elicitation in the context of LCA Environmental knowledge General knowledge on technical processes, causalities	Further structural and quantitative knowledge on specific internal and/or external technical processes ("know-how"), causalities Material and market characteristics LCA-methodology Environmental knowledge
Constraints	Limited resources Multi-criteria and multi-objective decision situation	Limited resources	Limited resources Expectations of the represented group(s)

carriers (*i.e.* participants) for mutual learning as well as harmonisation of mental models and values. Table 23-2 summarises the meeting of different goals, values, knowledge and personal or institutional constraints in an exemplary way. Emphasis is put on the decision-maker(s) – modeller(s) interaction; other participants are summarised as other stakeholders although their role in the modelling process can differ considerably.

Group model building brings a variety of people together, with different knowledge, goals, and values. It can even be desirable to have differing opinions together to avoid biased problem framing. In the case of destructive discussions, a facilitator can play a key role in group model building. Primarily concerned with procedure and process, the effective facilitator is neutral with regard to the content or the substance of the discusion. He or she is authentic and integer, has an inquiring attitude and is supposed to foster reflection and learning in a team. Vennix (1996) lists four critical skills for facilitators[547]: group process structuring skills including team building skills, skills to build consensus and commitment, intervention and conflict handling skills, communication skills, concentration (or focusing) skills, and, last but not least, thorough knowledge of LCA and model building to ask the right questions and to be able to translate what participants say into methodological terms.

Step 3: Goal and Scope Definition of the LCA

The next step in the conduct of an LCA is the goal and scope definition of the LCA. ISO/EN ISO 14041, chap. 5.2 states as part of the goal dependency:

- Intended application;
- Reasons for carrying out the study;
- Intended audience.

and as part of the scope definition (ISO/EN 14041, chap 5.3):

- Function, functional unit and reference flow;
- Initial system boundaries;
- Description of data categories;
- Criteria for initial inclusion of inputs and outputs;
- Data quality requirements;
- Type of critical review.

A crucial aspect is missing in this list, at least if the view of Hofstetter (1998) and Hofstetter *et al.* (2000) is shared that

> "It is the subjective elements that determine the view of the eco- and of the technosphere and shape the models representing them" (Hofstetter *et al.* 2000: 162).

The missing aspect is the clarification of the mental models and values on which the LCA will be based.

[547] Vennix 1996:151ff

As shown in this book, values are not restricted to the assessment step and dealt with by the choice of one or several impact assessment methods. Equally important is the discussion of:

- Sustainability goals with regard to material and energy uses within the technosphere, *i.e.* the definition of management rules for a sustainable use of the materials involved[548];
- Temporal preferences for the modelling of future processes[549];
- Valuation criteria for coproducts;
- Level of relevance for the cut-off of material and energy flows;
- Valuation criteria for recycling material with regard to its material value.

Discussing these points will automatically reveal the participants' mental models of relevant aspects related to the socio-economic system such as:

- The life cycle of the product itself;
- The limits of the technosphere and ecosphere;
- Knowledge on processes and technologies;
- Material and market characteristics of the most relevant materials involved;
- The range of the decision-makers' responsibility in view of the management rules for sustainable development;
- The role of environmental information for different actors for the achievement of sustainable development.

The outcome of this discussion will be decisive for methodological questions, such as:

- Criteria for setting system boundaries;
- Definition of modules, sub-modules and unit processes;
- Choice of allocation procedures for multi-output processes;
- Choice of allocation factors as a consequence of the mental model on the organisational principle of the socio-economic system;
- Allocation procedure(s) related to recycling (for different materials);
- Choice of an impact assessment method;
- Detection of relevant environmental issues that are not covered by the assessment method(s) chosen, such as workplace security, biodiversity, mechanical impacts, risk and toxicity considerations, etc (see also Chap. 3.3.4);
- Consideration of environmental opportunity 'costs'.

This third step of the group model-building process is challenging for all participants because structural, quantitative and operational knowledge about the technosphere and structural and operational methodological knowledge

[548] for reasons, see Chap. 3.3.4 and Chap. 3.3.5
[549] see Chap. 5.8.5

are shared and addressed, heading towards a consensus; further, values have to be articulated and discussed toward a consensus. Despite these difficulties, it is this step which is essential to achieve descriptiveness of LCA. For this reason, the involvement of all relevant stakeholders is required in step 3 according to the over-all goal of the project.

Various methodologies for stakeholder involvement are applicable, depending on the number and availability of participants, such as the Delphi-technique or a workshop. A workshop seems the most adequate method because interpersonal contact fosters mutual learning, motivation and commitment to the project.

For participatory processes, a variety of visualisation and diagramming techniques have proven to be valuable[550]. First, they can help team members to articulate their opinions accurately. Second, they help to bring tacit (causal) assumptions to the surface, which prevents miscommunication. Third, they clarify the different viewpoints; and fourth, they support the development of a common problem definition and consensus building[551].

LCA specific visualisation and diagramming methods can be process trees and/or relevance matrices as qualitative approaches to life cycle thinking[552], from which quantitative models can be set up (see Chap. 3.1.1).

Step 4: Data Collection and Inventory Analysis

In most cases, inventory data used in LCA is partly specific data gathered and/or processed for the purpose of the LCA; partly, public LCI data will be used for background data to lower cost and resource requirements of the project. Stakeholders, *e.g.* plant managers or technical staff, are involved in data collection procedures by questionnaires, telephone interviews, or plant visits.

During this step but also during steps 4 to 6, one or more feedback loops as internal critical review and clarification of further model-related problems seem advisable. Further workshops, workbooks or simple written feedback on process data sheets and calculations can be included. On the one hand, feedback is important for the validation of the data used. On the other hand, the consequence of the allocation procedures can be demonstrated and compared to the management rules for a sustainable use of the materials involved.

[550] see, *e.g.*, Cissé 1998; Rotmans and von Asselt 1996; Anonymous 1994e
[551] after Ulli-Beer 2002
[552] see Lewis 1996; Wenzel 1998:281ff

Step 5: Impact Assessment

In most cases, standard impact assessment methods like Eco-indicator 99[553], the Ecoscarcity-method[554] or the impact-oriented classification[555] will be applied according to the goal and scope definition (Step 3).

Step 6: Interpretation

Interpretation of the LCA with regard to goal and scope of the LCA will be made by the modeller(s) and the decision-maker(s). Other stakeholders can be invited to comment on the interpretation and conclusions drawn.

Step 7: External Review

A critical review shall be conducted as presented in ISO/EN 14040, chap. 7.3.3 if the use of the study is intended to support a comparative assertion that is disclosed to the public. External peer review should be conducted by stakeholders that did not participate in the group model building to avoid the preference of individual participants with their individual mental models and values.

Step 8: Approval of the Study

Based on the external review, final modifications and adaptations can be necessary. The final version should be approved by all participants of the group model-building process, *e.g.*, by signing the preface of the study.

Step 9: Implementation

If consensus on the LCA has been achieved, the recommended actions have to be considered in the light of further constraints of the decision situation, *i.e.* economic and social aspects. From a purely environmental perspective, the recommended actions can be considered sound, taking into account the limited descriptive power of the assessment method(s) applied as well as the modelling characteristics. Under these considerations, the recommended actions can be implemented.

[553] Goedkoop and Spriensma 2000
[554] Brand *et al.* 1998
[555] Heijungs *et al.* 1992

23.4 Pitfalls of participatory processes

Participatory processes have a lot of potential but also some pitfalls. As a consequence of the sensitive and often fragile nature of relations between different stakeholders, phenomena related to groupthink can occur, such as illusion of invulnerability, stereotyping of people outside the group, putting pressures on people inside the group who disagree with the majority opinion to remain silent, self-censorship and polarisation.

Further interpersonal issues in groups are trust, communication, and co-operation, as well as related performance and learning effects[556]. Psychological research highlights the fragile nature of trust and communication in creative problem solving and modelling for this purpose[557]. It is quite easy to fall into the closed and sterile mode of groupthink rather than to promote openness, dialectical testing of arguments, counterarguments, and learning. These considerations, once again, emphasise the skills of the process facilitator.

23.5 Evaluation of LCA as group model-building process

The group model-building process for the conduct of an LCA as outlined above has been derived and adapted from literature on group and organisational decision-making as well as on group model-building in system dynamics; the process has not been applied in a practice yet. The application and evaluation of the group model-building procedure for LCA as outlined above is one of the most important questions for future research (see Chap. 24).

Table 23-3. Aspects of strategic decision-making effectiveness (Akkermans and Vennix 1997:6)

Process effectiveness	Organisational platform	Model quality	Implementation results
Focus	Awareness	Completeness	Implementation of decision
Speed	Consensus	Thoroughness	Business performance
Involvement	Commitment	Theory-basedness	Insight
Communication	Ownership	Usability	Organisational learning
Willingness to co-operate	Confidence		

[556] see Kleindorfer *et al.* 1993:216 ff

[557] see Kleindorfer *et al.* 1993:221 and literature quoted there

Akkermans and Vennix (1997) provide a consistent framework for the evaluation of group model-building processes. They consider the following aspects relevant for the evaluation of group model-building processes with regard to strategic decision-making effectiveness (Table 23-3).

These aspects have to be addressed in relation to the (after Akkermans and Vennix 1997:7):

- Problem contingencies (problem scope, problem tangibility, data availability, problem urgency, and political sensitivity);
- Organisational contingencies (top management support, hierarchical diversity, problem ownership, group size, and working relations);
- Project design elements.

For the purpose of operationalisation, indicators need to be specified for each of the rather general aspects listed in Table 23-3. This specification could start off from Akkermans (1995).

23.6 Final remarks

One of the most fundamental principles in modelling and consulting should be that the problem determines the method. Standard procedures for LCA as well as group model building might look attractive at first sight but there is a danger of applying them in an unreflective manner. The method will prevail over the problem. Applying strict procedures is probably one of the best guarantees that the decision-makers' problem will *not* really be addressed.

A method prevailing over a problem is likely to occur at two levels: at project level and at model level. At project level, LCA is not necessarily the best method to address a specific problem or decision situation.

At model level, reflected selection of the approach to LCA, *i.e.* descriptive or marginal LCA, qualitative or quantitative approaches (see Chap. 3.1.1), a careful choice of attribution and allocation rules in dependency of – among others – the material and market characteristics of the materials involved with regard of their sustainable use, and a careful choice of the evaluation methods representing the decision-maker's value-system are essential. From a procedural perspective, the design of a group model-building process should be adapted to the particular situation to avoid ineffective work and endless discussions.

I hope that this book contributes to the understanding of the importance (and the pitfalls) of these issues and provides guidance for future LCA practice on content and on procedural level.

The final remark should be left to Heiskanen (2002):

"Of course, it is impossible to manage the life-cyclewide environmental impacts of all products, and there is no one with such power to wield. The most extensive life cycle study in the world could not conclusively encompass all environmental burdens related to a product, and the broader the chains of influence are extended, the greater the indeterminancies and uncertainties grow. As a rational, total project a life cycle approach is impossible. But as a cultural project, it appears vigorous. People in their different roles (as purchasers, designers and decision-makers) can push and pull at pieces of the product chain, and their efforts gain dignity and a sense of purpose from whole-system representations" (Heiskanen 2002:434).

Chapter 24

FUTURE RESEARCH NEEDS

Further research need has been detected for the following topics.

Modelling environmental impacts of product design-related decisions: The relevance of descriptive LCA for product design-related decision-making is rather limited. Marginal considerations related to decisions would depict the environmental consequences of product-related decision-making in a much better way. This means a totally different approach to LCA. Apart from the marginal accounting principle, also a new 'functional unit' would have to be defined: the decision situation with its different decision alternatives.

There has been little research and experience is sparse related to this type of LCA. A starting point is provided by Jungbluth (2000:220) who assesses the environmental impacts of buyers' decisions for their food consumption in dependency of different life styles (among others). This kind of research would preferably be done in close collaboration with a company involved in strategic product development.

Modelling mental models and value choices in the life cycle inventory step according to the cultural types used for the Eco-indicator 99-methodology: Subjective elements are not only part of the assessment step within LCA methodology. They also guide the setting up of the life cycle model within the inventory step. For a further development within the framework of Eco-indicator 99[558], it has to be investigated if differing mental models and value choices necessary to set up the product system can be attributed to

[558] Goedkoop and Spriensma 2000; see also Hofstetter 1998; Hofstetter *et al.* 2000

different cultural types. This would allow one to set up life cycle models in consistency with the mental models and values for each of the cultural types.

The consideration of time-dependent aspects related to reuse & recycling within Eco-indicator 99-methodology could also only be discussed in an exemplary way (Chap. 5.8.5). Handling ignorance on the future expresses the decision-maker's attitude towards risk. Attitude towards risk is a criterion for the distinction of different cultural types, on which Eco-indicator 99-methodology is based. For consistent modelling with Eco-indicator 99, the integration of the decision-maker's attitude towards risk on decisions related to reuse & recycling should further be investigated (see also footnote 241).

Although these intents have to be considered with scepticism, the development of suitable schemes cannot *a priori* be excluded (see Chap. 5.8.5).

Framework for stakeholder involvement in LCA (*see Chapter 23*)**:** The involvement of the relevant stakeholders into methodological questions is crucial for the 'descriptiveness' of an LCA and thus for the acceptance of LCA results and recommendations derived from it[559]. A common sense must be achieved among the different stakeholders concerning the material and market characteristics as well as concerning the management rules for a sustainable use of the materials involved. Based on this, methodological decisions have to be made and shared by the stakeholders.

It has been realised that steering committees for comparative LCAs composed of representatives of differing material industries are not able to understand the consequences of methodological decisions for their products for various reasons.

Nevertheless, promising experiences have been made in some cases leading different, conflicting stakeholders to a common sense in methodological questions related to LCA and sustainability assessments of products[560]. Important experiences with stakeholder involvement in modelling processes for decision-making have also been made with system dynamics[561].

Such experiences have been compiled and a framework for stakeholder involvement in LCA has been elaborated in Chapter 23. This framework should be applied, evaluated and refined in different LCA-projects.

[559] there exists some knowledge on stakeholder involvement in system dynamics modelling; see, *e.g.*, Vennix 1996, Vennix *et al.* 1996, Akkermans and Vennix 1997, Vásquez *et al.* 1996.

[560] see, *e.g.*, Plinke *et al.* 1999

[561] Vennix 1996; Vennix and Gubbels 1992; Akkermans 1995; Akkermans and Vennix 1997; Ulli-Beer, S. 2002; Vázquez *et al.* 1996;

Influence of mental models and values in the peer review process of LCA: The standard ISO/EN 14040, chap. 7.2, plans a critical review by interested parties or by external experts to enhance the credibility of an LCA. Common sense among interested parties or external experts on methodological questions and common sense on mental model of material and market characteristics and on values cannot be assumed. It should be investigated, *e.g.*, by text analysis of review reports on one LCA, how mental models and values of reviewers influence the results of the peer review process.

Consequences for environmental accounting schemes of companies: Environmental accounting schemes of companies commonly can be divided into an on-site part, accounting all the direct environmental interventions occurring within the gates of a plant or company. The second part covers all the up-stream and down-stream processes of the products bought and sold with all indirect environmental interventions[562].

While the on-site part can geographically be situated and thus be demarcated, the complementary part faces similar problems of setting system boundaries and attribution like product LCA as treated in this book. The research question of this book can be reformulated for environmental accounting schemes of companies: How does the complementary part of an environmental accounting scheme have to be modelled to provide environmentally optimal decision support? This question is important as consensus exists to apply the cut-off procedure for the treatment of recycling and other end-of-life options in environmental accounting schemes[563].

This book concludes that it is highly questionable if the cut-off procedure provides rational recommendations for environmental improvement for each of the actors of the socio-economic system and if it does not lead to suboptimal results if re-evaluated from a meta-perspective.

It should be assumed instead that material-specific accounting principles for the complementary part of environmental accounting schemes lead to more consistent action recommendations for each actor within the socio-economic system if reassessed from a meta-perspective. In this sense, the adequacy of the cut-off procedure for environmental accounting schemes must be substantiated.

[562] see, *e.g.*, Braunschweig and Müller-Wenk 1993:57
[563] Schaltegger and Sturm 1992:157/187ff; Braunschweig and Müller-Wenk 1993:73

Acronyms

AA	Aluminium Association Inc.
ASTM	American Society for Testing and Materials
Bu	Buche (Beech)
CHF	Swiss Franc (1.5 CHF ~ 1 Euro; 2004)
CML	Centrum voor Milieukunde (CML), Rijksuniversiteit Leiden
d.t.	dross treatment
DIN	Deutsche Industrie Norm
EAA	European Aluminium Association
EC	eddy current
EDIB	Environmental Data Inventory of Building materials; data base at the Wood Department at EMPA
EMPA	Eidgenössische Materialprüfungs- und Forschungsanstalt
EN	Euro Norm
EPDM	ethylene-propylene elastomere
EPS	Environmental Priority Strategies; basis for an impact assessment method originating from Sweden
ESU	Energie-Stoffe-Umwelt; former research group at ETH Zurich
ewh	engine working hours
ETH	Eidgenössische Technische Hochschule
H_u	lower heating value
IL	Industrieholz lang (Swiss sorting category for industrial wood)
IS	Industrieholz schwach (Swiss sorting category for industrial wood)
ISO	International Standard Organisation
JIS	Japanese Industrial Standards
LC	life cycle

LCA	life cycle assessment
LCI	life cycle inventory
LME	London Metal Exchange
LP	linear programming
LSA	landscape architecture
mc	moisture content
MEA	material and energy aspect (of wood)
n.k	not known
no d.t.	without dross treatment
MR	management rule, referring to management rules for the environmental dimension of sustainable development (Chapter 3.3)
mth	month
odm	oven-dried matter (mc = 0%)
Op-cost (MEA)	allocation procedure considering 'environmental opportunity costs' and substitutions related to the material and energy aspect of wood
PA	polyamides
PCDD/F	polychlorinated dibenzodioxines
PE (HD)	polyethylene (high density)
PVC	polyvinyl chloride
SC-PA	strict coproduct allocation
SETAC	Society of Environmental Toxicology and Chemistry
SIA	Schweizerischer Ingenieur- und Architekten-Verein
Sub-Con	allocation procedure considering substitutions related to the material aspect of wood, *i.e.* a concrete element
Sub-En	allocation procedure considering substitutions related to the energy aspect of wood
u.b.	under bark; volume measure of harvested timber exclusive of bark (m^3)
UCPTE	Union de la Coordination de la Production et du Transport de l'Electricité
VCS	value-corrected substitution
WF	Frank Werner
WI	waste incineration

Acronyms of metals used in the text and units according to the International System of Units are not listed

Glossary

The following definitions are based on the series of standards ISO 14040, Frischknecht (1998), Guinée *et al.* (2001), Heijungs (1997), and others:

Attribution: a) Determination of material and energy flows out of the technosphere that are causally linked to the product under study, and b) determination of the environmental impacts caused by these material and energy flows within ecosphere; -> *allocation*

Allocation: Partitioning of the input and output flows of a unit process to the coproducts produced either in fixed or variable ratio; a specific -> *attribution*

Avoided-burden approach: Procedure that enlarges system boundaries of a multi-functional product system and subtracts the environmental interventions of an equi-functional single-functional product system; -> *system expansion*

Benefit: usefulness; in micro-economics, consumer's preferences are considered the expression of different ordinal personal benefits; market price is considered an expression of the marginal benefit of a product

Closed-loop recycling: Recycling where the material is used for the same application or where the material's inherent properties of the output material are not changed compared to the input material (open-loop recycling considered closed-loop); -> *open-loop recycling*

Mental model: Knowledge representation, *i.e.* a conceptualisation of aspects of our world

Coproduct: Any of two or more physical or symbolic objects that leave a unit process and whose values in monetary terms are positive. A coproduct is a commercial commodity

Datum/data: numeric value(s) of a feature or features

Descriptive LCA: LCA assessing an average product using proportional attribution and average data. Note that both, descriptive and marginal LCA have to be 'descriptive' of the problem environment; -> *marginal LCA*

Ecosphere: Natural environment and its constituent processes

Environmental intervention: Matter or energy entering or leaving the product system under study that a) has been extracted from ecosphere without previous human transformation (*e.g.*, timber, water, iron ore, coal), or b) is emitted or discarded into the environment without subsequent human transformation (*e.g.*, CO_2, noise emissions, wastes discarded; -> *environmental flow*

Elementary flow: -> *environmental intervention*

Functional unit: Quantified performance of a product system as reference in LCA

Joint coproduction process: Multifunctional process providing more than one product in a fixed ratio, *i.e.* the ration of the products cannot be influenced; -> *shared coproduction process*

Information: communication; news; logically structured data

LCA-model: the overall model set up during life cycle assessment covering the product system and the models of the environmental cause-effect chain. It describes the potential environmental implications of a functional unit

Life cycle: Consecutive and interlinked stages of a product system, from raw material acquisition or generation of natural resources, production and use to recycling or final disposal

Life cycle assessment (*LCA*): Compilation and evaluation of the inputs, outputs and the potential environmental impacts of a functional unit throughout its life cycle

Life cycle impact assessment: Phase of the life cycle impact assessment aimed at understanding and evaluating the magnitude and significance of the potential environmental impacts of a functional unit

Life cycle inventory (*LCI*): Spatially and temporally undifferentiated summary of the environmental interventions attributed to a functional unit throughout its life cycle

Life cycle inventory analysis: Phase of life cycle assessment involving compilation and quantification of environmental interventions related to the life cycle of a functional unit. The life cycle inventory results

Marginal LCA: Change-oriented, prospective LCA assessing the environmental consequences of induced changes in a system (marginal attribution), *e.g.*, by decisions; uses marginal data; -> *descriptive LCA*

Model: material or ideal (re-)production of an object by means of analogies realised by a cognitive subject

Open-loop recycling: Recycling where the material's inherent properties of the output material are changed compared to the input material; -> *closed-loop recycling*

Product system: Collection of materially and energetically connected unit processes which performs one or more defined functions; model of the life cycle of a product; its summary of the environmental interventions is called life cycle inventory (LCI)

Shared coproduction process: Multifunctional process providing more than one product in an unfixed ratio, *i.e.* the ratio of the products can be influenced; -> *joint coproduction process*

Socio-economic system: total of human activities covering production, consumption and disposal of products

System boundary: Interface between the product system and a) the rest of technosphere (other product systems), and b) ecosphere

System expansion: Procedure which enlarges system boundaries of a mono-functional product system in view of the comparability with alternative multi-functional systems or to avoid allocation of multifunctional processes; -> *avoided-burden approach*

Technosphere: -> *socio-economic system*

Unit process: Smallest portion of a product system for which data is collected when performing a life cycle assessment

Value: ethical category as socially accepted, internalised idea of the desirable from which action maxims can be derived; on an individual basis, a criterion of assessment to which preference action can be traced back; the values of a person are organised in a (subjective) value system; -> *valuesphere*

Valuesphere: The decision-maker's value system as disposition

Whole-system model: model set using system expansion (thus avoiding allocation) covering various product systems with various functions

Appendix

Appendix A1: **LME prices of the primary aluminium and alloyed aluminium contracts 1993-1999/5**

Appendix A2: **Impact assessment of the two Al windows regarding the potential contributions to the greenhouse effect**

Appendix A3: **Detailed data: expenditures for the production of beech wood, yield and market price (yield class II, intense thinning)**

Appendix A4: **Inventory data for forestry processes for the production of beech railway sleepers, allocation based on volume and revenues**

Appendix A5: **Inventory data for forestry processes for the production of railway sleepers, allocation based on mass (odm)**

Appendix A1: LME prices of the primary aluminium and alloyed aluminium contracts 1993-1999/5

Table A1-1. LME prices of the primary aluminium and alloyed aluminium contracts 1993-1999/5 (Lysheim 1999) (US $ per t)

	1993		1994		1995		1996		1997		1998		1999	
	cash	3 mth	cash	3 mth	cash	3 mth	cash	3 mth	cash	3 mth	cash	3 mth	cash	3 mth
Average (bid/ask) LME primary Al														
Jan	1207	1229	1175	1192	2061	2095	1589	1606	1576	1604	1486	1509	1218	1224
Feb	1202	1223	1270	1291	1916	1956	1592	1619	1580	1610	1466	1484	1187	1207
Mar	1151	1174	1289	1312	1805	1839	1612	1643	1632	1663	1438	1460	1182	1204
Apr	1109	1131	1279	1304	1849	1855	1587	1620	1561	1594	1418	1446	1278	1300
May	1124	1147	1323	1349	1763	1771	1589	1622	1625	1646	1365	1392	1323	1350
Jun	1165	1189	1401	1429	1780	1801	1483	1519	1568	1590	1307	1337		
Jul	1202	1226	1492	1515	1860	1876	1459	1495	1592	1611	1309	1332		
Aug	1172	1195	1455	1484	1888	1914	1463	1499	1711	1686	1311	1331		
Sep	1115	1135	1569	1593	1761	1792	1407	1443	1611	1624	1342	1370		
Oct	1087	1108	1698	1718	1674	1710	1336	1368	1608	1624	1304	1333		
Nov	1040	1061	1893	1906	1654	1690	1450	1473	1599	1626	1295	1310		
Dec	1094	1114	1878	1907	1657	1688	1500	1530	1531	1555	1249	1259		
Q1	1187	1209	1245	1265	1927	1963	1598	1623	1596	1626	1463	1484	1196	1212
Q2	1133	1156	1334	1361	1797	1809	1553	1587	1585	1610	1363	1392	1301	1325
Q3	1163	1185	1505	1531	1836	1861	1443	1479	1638	1640	1321	1344		
Q4	1074	1094	1823	1844	1662	1696	1429	1457	1579	1602	1283	1301		
H1	1160	1182	1290	1313	1862	1886	1575	1605	1590	1618	1413	1438	1238	1257
H2	1118	1140	1664	1687	1749	1778	1436	1468	1609	1621	1302	1323		
Year	1139	1161	1477	1500	1806	1832	1506	1536	1600	1619	1358	1380	1238	1257
Average (bid/ask) LME secondary Al														
Jan	1108	1128	1016	1040	1964	2006	1395	1431	1491	1500	1330	1353	1024	1052
Feb	1039	1059	1127	1147	1880	1919	1357	1392	1497	1520	1291	1317	1023	1047
Mar	1013	1036	1241	1252	1813	1823	1364	1403	1523	1547	1271	1298	1059	1081
Apr	980	1003	1314	1315	1761	1778	1346	1384	1454	1483	1284	1312	1161	1181
May	988	1009	1331	1337	1655	1666	1327	1365	1482	1505	1264	1277	1232	1249
Jun	1019	1042	1413	1421	1626	1651	1254	1290	1447	1471	1224	1221		
Jul	1049	1073	1496	1514	1666	1699	1244	1279	1425	1454	1242	1199		
Aug	1016	1039	1526	1536	1693	1728	1258	1293	1476	1503	1148	1171		
Sep	984	1005	1600	1617	1564	1604	1223	1253	1427	1455	1152	1181		
Oct	964	988	1703	1730	1435	1476	1211	1234	1443	1464	1112	1146		
Nov	933	956	1842	1877	1393	1432	1294	1323	1470	1479	1083	1115		
Dec	960	984	1826	1865	1424	1463	1347	1372	1396	1417	1045	1076		
Q1	1053	1074	1128	1146	1886	1916	1372	1409	1504	1522	1297	1323	1035	1060
Q2	996	1018	1353	1358	1681	1698	1309	1346	1461	1486	1257	1270	1197	1215
Q3	1016	1039	1541	1556	1641	1677	1242	1275	1443	1471	1181	1184		
Q4	952	976	1790	1824	1417	1457	1284	1310	1436	1453	1080	1112		
H1	1025	1046	1240	1252	1783	1807	1341	1378	1482	1504	1277	1296	1100	1122
H2	984	1008	1666	1690	1529	1567	1263	1292	1440	1462	1130	1148		
Year	1004	1027	1453	1471	1656	1687	1302	1335	1461	1483	1204	1222	1100	1122

Q1: first three month; H1: first half year; Q2: second three month ; H2: second half year; Q3: third three month; Q4: fourth three month

Appendix A2: Impact assessment of the two Al windows regarding the potential contributions to the greenhouse effect

Greenhouse gas potential of the two Al window variants, weighted after Heijungs *et al.* (1992) and Houghton *et al.* (1996) (for details, see Chap 12)

Figures A2-1 and A2-2 illustrate the results; the corresponding data is presented in Table A2-1.

Table A2-1. Results of the impact assessment of the aluminium case study (greenhouse gas potential) (kg CO_2-equivalent)

Allocation Procedure	cut-off	cut-off	closed-loop	closed-loop	VCS incl.	VCS
Including dross treatment/no	incl. d.t.	no d.t.	incl. d.t.	no d.t.	d.t.	no d.t.
Al-window						
Value-corrected substitution[564]					-417.9	-393.4
Frame material	343.9	344.6	137.5	138.1	490.8	490.8
Surface treatment	15.3	15.3	15.3	15.3	15.3	15.3
Window production	2.4	2.4	2.4	2.4	2.4	2.4
Fittings Al	25.9	25.9	8.0	8.1	37.6	37.6
Sealing	7.1	7.1	7.1	7.1	7.1	7.1
Thermal insulation glazing/LCA	123.7	123.7	123.7	123.7	123.7	123.7
Use phase	238.0	238.0	238.0	238.0	238.0	238.0
Transport window	6.3	6.3	9.5	9.5	9.5	9.2
Recycling process scrap	1)	1)	1)	1)	0.9	0.9
Recycling old scrap					33.4	31.9
Total	762.5	763.2	541.6	542.1	540.8	563.6
Al(Zn)-window						
Value-corrected substitution[564]					-351.0	-330.6
Frame material	343.9	344.6	137.5	138.1	490.8	490.8
Surface treatment	15.3	15.3	15.3	15.3	15.3	15.3
Window production	2.4	2.4	2.4	2.4	2.4	2.4
Fittings Al(Zn)	16.3	16.3	13.9	13.9	18.3	18.3
Sealing	7.1	7.1	7.1	7.1	7.1	7.1
Thermal insulation glazing/LCA	123.7	123.7	123.7	123.7	123.7	123.7
Use phase	238.0	238.0	238.0	238.0	238.0	238.0
Transport Window	6.5	6.5	10.0	10.0	10.0	9.6
Recycling process scrap	1)	1)	1)	1)	0.9	0.9
Recycling old scrap					34.1	31.2
Total	753.1	753.8	547.9	548.5	589.5	606.7

1) considered closed-loop; added to the recycled metal content for the frame material

564 'value-corrected substitution' includes not only material recovered from 'frame materials' but also from 'fittings'.

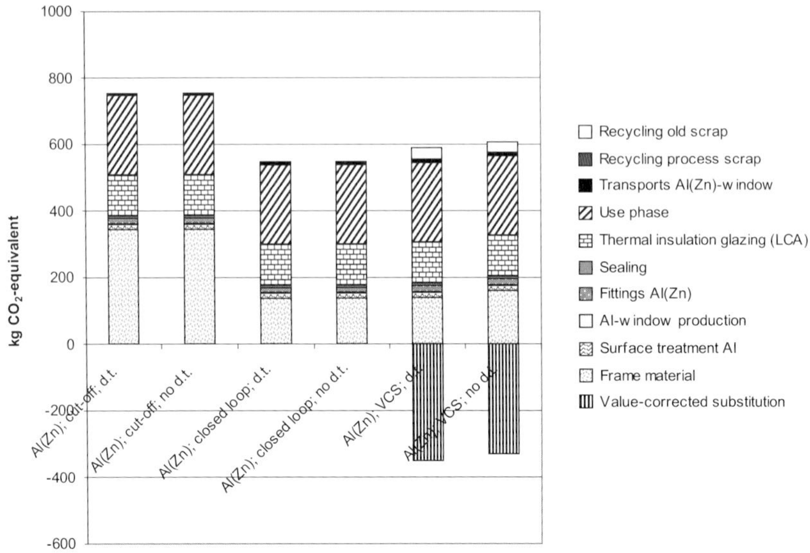

Figure A2-1. Impact assessment of the Al (Zn)-window, greenhouse gas potential
For graphical reasons, the credit given for the 'value-corrected substitution' (the negative part in Figure A2-1) is subtracted from the 'frame material'. The two absolute amounts add up to the total impacts of the 'frame material' at the input side.

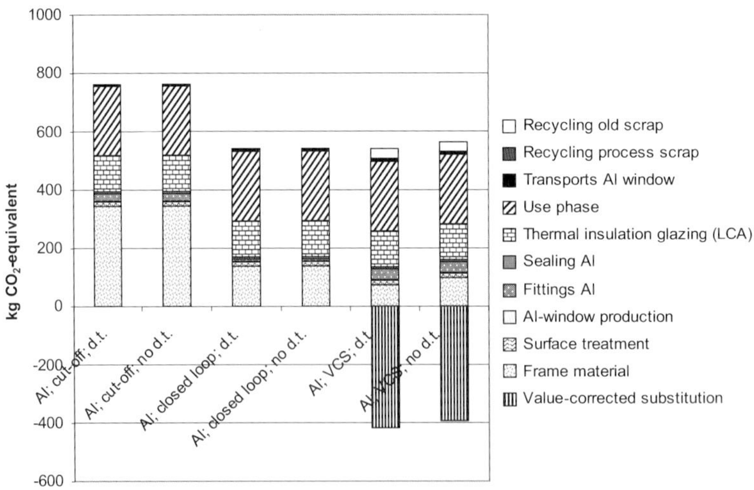

Figure A2-2. Impact assessment of the Al-window, greenhouse gas potential
For graphical reasons, the credit given for the 'value-corrected substitution' (the negative part in Figure A2-2) is subtracted from the 'frame material'. The two absolute amounts add up to the total impacts of the 'frame material' at the input side.

Appendix A3: Detailed data: expenditures for the production of beech wood, yield and market price (yield class II, intense thinning)

Table A3-1. Expenditures for the production of beech wood, yield and revenues (yield class II, intense thinning) (after Schweinle 2000; Schweinle 2001; Frühwald *et al.* 1996: 31; Anonymous 2000b; Udry 2001)

	Nursery, stand establ.	Tending	Young-growth tending	Clearing (selective)	Forestry roads	Expenditures without yield	Thinnings	Cuttings	Total
Expenditures per hectare									
Plants/ha	8000								
Capacity plants/h	500								
Expenditure ewh/ha					0.12				
Diesel fuel l/ewh	7				3		6	6	
Diesel fuel l/ha	112				0.36	112	451	434	997
Expenditure ewh/ha		14	15	15					
Expenditure ewh/m^3 (u.b.) Bu IS/IL [1]							0.300	0.300	
Expenditure ewh/m^3 (u.b.) Bu L [2]							0.125	0.125	
Gasoline for chainsaw l/ewh		2.4	2.4	2.4			2.000	2.000	
Gasol. f. chainsaw l/m^3 (u.b.) Bu IS/IL							0.600	0.600	
Gasoline for chainsaw l/m^3 (u.b.) Bu L							0.250	0.250	
Gasoline for chainsaw l/ha		33.6	36	36		106	150	145	401
Ratio oil/gasoline [3]		0.31	0.31	0.31			0.31	0.31	
Oil for chainsaw l/m^3 (u.b.) Bu IS/IL							0.19	0.19	
Oil for chainsaw l/m^3 (u.b.) Bu L							0.08	0.08	
Oil for chainsaw l/ha		10.5	11.25	11.25		33	47	45	125
Yield per hectare m^3 (u.b.)									
Industrial wood (m^3 (u.b.))							223	120	343
Roundwood (m^3 (u.b.))							66		66
Roundwood for sleepers(m^3 (u.b.))								291	291
Total m^3 (u.b.)							289	411	700
Revenues CHF per hectare									
Revenues from industrial wood (16.2 CHF/m^3 u.b.) [4]							3612	1941	5553
Revenues from roundwood (52.1 CHF/m^3 u.b.)							3428		3428
Revenues from roundw. for sleepers (52.1 CHF/m^3 u.b.)								15146	15146
Total CHF/ha							7040	17087	24127

[1] Bu IS 0.4-0.45 ewh/m^3; Bu IL 0.2-0.25 ewh/m^3 ; average assumed: 0.3 ewh/m^3

[2] Bu L 0.1-0.15 ewh/m^3; average assumed: 0.125 ewh/m^3

[3] according to Wegener *et al.* 1994:34

[4] IS + IL (cannot be distinguished)

Appendix A4: Inventory data for forestry processes for the production of beech railway sleepers, allocation based on volume and revenues

Inventory data for forestry processes for the production of beech wood railway sleepers allocated based on volume or revenues, per m^3 wet wood and per kg oven-dried wood. For input data allocated based on mass, see Appendix A5.

Table A4-1. Inventory data for forestry processes for the production of beech wood railway sleepers, per m^3 wet wood; allocated based on volume and revenues (mc = 80% for industrial wood; mc = 70% for roundwood)

per m^3 wet wood (ρ = 578 kg/m^3)	Diesel, burned in building machine MJ	Petrol, unleaded, at regional storage CH kg	Heavy fueloil, at refinery EU kg	Chainsaw, with catalytic converter l	Oil (biol. degr.) into soil gr	Diesel, in building machine MJ	Petrol, unleaded, at regional storage CH kg	Heavy fueloil, at refinery EU kg	Chainsaw, with catalytic converter l	Oil (biol. degr.) into soil gr
	Industrial wood from thinning (IW-T)					*Industrial wood from cutting (IW-C)*				
Over-all; volume										
Over-all; revenues										
Step-by-Step; volume (I)	72.6	0.664	0.260	0.886	26.0	39.3	0.264	0.103	0.352	10.3
Step-by-Step; reven.(I)	48.2	0.441	0.173	0.588	17.3	15.3	0.103	0.040	0.137	4.0
Step-by-Step; vol. (II)	64.1	0.503	0.197	0.671	19.7	45.3	0.377	0.148	0.503	14.8
Step-by-Step; reven. (II)	41.4	0.312	0.122	0.416	12.2	18.1	0.156	0.061	0.208	6.1
All-to-sleeper										
	Industrial wood (IW-T+C)					*Roundwood*				
Over-all; volume	0.156	1.26E-03	4.95E-04	1.68E-03	4.95E-02	0.156	1.26E-03	4.95E-04	1.68E-03	4.95E-02
Over-all; revenues	0.236	1.91E-03	7.47E-04	2.54E-03	7.47E-02	0.236	1.91E-03	7.47E-04	2.54E-03	7.47E-02
Step-by-Step; vol.(I)	0.116	7.77E-04	3.04E-04	1.04E-03	3.04E-02	0.116	7.77E-04	3.04E-04	1.04E-03	3.04E-02
Step-by-Step; reven.(I)	0.145	9.73E-04	3.81E-04	1.30E-03	3.81E-02	0.145	9.73E-04	3.81E-04	1.30E-03	3.81E-02
Step-by-Step; vol.(II)	0.133	1.11E-03	4.35E-04	1.48E-03	4.35E-02	0.133	1.11E-03	4.35E-04	1.48E-03	4.35E-02
Step-by-Step; reven. (II)	0.171	1.48E-03	5.78E-04	1.97E-03	5.78E-02	0.171	1.48E-03	5.78E-04	1.97E-03	5.78E-02
All-to-sleeper										
	Sleepers									
Over-all; volume	0.156	1.26E-03	4.95E-04	1.68E-03	4.95E-02					
Over-all; revenues	0.236	1.91E-03	7.47E-04	2.54E-03	7.47E-02					
Step-by-Step; volume (I)	0.116	7.77E-04	3.04E-04	1.04E-03	3.04E-02					
Step-by-Step; reven.(I)	0.145	9.73E-04	3.81E-04	1.30E-03	3.81E-02					
Step-by-Step; vol. (II)	0.133	1.11E-03	4.35E-04	1.48E-03	4.35E-02					
Step-by-Step; reven.(II)	0.171	1.48E-03	5.78E-04	1.97E-03	5.78E-02					
All-to-sleeper	0.376	3.04E-03	1.19E-03	4.05E-03	1.19E-01					

Table A4-2. Inventory data for forestry processes for the production of beech wood railway sleepers, per kg wood (odm); allocated based on volume and revenues (mc = 80% for industrial wood; mc = 70% for roundwood)

per kg wood (mc = 0%) (ρ = 578 kg/m^3)	Diesel, burned in building machine MJ	Petrol, unleaded, at regional storage CH kg	Heavy fueloil, at refinery EU kg	Chainsaw, with catalytic converter l	Oil (biol. degr.) into soil gr	Diesel, in building machine MJ	Petrol, unleaded, at regional storage CH kg	Heavy fueloil, at refinery EU kg	Chainsaw, with catalytic converter l	Oil (biol. degr.) into soil gr
	Industrial wood from thinning (IW-T)					*Industrial wood from cutting (IW-C)*				
Over-all; volume										
Over-all; revenues										
Step-by-Step; volume (I)	0.226	2.07E-03	8.10E-04	2.76E-03	8.10E-02	0.123	8.23E-04	3.22E-04	1.10E-03	3.22E-02
Step-by-Step; reven.(I)	0.150	1.37E-03	5.38E-04	1.83E-03	5.38E-02	0.048	3.20E-04	1.25E-04	4.27E-04	1.25E-02
Step-by-Step; vol. (II)	0.200	1.57E-03	6.14E-04	2.09E-03	6.14E-02	0.141	1.18E-03	4.60E-04	1.57E-03	4.60E-02
Step-by-Step; reven. (II)	0.129	9.73E-04	3.81E-04	1.30E-03	3.81E-02	0.056	4.85E-04	1.90E-04	6.47E-04	1.90E-02
All-to-sleeper										
	Industrial wood (IW-T+C)					*Roundwood*				
Over-all; volume	0.165	1.34E-03	5.24E-04	1.78E-03	5.24E-02	0.156	1.26E-03	4.95E-04	1.68E-03	4.95E-02
Over-all; revenues	0.078	6.28E-04	2.46E-04	8.37E-04	2.46E-02	0.236	1.91E-03	7.47E-04	2.54E-03	7.47E-02
Step-by-Step; vol.(I)						0.214	1.95E-03	7.65E-04	2.61E-03	7.65E-02
Step-by-Step; reven.(I)						0.457	4.18E-03	1.64E-03	5.57E-03	1.64E-01
Step-by-Step; vol.(II)						0.189	1.48E-03	5.80E-04	1.97E-03	5.80E-02
Step-by-Step; reven. (II)						0.392	2.96E-03	1.16E-03	3.94E-03	1.16E-01
All-to-sleeper										
	Sleepers									
Over-all; volume	0.156	1.26E-03	4.95E-04	1.68E-03	4.95E-02					
Over-all; revenues	0.236	1.91E-03	7.47E-04	2.54E-03	7.47E-02					
Step-by-Step; volume (I)	0.116	7.77E-04	3.04E-04	1.04E-03	3.04E-02					
Step-by-Step; reven.(I)	0.145	9.73E-04	3.81E-04	1.30E-03	3.81E-02					
Step-by-Step; vol. (II)	0.133	1.11E-03	4.35E-04	1.48E-03	4.35E-02					
Step-by-Step; reven.(II)	0.171	1.48E-03	5.78E-04	1.97E-03	5.78E-02					
All-to-sleeper	0.376	3.04E-03	1.19E-03	4.05E-03	1.19E-01					

Appendix A5: Inventory data for forestry processes for the production of railway sleepers, allocation based on mass (odm)

Table A5-1. Inputs of fossil fuels per hectare for the production of beech wood, based on mass per kg odm (I. yield class, intense thinning) (after Schweinle 2000; Schweinle 2001; Frühwald *et al.* 1996: 31; Anonymous 2000b; Udry 2001)

Beech wood	Nursery, stand establ.	Tending	Young-growth tending	Clearing (selective)	Forestry roads	Expenditures without yield	Thinnings	Cuttings	Total
Inputs per hectare									
Diesel fuel (l/ha)	112				0.36	112	451	434	997
Gasoline for chainsaw (l/ha)		33.6	36.0	36.0		106	150	145	401
Oil for chainsaw (l/ha)		10.5	11.25	11.25		33	47.0	45.0	125
Yield per hectare									
Industrial wood (kg odm)							71608	38533	110141
Roundwood (kg odm)							22440		22440
Roundwood for sleepers (kg)								98940	98940
Total kg odm							94048	137473	231521

Table A5-2. Collection of the inputs into the forestry processes of beech production and its allocation according to different mental model of forestry planning and management; allocation based on mass (odm) (mc = 80% for industrial wood; mc = 70% for roundwood); per kg wood (odm)

per kg wood (odm)	IW-T	IW-C	IW-T+C	RW	S
Diesel fuel into tractor (l/m^3 u.b.)					
Over-all; mass			4.31E-03	4.31E-03	4.31E-03
Step-by-Step; mass (I)	5.99E-03	3.16E-03		5.99E-03	3.16E-03
Step-by-Step; mass (II)	5.28E-03	3.64E-03		5.28E-03	3.64E-03
Gasoline into chainsaw (l/m^3 u.b.)					
Over-all; mass			1.73E-03	1.73E-03	1.73E-03
Step-by-Step; mass (I)	2.72E-03	1.05E-03		2.72E-03	1.05E-03
Step-by-Step; mass (II)	2.05E-03	1.51E-03		2.05E-03	1.51E-03
Oil into chainsaw (l/ m^3 u.b.)					
Over-all; mass			5.41E-04	5.41E-04	5.41E-04
Step-by-Step; mass (I)	8.51E-04	3.29E-04		8.51E-04	3.29E-04
Step-by-Step; mass (II)	6.42E-04	4.71E-04		6.42E-04	4.71E-04

IW-T: Industrial wood from thinnings; IW-C: Industrial wood from cuttings; IW-T+C: Industrial wood from thinnings and cuttings; RW: Roundwood from thinnings; S: Roundwood for railway sleepers from cuttings

Table A5-3. Collection of the inputs into the forestry processes of beech production and its allocation according to different mental model of forestry planning and management; allocation based on mass (odm); per m^3 wet wood

per kg wet wood (mass based)	IW-T	IW-C	IW-T+C	RW	S
Diesel fuel into tractor (*l/m³ u.b.*)					
Over-all; mass			1.38	1.46	1.46
Step-by-Step; mass (I)	1.92	1.01		2.04	1.07
Step-by-Step; mass (II)	1.70	1.17		1.80	1.24
Gasoline into chainsaw (*l/m³ u.b.*)					
Over-all; mass			0.56	0.59	0.59
Step-by-Step; mass (I)	0.87	0.34		0.93	0.36
Step-by-Step; mass (II)	0.66	0.48		0.70	0.51
Oil into chainsaw (*l/ m³ u.b.*)					
Over-all; mass			0.17	0.18	0.18
Step-by-Step; mass (I)	0.27	0.11		0.29	0.11
Step-by-Step; mass (II)	0.21	0.15		0.22	0.16

IW-T: Industrial wood from thinnings; IW-C: Industrial wood from cuttings; IW-T+C: Industrial wood from thinnings and cuttings; RW: Roundwood from thinnings; S: Roundwood for railway sleepers from cuttings

Tables *A5-4* and *A5-5* summarise the inventory data for forestry processes for the production of beech wood railway sleepers allocated based on mass, per m^3 wet wood and per kg oven-dried wood, as inventoried in the LCA software.

Table A5-4. Inventory data for forestry processes for the production of beech wood railway sleepers, per m^3 wet wood

per m^3 wet wood (ρ = 578 kg/m^3)	Diesel in Bauma-schine MJ	Benzin bleifrei ab Regional-lager CH kg	Heizöl Petro ab Raffiner ie Euro kg	Motorsäge Vergaser o. Kat l	Öl (biol abb.) in Boden gr	Diesel in Bauma-schine MJ	Benzin bleifrei ab Regional-lager CH kg	Heizöl Petro ab Raffinerie Euro kg	Motorsäge Vergaser o. Kat l	Öl (biol abb.) in Boden gr
	Industrial wood from thinning (IW-T)					*Industrial wood from cutting (IW-C)*				
Over-all; mass basis										
Step-by-Step; mass (I)	71.6	0.655	0.257	0.874	25.7	37.7	0.253	0.099	0.338	9.9
Step-by-Step; mass (II)	63.2	0.495	0.194	0.660	19.4	43.5	0.363	0.142	0.484	14.2
	Industrial wood (IW-T+C)					*Roundwood*				
Over-all; mass basis						54.5	0.441	0.173	0.588	17.3
Step-by-Step; mass (I)						75.9	0.694	0.272	0.925	27.2
Step-by-Step; mass (II)						66.9	0.524	0.205	0.699	20.5
	Sleepers									
Over-all; mass basis	54.5	0.441	0.173	0.588	17.3					
Step-by-Step; mass (I)	40.0	0.268	0.105	0.358	10.5					
Step-by-Step; mass (II)	46.1	0.385	0.151	174.3	15.1					

Table A5-5. Inventory data for forestry processes for the production of beech wood railway sleepers, per kg wood (odm); allocated based on mass odm (mc = 80% for industrial wood; mc = 70% for roundwood)

per kg wood (mc = 0%) (ρ = 578 kg/m^3)	Diesel in Baumaschine MJ	Benzin bleifrei ab Regionallager CH kg	Heizöl Petro ab Raffinerie Euro kg	Motorsäge Vergaser o. Kat l	Öl (biol abb.) in Boden gr	Diesel in Baumaschine MJ	Benzin bleifrei ab Regionallager CH kg	Heizöl Petro ab Raffinerie Euro kg	Motorsäge Vergaser o. Kat l	Öl (biol abb.) in Boden gr
	Industrial wood from thinning (IW-T)					*Industrial wood from cutting (IW-C)*				
Over-all; mass basis										
Step-by-Step; mass (I)	0.223	2.04E-03	7.99E-04	2.72E-03	7.99E-02	0.118	7.89E-04	3.09E-04	1.05E-03	3.09E-02
Step-by-Step; mass (II)	0.197	1.54E-03	6.04E-04	2.05E-03	6.04E-02	0.136	1.13E-03	4.43E-04	1.51E-03	4.43E-02
	Industrial wood (IW-T+C)					*Roundwood*				
Over-all; mass basis	0.160	1.30E-03	5.08E-04	1.73E-03	5.08E-02	0.160	1.30E-03	5.08E-04	1.73E-03	5.08E-02
Step-by-Step; mass (I)						0.223	2.04E-03	7.99E-04	2.72E-03	7.99E-02
Step-by-Step; mass (II)						0.197	1.54E-03	6.04E-04	2.05E-03	6.04E-02
	Sleepers									
Over-all; mass basis	0.160	1.30E-03	5.08E-04	1.73E-03	5.08E-02					
Step-by-Step; mass (I)	0.118	7.89E-04	3.09E-04	1.05E-03	3.09E-02					
Step-by-Step; mass (II)	0.136	1.13E-03	4.43E-04	5.13E-01	4.43E-02					

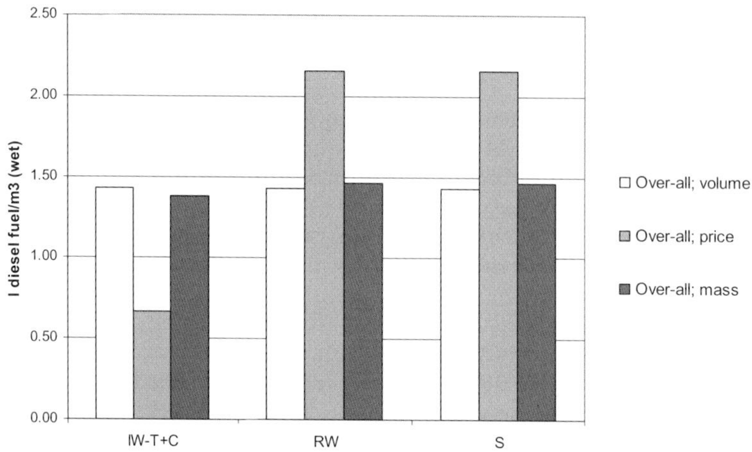

Figure A5-1. Influence of different allocation factors (volume, revenues, mass mc = 0) on the allocation of diesel fuel to industrial wood (IW-T+C) and to two fraction of equivalent roundwood (RW and S)

Figure A5.1 illustrates the consequences of the choice of an allocation factor on the attribution of inputs to forestry, shown on the over-all procedure considered to be the most suitable for forestry processes. Possible allocation factors are volume (wet), revenues, and mass (mc = 0%). Data is taken from Table 18-3 and Tables A5-2 and A5-3; results are representative for all inputs to forestry processes (diesel fuel, gasoline, engine oil).

For beech with similar moisture content of industrial wood (mc = 80%) and roundwood (mc = 70%), the difference between volume- and mass-based allocation is not very significant. However, this is not true for other, mainly softwood species such as spruce and pine with significantly higher moisture content of industrial wood (mc = 140%) than roundwood (mc = 70%)[565].

[565] data on moisture content taken from Frühwald *et al.* 1996:31

References

References

Achleitner, P. (1985): *Sozio-politische Strategien multinationaler Unternehmungen.* Diss., Hochschule St. Gallen, Paul Haupt Verl., Bern.

Ahbe, S., A. Braunschweig and R. Müller-Wenk (1990): *Methodik für Ökobilanzen auf der Basis ökologischer Optimierung.* Schriftenreihe Umwelt SRU 133, Bundesamt für Umwelt, Wald und Landschaft (BUWAL), Bern.

Akkermans, H.A. (1995): *Modelling with Managers; Participative Business Modelling for Effective Strategic Decision-Making.* Diss., University of Technology, Einhoven.

Akkermans, H.A. and J.A.M. Vennix (1997): Client's Opinions on Group Model-Building: an Exploratory Study. In: *System Dynamics Review*, ***13***(1):3-31.

Anderson, J.R. (1985): *Cognitive Psychology and Its Implications.* 2nd ed., W.H. Freeman and Company, New York.

Anonymous (1977): *Unified Numbering System for Metals and Alloys and Cross Index of Chemically Similar Specifications.* 2nd ed., Society of Automotive Engineers, Warrendale/PE.

Anonymous (1991): VDI-2243: *Konstruieren recyclinggerechter technischer Produkte.* VDI-Verl., Düsseldorf.

Anonymous (1992): *Abfallkonzept für die Schweiz; Ziele, Massnahmen, Wirkung.* Schriftenreihe Umwelt SRU 173, Bundesamt für Umwelt, Wald und Landschaft (BUWAL), Bern.

Anonymous (1993): *The LCA Sourcebook; A European Business Guide to Life-Cycle Assessment.* SustainAbility, London.

Anonymous (1994a): *Alterungsverhalten von Bauteilen und Unterhaltskosten; Grundlagendaten für den Unterhalt und die Erneuerung von Wohnbauten.* Impulsprogramm Bau (IP Bau), Bundesamt für Konjunkturfragen (BKF), Bern.

Anonymous (1994b): *Anleitung zur Erstellung einer Produkt-Ökobilanz; Theoretischer Teil.* Frauenhofer-Institut für Lebensmitteltechnologie und Verpackung, München.

Anonymous (1994c): Market Outlook for Major Energy Products, Metals, and Minerals. The World Bank, Washington/D.C.

Anonymous (1994d): *Umweltgutachten 1994.* Metzler-Poeschel, Stuttgart.

Anonymous (ed.) (1994e): *Annotated Bibliography on Gender, Rapid Rural Appraisal and Participatory Rural Appraisal.* Bibliography No 1, Institute for Development Studies, Brighton.

Anonymous (1995a): *DGfH-Merkblatt: Stoffliche Verwertung von Holz und Holzwerkstoffen.* Deutsche Gesellschaft für Holzforschung e.V. (DGfH), München.

Anonymous (1995b): *Nutzungszeiten von Gebäuden und Bauteilen.* Amt für Bundesbauten (AfB), Bern.

Anonymous (1996a): Databases for Materials Selection. In: *Advanced Materials & Processes,* **149**(6):35-37.

Anonymous (1996b): *DGfH-Merkblatt: Energetische Verwertung von Altholz.* Deutsche Gesellschaft für Holzforschung (DGfH), München.

Anonymous (1996c): *Ecological Profile Report for the European Aluminium Industry.* European Aluminium Association (EAA), Brussels.

Anonymous (1996d): Gebrauchtholzverwertung in der Holzwerkstoffindustrie; Qualitätsbestimmungen für die Annahme zur stofflichen Verwertung. In: *Holz-Zentralblatt,* **122**(75):1213-14.

Anonymous (1996e): *Life Cycle Assessment: What It Is and How to Do It.* Cleaner Production Series, United Nations Environment Programme; Industry and Environment (UNEP), Paris.

Anonymous (1996f): *Nachhaltige Entwicklung in der Schweiz.* Interdepartementaler Ausschuss Rio (IDARio), Bundesamt für Umwelt, Wald und Landschaft (BUWAL), Bern.

Anonymous (1996g): Supply of Recovered Wood and Paper is an Impetus for Recycling. In: *Industrial Uses,* **6**(96/9):30-31.

Anonymous (1997a): *Aluminium und Umwelt im Fenster- und Fassadenbau; Werkstoffkreislauf A/U/F.* Aluminium und Umwelt im Fenster- und Fassadenbau (A/U/F), Frankfurt a.M.

Anonymous (1997b): *Kyoto Protocol to the United Nations Framework Convention on Climate Change.* United Nations Framework Convention on Climate Change (UNFCCC),
http://www.unfccc.de/resource/docs/convkp/kpeng.html.

Anonymous (1997c): Recycling von Aluminium; Der Kreis ist geschlossen. In: *Aluminium Kurier,* **1997**(3):8-9.

Anonymous (1998a): *Entsorgung von Abfällen in Zementwerken; Richtlinie.* Bundesamt für Umwelt, Wald und Landschaft (BUWAL), Bern,
http://www.buwal.ch/abfall/docu/pdf/d_zementwerkrichtlinie.pdf.

Anonymous (1998b): *Export und Import von Altholz; Empfehlungen und Hinweise für den Exporteur und die Zollorgane, Stand 16.08.98.* Bundesamt für Umwelt, Wald und Landschaft (BUWAL), Bern.

Anonymous (1998c): *General Declaration and Resolutions Adopted.* Ministry of Agriculture, Rural Development and Fishery of Portugal, Lisbon.

Anonymous (1998d): *Schweizerische Vereinigung für Holzenergie; Schwerpunkt Altholz.* Schweizerische Vereinigung für Holzenergie, Maschwanden.

Anonymous (1999a): *Gewinnung von Substraten für Gartenbau und Landwirtschaft aus Spanplatten- und MDF-Resten.* WKI-Kurzbericht Nr. 20/1999, Braunschweig.

Anonymous (1999b): Qualitätsanforderungen an den Rohstoff Holz; Wald, Rest- und Gebrauchtholz für den stofflichen Einsatz in Holzwerkstoffen. In: *Holz-Zentralblatt,* **125**(7):70-71.

Anonymous (1999c): *State of European Forests and Forestry.* Timber Section, UN/ECE Trade Division, Geneva,
http://www.unece.org/trade/timber/docs/sp/eur-for/sp16.htm#Overview

Anonymous (1999d): *Statistik der erneuerbaren Energieträger.* Bundesamt für Energiewirtschaft (BEW), Bern.

Anonymous (1999e): *Towards the Global Use of Life Cycle Assessment.* Sustainable Consumption Series, United Nations Environment Programme; Industry and Environment (UNEP), Paris.

Anonymous (2000a): Rohwarenmärkte im Lauf der Zeit; Ein Abriss von der Antike zu modernen Futures-Börsen. In: *Neue Zürcher Zeitung* (*NZZ*), **221**(1):B 9.

Anonymous (2000b): *Wald und Holz in der Schweiz; Jahrbuch 2000.* Statistik der Schweiz, Bundesamt für Umwelt, Wald und Landschaft (BUWAL), Bundesamt für Statistik (BFS), Neuenburg.

Anonymous (2000c): *Waldpolitik Bund; Schwerpunkte des Eidg. Departementes für Umwelt, Verkehr, Energie und Kommunikation.* Dokument 3/2000 zur Vernehmlassung, Eidgenössische Forstdirektion, Bern.

Anonymous (2000d): Wenn der virtuelle Hammer fällt...; Die Auktion und ihr Aufstieg im Internet. In: *Neue Zürcher Zeitung* (*NZZ*), **221**(175):23.

Anonymous (2001a): Die Abfallberge türmen sich unvermindert; Keine Entkoppelung von Wirtschaftsentwicklung und Abfallaufkommen. In: *Neue Zürcher Zeitung*, **265**(100):15.

Anonymous (2001b): Schweiz bleibt Weltmeister im Aludosensammeln. In: *Neue Zürcher Zeitung*, **265**(135):64.

Ashby, M.F. (1992): *Materials Selection in Mechanical Design.* Pergamon Press, Oxford, New York, Seoul, Tokyo.

Ashby, W.R. (1971): *An Introduction to Cybernetics.* 3rd ed., Chapman & Hall Ltd. and University Paperbacks, London.

Askeland, D.R. (1996): *Materialwissenschaften; Grundlagen, Übungen, Lösungen.* Spektrum Akad. Verl., Heidelberg, Berlin, Oxford.

Ayres, R.U. (1999): The Second Law, the Fourth Law, Recycling and Limits to Growth. In: *Ecological Economics*, **29**:473-483.

Ayres, R.U. and U.E. Simonis (1994): *Industrial Metabolism; Restructuring for Sustainable Development.* United Nations University Press, Tokyo, New York, Paris.

Azapagic, A. and R. Clift (1994): Allocation of Environmental Burdens by Whole-System Modelling; The Use of Linear Programming. In: G. Huppes and F. Schneider (eds.): *Proceedings of the European Workshop Allocation in LCA.* Centrum voor Milieukunde (CML), Leiden, Society of Environmental Toxicology and Chemistry (SETAC), Brussels, pp. 54-60.

Azapagic, A. and R. Clift (1999a): Allocation of Environmental Burdens in Coproduct Systems: Product-related Burdens (Part 1). In: *International Journal for Life Cycle Assessment*, **4**(6):357-369.

Azapagic, A. and R. Clift (1999b): Allocation of Environmental Burdens in Multiple-Function Systems. In: *Journal of Cleaner Production*, **7**:101-109.

Banks, F.E. (1979): *Bauxite and Aluminium; An Introduction to the Economics of Nonfuel Minerals.* Lexington Books, D.C. Heath and Company, Lexington, Toronto.

Banks, F.E. (1980): Prices and Price Formation in the Market for Non-Fuel-Minerals. In: H. Siebert (ed.): *Erschöpfbare Ressourcen; Verhandlungen auf der Arbeitstagung des Vereins für Socialpolitik der Gesellschaft für Wirtschafts- und Sozialwissenschaften in Mannheim 1979.* Neue Folge Band 108, Verl. Duncker & Humbolt, Berlin, pp. 475-496.

Baumgarten, H.L. (1986): Techno-ökonomische Probleme der Erzeugung und Verwendung von Altpapierstoff. In: K.J. Thomé-Kozmiensky (ed.): *Recycling von Holz, Zellstoff und Papier; Beiträge des 5. Internationalen Recycling Congresses.* EF-Verl. für Energie- und Umwelttechnik, Berlin, pp. 80-93.

Beck, A. and S. Bosshart (1995): *Umweltanalyseinstrumente im Vergleich.* Abteilung Umweltnaturwissenschaften, Eidgenössische Techische Hochschule (ETH), Zürich.

Beck, A., M. Scheringer and K. Hungerbühler (2000): Fate Modelling Within LCA; The Case of Textile Chemicals. In: *International Journal for Life Cycle Assessment*, **5**(6):335-344.

Berger, P.L. and T. Luckmann (1980): *Die gesellschaftliche Konstruktion der Wirklichkeit – eine Theorie der Wissenssoziologie*. Fischer Taschenbuch Verl., Frankfurt a.M.

Bhamra, T.A., S. Evans, T.C. McAloone, M. Simon, S. Poole and A. Sweatman (1999): Integrating Environmental Decisions into the Product Development Process; Part 1: The Early Stage. In: *Proceedings of the First International Symposium on Environmentally Conscious Design and Inverse Manufacturing*, 1999, Tokyo, pp. 329-337.

Binder, M. (1998): *Produktkostenmanagement in Entwicklung und Konstruktion*. Gabler Verl., Deutscher Universitätsverl., Wiesbaden.

Birch, G. (2000): Trotz tiefer Performance nicht unbeliebt; Deutliche Zunahme der Anzahl Rohstoff-Fonds. In: *Neue Zürcher Zeitung (NZZ)*, **221**(20):B11.

Bird, A. (1990): Price Determination in the Market for Aluminium. In: L.A. Winters and D. Sapsford (eds.): *Primary Commodity Prices*. Cambridge University Press, Cambridge, pp. 134-156.

Bishop, R. (1978): Endangered Species and Uncertainty; the Economics of a Safe Minimum Standard. In: *American Journal of Agricultural Economics*, **60**:10-13.

Boguski, T.K., R. Hunt and W.E. Franklin (1994): General Mathematical Models for LCI Recycling. In: *Resources, Conservation and Recycling*, **12**:147-163.

Boguski, T.K., R.G. Hunt, J.M. Cholakis and W.E. Franklin (1996): LCA Methodology. In: M.A. Curran (ed.): *Environmental Life-Cycle Assessment*. McGraw-Hill, New York, chapter 2.

Borg, M., J. Paulsen and W. Trinius (2001): Proposal of a Method for Allocation in Building-Related Environmental LCA Based on Economic Parameters. In: *International Journal for Life Cycle Assessment*, **6**(6):219-230.

Bouchard, T. (1969): A Comparison of Two Group Brainstorming Procedures. In: *Journal of Applied Psychology*, **56**:418-421.

Boulding, K.E. (1966): *Economic Analysis*. Harper & Row, New York.

Bower, B.T. (1977): Economic Dimensions of Waste Recycling and Re-use: Some Definitions, Facts, and Issues. In: D.W. Pearce and I. Walter (eds.): *Resource Conservation: Social and Economic Dimensions of Recycling*. Longman Publ., London, pp. 1-19.

Bowyer, J. (1995): Wood and Other Raw Materials for the 21st Century. In: *Forest Products Journal*, **45**(2):17-24.

Brand, G., A. Scheidegger, O. Schwank and A. Braunschweig (1998): *Bewertung in Ökobilanzen mit der Methode der ökologischen Knappheit, Ökofaktoren 1997*. Schriftenreihe Umwelt SRU 297, Bundesamt für Umwelt, Wald und Landschaft (BUWAL), Bern.

Brassel, P. and U.-B. Brändli (1999): *Schweizerisches Landesforstinventar. Ergebnisse der Zeitaufnahme 1993-1995*. Haupt-Verl., Bern, Stuttgart, Wien.

Braungart, M.R. and W.A. McDonough (1999): Die nächste industrielle rEvolution. In: *Politische Ökologie*, **62**:18-22.

Braunschweig, A. and R. Müller-Wenk (1993): *Ökobilanzen für Unternehmungen: eine Wegleitung für die Praxis*. Verl. Paul Haupt, Bern, Stuttgart, Wien.

Brezet, H., A. Stevels and J. Rombouts (1999): LCA for EcoDesign; The Dutch Experience. In: *Proceedings of the First International Symposium on Environmentally Conscious Design and Inverse Manufacturing*, 1999, Tokyo, pp. 36-40.

Brouwer, J.M. and E.W. Lindeijer (1993): *Environmental Judgement of Batteries for Photovoltaic Systems* (in Dutch). IVAM Report 72, Institute for Environmental Studies, Vrije Universiteit, Amsterdam.

Brundtland, G.H., M. Khalid, S. Agnelli, S. Al-Athel, B. Chidzero, L.M. Fadika, V. Hauff, I. Lang, M. Shijun, M. Marino de Botero, N. Singh, P. Nogueira-Neto, S. Okita, S.S. Ramphal, W.D. Ruckelshaus, E. Sahnoun, E. Salim, B. Shaib, V. Sokolov, J. Stanovnik and M. Strong (1987): *Our Common Future*. Oxford University Press, Oxford, London.

Brunswik, E. (1950): *The conceptual framework of psychology*. University of Chicago Press, Chicago

Buhé, C., G. Achard, J.F. Le Téno and J.L. Chevalier (1997): Integration of the Recycling Processes to the Life Cycle Analysis of Construction Products. In: *Resources, Conservation and Recycling*, **20**(4):227-243.

Buser, B. (1998): Die Bauteilbörse. In: *Schweizerischer Ingenieur und Architekt*, **13**(1998):203ff.

Buxmann, K. (1997): Ökobilanzen, Methodik nach ISO 14040ff. In: J.-F. Hake and A. Möckel (eds.): *Proceedings of the Workshop Aluminium in Ökobilanzen – Methodik und Anwendungen*, Bad Honnef. Interner Bericht FZJ-STE-IB11/97, Forschungszentrum Jülich, Jülich, Vortrag A1.

Buxmann, K. (1998a): Ökobilanz und Recycling von Aluminium-Press- und -Walzprodukten im Baubereich. In: *Faszinierendes Bauen mit Aluminium*, Tagungsdokumentation zum Aluminium-Symposium vom 26. November 1998 in Luzern. Aluminium-Verband der Schweiz, Schweizerischen Zentralstelle für Fenster- und Fassadenbau (SZFF), Luzern, Dietikon, pp. 26-29.

Buxmann, K. (1998b): *Personal communication*. Alusuisse Technology & Management, Chippis.

Carlsen, E.H. (1980): Aluminium Recycling Coefficients. In: *Business Economics*, **15**(1):41-4.

Caron, R.N. (1997): Effects of Composition, Processing, and Structure on Properties of Nonferrous Alloys. In: G.E. Dieter, S.R. Lampman, G.M Davidson and B.R. Sanders (eds.): *Materials Selection and Design*. ASM Handbook, Vol. 20, ASM International, Materials Park/OH, pp. 383-415.

Cerutti, H. (2000): Der wackelnde Gartentisch. In: *NZZ-Folio*, **2000**(10):17.

Chadwick, M.J. (1997): Industrial Ecocycles; Rate Adjustments and Dematerialzation. In: *Philosophical Transactions of the Royal Society of London, Series A*, **355**:1439-1447.

Chen, R.W., D. Navin-Chandra, I. Nair, F. Prinz and I.L. Wadehra (1995): ImSelection – An Approach for Material Selection that Integrates Mechanical Design and Life Cycle Environmental Burdens. In: *Proceedings of the International Symposium on Electronics and the Environment* (*ISEE*), 1995, Orlando/FL, pp. 68-75.

Chenery, H.B. and P.G. Clark (1959): *Interindustry Economics*. John Wiley & Sons, London.

Ciriacy-Wantrup, S.V. (1952): *Resource Conservation; Economics and Policies*. University of California Press, Berkeley/CA.

Cissé, G. (1998): *Partnership in Local Resource Management; The Stakeholder Approach*. Discussion Forum North-South, Basic Documents No 1, University of Bern, Centre for General Ecology (IKAÖ), Bern.

Clark, J.P., R. Roth and F.R. Field III (1997): Techno-Economic Issues in Materials Selection. In: G.E. Dieter, S.R. Lampman, G.M Davidson and B.R. Sanders (eds.): *Materials Selection and Design*. ASM Handbook, Vol. 20, ASM International, Materials Park/OH, pp. 255-265.

Clénin, P. (2001): Nachhaltigkeit bis am Schluss...; Das Holz des Schweizer Expo-2000-Pavillons wurde in fünf Länder verkauft. In: *Schweizer Holzzeitung*, **113**(15):6.

Cleveland, C.J. and D.I. Stern (1993): Productive and Exchange Scarcity; An Empirical Analysis of the U.S. Forest Industry. In: *Canadien Journal of Forestry Research* **23**:1537-1549.

Coates, J.F. (1994): *Normative Forecasting*. United Nations University, Washington/D.C.

Consoli, F., D. Allen, I. Boustead, J.A. Fava, W.E. Franklin, A.A. Jensen, N. de Oude, R. Parrish, R. Perriman, D. Postlethwaite, B. Quay, J. Seguin and B.W. Vigon (1993): *Guidelines for Life-Cycle Assessment: A Code of Practice*. 1st ed., Society of Environmental Toxicology and Chemistry (SETAC), Brussels, Washington/D.C.

Constanza, R. (1981): Embodied Energy, Energy Analysis and Economics. In: H.E. Daly and A.F. Umana (eds.): *Energy, Economics and the Environment: Conflicting Views of an Essential Relationship*. AAAS Selected Symposium No 64, WestviewPress/CO, pp. 119-145.

Corbat, P. (2001): *Personal communication*. Roethlisberger S.A., Glovelier.

Dalkey, N. (1969): An Experimental Study of Group Opinion: The Delphi Method. In: *Futures*, **1**:408-426.

Daly, H.E. (1974): The Economics of the Steady State. In: *American Economic Review*, **64**(2):15-21.

Danthine, J.-P. (1978): Information, Futures Prices, and Stabilizing Speculation. In: *Journal of Economic Theory*, **17**:79-98.

Dasgupta, P.S. and G.M. Heal (1979): *Economic Theory and Exhaustible Resources*. James Nisbet, Welwyn and Cambridge University Press, Cambridge/MA

Datsko, J. (1997): *Materials Selection for Design and Manufacturing; Theory and Practices*. Marcel Dekker, Inc., New York, Basel, Hong Kong.

de Bakker, F.G.A., O.M. Fisscher and A.J.P. Brack (2002): Organizing Product-oriented Environmental Management From a Firm's Perspective. In: *Journal of Cleaner Production*, **10**(2002):455-464.

de Feyter, S. (1995): Handling of the Carbon Balance of Forests in LCA. In: A. Frühwald and B. Solberg (eds.): *Life-Cycle Analysis – A Challenge for Forestry and Forest Industry*. EFI Proceedings No 8, European Forest Institute, Hamburg, pp. 33-39.

de Fries, H. (1986): Der internationale Altpapierhandel. In: K.J. Thomé-Kozmiensky (ed.): *Recycling von Holz, Zellstoff und Papier*; Beiträge des 5. Internationalen Recycling Congresses. EF-Verl. für Energie- und Umwelttechnik, Berlin, pp. 46-57.

Delbecq, A.L., A.H. Van de Ven and D.H. Gustafson (1975): *Group Techniques for Program Planning: A Guide to Nominal Group and Delphi Processes*. Scott Foresman, Glenview/IL.

de Montmollin, A. and D. Altwegg (1999): *Nachhaltige Entwicklung in der Schweiz; Materialien für ein Indikatorensystem; Eine Pilotstudie unter Verwendung der Kommission für nachhaltige Entwicklung der UNO (CSC)*. Statistik der Schweiz, 2 Raum und Umwelt, Bundesamt für Umwelt, Wald und Landschaft (BUWAL), Bundesamt für Statistik (BfS), Neuchâtel.

Diehl, E.W. (1994): Managerial Microworlds as Learning Support Tools. In: J.D.W. Morecroft and J.D. Sterman (eds.): *Modelling for Learning Organizations*. Productivity Press, Portland/OR, pp. 327-337.

Diehl, M. and W. Stroebe (1987): Productivity Loss in Brainstorming Groups: Toward the Solution of a Riddle. In: *Journal of Personality and Social Psychology*, **53**:497-509.

Dieter, G.E. (1997a): Overview of the Materials Selection Process. In: G.E. Dieter, S.R. Lampman, G.M Davidson and B.R Sanders (eds.): *Materials Selection and Design*. ASM Handbook, Vol. 20, ASM International, Materials Park/OH, pp. 243-254.

Dieter, G.E. (1997b): Relationship Between Materials Selection and Processing. In: G.E. Dieter, S.R. Lampman, G.M Davidson and B.R. Sanders (eds.): *Materials Selection and Design*. ASM Handbook, Vol. 20, ASM International, Materials Park/OH, pp. 297-308.

Dinkelbach, W. and O. Rosenberg (1996): *Erfolgs-und umweltorientierte Produktionstheorie*. Heidelberger Lehrtexte Wirtschaftswissenschaften, 2. Aufl., Springer-Verl., Berlin, Heidelberg.

Dix, B., E. Roeffael and G. Bär (1997): Zur Mitverwendung von Altpapier in mitteldichten Faserplatten. In: A. Kharazipour and E. Roeffael (eds.): *Recyclingkonzepte in der Holzwerkstoffindustrie*. Institut für Holzbiologie und Holztechnologie, Göttingen, pp. 45-63.

Dixit, A., P. Hammond and M. Hoel (1980): On Hartwick's Rule for Regular Maximin Paths of Capital Accumulation and Resource Depletion. In: *Review of Economic Studies*, **47**:551-556.

Doka, G. (2000): *Ökoinventar der Entsorgungsprozesse von Baumaterialien; Grundlagen zur Integration der Entsorgung in Ökobilanzen von Gebäuden*. Untersuchung im Rahmen des IEA BCS Annex 31: Energy Related Environmental Impact of Buildings, Bundesamt für Energiewirtschaft (BEW) und Zentrum für Energie und Nachhaltigkeit im Bauwesen (ZEN), Bern, Dübendorf.

Dörner, D. (1976): *Problemlösen als Informationsverarbeitung*. Kohlhammer Standards Psychologie. Teilgebiet 4: Denkpsychologie, Kohlhammer Verl., Stuttgart.

Dosi, G., F. Malerba and L. Orsenigo (1994): Evolutionary Regimes and Industrial Dynamics. In: L. Magnusson (ed.): *Evolutionary and neo-Schumpeterian Approaches to Economics*. Kluwer Academic Publ., Boston, Dordrecht, London, pp. 203-229.

Duncker, K. (1974): *Zur Psychologie des produktiven Denkens*. 3rd ed., Springer Verl., Berlin.

Dupré, F. (1986): Aufbereitung von Holzabfällen aus Baurestmassen als Brennstoff für Kleinfeuerungen. In: K.J. Thomé-Kozmiensky (ed.): *Recycling von Holz, Zellstoff und Papier; Beiträge des 5. Internationalen Recycling Congresses*. EF-Verl. für Energie- und Umwelttechnik, Berlin, pp. 143-147.

Dyllick, T., F. Belz and U. Schneidewind (1997): *Ökologie und Wettbewerbsfähigkeit*. Carl Hanser Verl. und Verl. Neue Zürcher Zeitung, München, Wien und Zürich.

Ehrlenspiel, K. (1995): *Integrierte Produktentwicklung; Methoden für Prozessorganisation, Produkterstellung und Konstruktion*. Carl Hanser Verl., München, Wien.

Ekvall, T. (1994): Principles for Allocation at Multi-Output Processes and Cascade Recycling. In: G. Huppes and F. Schneider (eds.): *Proceedings of the European Workshop Allocation in LCA*. Centrum voor Milieukunde (CML), Leiden, Society of Environmental Toxicology and Chemistry (SETAC), Brussels, pp. 91-101.

Ekvall, T. (1999): *System Expansion and Allocation in Life Cycle Assessment with Implications for Wastepaper Management*. AFR Report 245, Department of Technical Environmental Planning, Chalmers University of Technology, Göteborg.

Ekvall, T. (2000): A Market-based Approach to Allocation at Open-loop Recycling. In: *Resources, Conservation and Recycling*, **29**(1-2):91-109.

Ekvall, T. and G. Finnveden (1999): Allocation in ISO 14041- A Critical Review. In: T Ekvall (ed.): *System Expansion and Allocation in Live Cycle Assessment, with Implications for Wastepaper Management*. AFR Report 245, Chalmers University of Technology, Göteborg, Part VI, pp. 1-19.

Ekvall, T. and A.M. Tillman (1997): Open-Loop Recycling: Criteria for Allocation Procedures. In: *International Journal for Life Cycle Assessment*, **2**(3):155-162.

Eldag, H. (1980): Wood and Wood Products. In: D.G. Altenpohl (ed.): *Materials in World Perspective; Assessment of Resources, Technologies and Trends for Key Materials Industries*. Springer-Verl., Berlin, Heidelberg, New York, pp. 105-117.

Elkington, J. and J. Hailes (1993): The LCA Sourcebook. A European Business Guide to Life-Cycle Assessment. *SustainAbility, London 1993*

Engelhardt, W.H. and B. Günter (1996): *Investitionsgüter-Marketing;* Kurseinheiten 1-3. FernUniv. Gesamtschule in Hagen, Hagen.

Erlandson, M. (1996): *Methodology for Environmental Assessment of Wood-Based Products*. Rapport I 9608070, Trätek, Stockholm.

Esser, P. and D.J. Robson (1999a): *Life-Sys Wood; Consistent Life Cycle Analysis of Wood Products; Contract No FAIR-CT95-072; Final Consolidated Report*. TNO Centre for Timber Research, Delft.

Esser, P. and D.J. Robson (1999b): *Life-Sys Wood; Consistent Life Cycle Analysis of Wood Products; Contract No FAIR-CT95-072; Final Consolidated Report – Appendix 1*. TNO Centre for Timber Research, Delft.

Fallscheer, F.C. (1997): *Ökoeffizienter Einsatz von Primär- und Recyclingstoffen für die Papier- und Kartonherstellung*. Diss., Eidgenössische Techische Hochschule (ETH), Zürich.

Fava, J.A. and R. Denison (eds.) (1991): *Workshop Report: Life Cycle Assessment, Inventory, Classification, Valuation, Data Bases*. SETAC Europe, Brussels.

Fava, J.A., R. Denison, B. Jones, M.A. Curran, B.W. Vigon, S. Selke and J. Barnum (eds.) (1991): *A Technical Framework for Life Cycle Assessment*. SETAC and SETAC Foundation for Environmental Education, Washington/D.C.

Fava, J.A., J. Senner and N. Kirkpatrick (1996): *Life Cycle Inventory Analysis; User's Guide; Enhanced Methods and Applications for the Products of the Forest Industry; The International Working Group*. American Forest & Paper Association, Inc., Washington/D.C.

Felton, C.C. (1996): The Recycling Potential of Preservative-Treated Wood. In: *Forest Products Journal*, **46**(7/8):37-46.

Feyerabend, P. (1983): *Wider den Methodenzwang*. 3. Aufl. 1991, Suhrkamp Verl., Frankfurt a. M.

Fischlin, A. (2000): Die Klimapolitik bringt die Wissenschaft an ihre Grenzen; die Herausforderung des Kyoto-Protokolls für die Ökologie. In: *Neue Zürcher Zeitung* (*NZZ*), **221**(262):11.

Fleischer, G. (1994): The Allocation of Open-Loop-Recycling. In: G. Huppes and F. Schneider (eds.): *Proceedings of the European Workshop Allocation in LCA*. Centrum voor Milieukunde (CML), Leiden, Society of Environmental Toxicology and Chemistry (SETAC), Brussels, pp. 61-63.

Fleischer, G. and W.-P. Schmidt (1996): Functional Unit for Systems Using Natural Raw Materials. In: *International Journal for Life Cycle Assessment*, **1**(1):23-27.

Forrester, J.W. (1971): *World Dynamics*. Wright-Allen Press Inc., Cambridge/MA.

Foerster, H. (1984): Das Konstruieren einer Wirklichkeit. In: P. Watzlawick (ed.): *Die erfundene Wirklichkeit*. Piper Verl., München, pp. 39-60.

Förster, R. and P.W. Gilgen (1997): How to Deal with Open-Loop Recycling Processes Within LCA; Example of BUWAL SRU 250. In: *Proceedings of the Recovery, Recycling, Reintegration 1997* (*R'97*), 1997, Geneva (draft).

Förster, R., U. Stahel and A. Scheidegger (1998): *Zuordnung der Ökofaktoren 97 und des Eco-Indicator 95 zu Schweizer Ökoinventaren, Standardisierte und kommentierte Listen*. Schweizerische Vereinigung für ökologisch bewusste Unternehmungsführung (öbu), Zürich.

Fraanje, P.J. (1997): Cascading of Pine Wood. In: *Resources, Conservation and Recycling*, **19**(1):21-28.

Frank, W. (1990): *Die Abfallwirtschaft als Teil der Rohstoffwirtschaft*. VDI-Verl., Düsseldorf.

Frankl, P. and F. Rubik (1998): *Application Patterns of Life Cycle Assessment in German, Italian, Swedish and Swiss Companies, Comparative Results and Conclusions*. Schriftenreihe des Instituts für Ökologie und Wirtschaft 130/98, Institut für Ökologie und Wirtschaft, Berlin/Heidelberg.

Frech, W. and K. Schemme (1993): Verschrottungsgerechtes Konstruieren. In: E. Hornbogen, R. Bode and P. Donner (eds.): *Recycling*. Springer-Verl., Berlin, pp. 74-86.

Freeman, A.M. (1993): *The Measurement of Environmental and Resource Values; Theory and Methods*. Resources for the Future, Washington/D.C.

Freeman, R.E. (1984): *Strategic Management – A Stakeholder Approach*. Pitman, Boston.

Frei, M. (1998): *Die öko-effektive Produktentwicklung; Der Beitrag des Umweltmanagements zur Entwicklung umweltgerechter Produkte*. Diss. ETH Nr. 12593, Betriebswissenschaftliches Institut (BWI), Eidgenössische Techische Hochschule (ETH), Zürich.

Frei, M. and R. Züst (1998): Die öko-effektive Produktentwicklung; Systematische Berücksichtigung ökologischer Aspekte in der Anforderungsliste. In: R. Züst and A. Schlatter (eds.): *Eco-Performance; Beiträge zum betrieblichen Umweltmanagement*. Verl. Eco-Performance, Zürich, pp. 123-133.

Frischknecht, R. (1994): Allocation – an Issue of Valuation? In: G. Huppes and F. Schneider (eds.): *Proceedings of the European Workshop Allocation in LCA*. Centrum voor Milieukunde (CML), Leiden, Society of Environmental Toxicology and Chemistry (SETAC), Brussels, pp. 122-131.

Frischknecht, R. (1998): *Life Cycle Inventory Analysis for Decision-Making; Scope-dependent Inventory System Models and Context Specific Joint Product Allocation*. Diss. ETH Nr. 12599, Eidgenössische Techische Hochschule (ETH), Zürich.

Frischknecht, R., P. Suter, U. Bollens, S. Bosshart, M. Ciot, L. Ciseri, G. Doka, R. Hischier, A. Martin, R. Dones and U. Gantner (1996): *Ökoinventare von Energiesystemen; Grundlagen für den ökologischen Vergleich von Energiesystemen und den Einbezug von Energiesystemen in Ökobilanzen für die Schweiz*. 3rd ed., Bundesamt für Energiewirtschaft (BEW), Bern.

Fritsch, B. (1980): Über die partielle Substitution von Energie, Ressourcen und Wissen. In: H. Siebert (ed.): *Erschöpfbare Ressourcen*; *Verhandlungen auf der Arbeitstagung des Vereins für Socialpolitik der Gesellschaft für Wirtschafts- und Sozialwissenschaften in Mannheim 1979*. Neue Folge Band 108, Verl. Duncker & Humbolt, Berlin, pp. 339-383.

Frosch, R.A., W.C. Clark, J. Crawford, A. Sagar, F.T. Tschang and A. Webber (1997): The Industrial Ecology of Metals: a Reconnaissance. In: *Philosophical Transactions of the Royal Society of London, Series A*, **355**:1335-1347.

Frühwald, A., C.M. Pohlmann and G. Wegener (2001): *Holz, Rohstoff der Zukunft; nachhaltig verfügbar und umweltgerecht*. Informationsdienst Holz, Deutsche Gesellschaft für Holzforschung (DGfH), München.

Frühwald, A. and G. Wegener (1996): *Querschnittsanalyse zur ökologischen Bilanzierung von Holzprodukten in Deutschland, Europa und Nordamerika*. Studie und Kurzfassung. Ordinariat für Holztechnologie der Univ. Hamburg, Hamburg.

Frühwald, A., G. Wegener, M. Scharai-Rad, B. Zimmer and J. Hasch (1996): *Grundlagen für Ökoprofile und Ökobilanzen in der Forst- und Holzwirtschaft*. Ordinariat für Holztechnologie der Univ. Hamburg, Institut für Holzforschung der Univ. München, in Zusammenarbeit mit Bundesforschungsanstalt für Forst- und Holzwirtschaft, Hamburg.

Gabathuler, H. (1997): The CML-Story; How Environmental Sciences Entered the Debate on LCA. In: *International Journal for Life Cycle Assessment*, **2**(4):187-194.

Gautschi, H.-P. (2001): *Die betriebswirtschaftliche Untersuchung in der Sägereiindustrie; Branchenspiegel; Ergebnisse des Jahres 2000*. Holzindustrie Schweiz, Affoltern a.A.

Georgescu-Roegen, N. (1971): *The Entropy Law and the Economic Process*. Harvard University Press, Cambridge, London.

Georgescu-Roegen, N. (1977): Matter Matters, too. In: N.N. Wilson (ed.): Prospects for Growth – Changing Expectations for the Future. Praeger Publ., New York, London, pp. 293-313.

Georgescu-Roegen, N. (1979): Energy Analysis and Economic Valuation. In: *Southern Economic Journal*, **45**(4):1023-1058.

Ghodsypour, S.H. and C. O'Brien (1998): A Decision Support System for Supplier Selection Using an Integrated Analytic Hierarchy Process and Linear Programming. In: *International Journal of Production Economics*, **56/57**:199-212.

Giachetti, R.E. (1998): A Decision Support System for Material and Manufacturing Process Selection. In: *Journal of Intelligent Manfacturing*, **9**:265-276.

Giarini, O. (1980): *Dialogue on Wealth and Welfare; An Alternative View of World Capital Formation, A Report to the Club of Rome*. Pergamon Press, Oxford.

Gibson-Jarvie, R. (1976): *The London Metal Exchange*. Woodhead-Faulkner, Cambridge, Metallgesellschaft, Frankfurt a.M.

Gielen, D. (1995): Wood for Energy or Materials Applications – Integrated Energy and Materials System Optimisation for CO_2 Reduction. In: A. Frühwald and B. Solberg (eds.): *Life-Cycle Analysis – A Challenge for Forestry and Forest Industry*. EFI Proceedings No 8, European Forest Institute, Hamburg, pp. 149-168.

Glaserfeld, E. (1984): Einführung in den Radikalen Konstruktivismus. In: P. Watzlawick (ed.): *Die erfundene Wirklichkeit*. Piper Verl., München, pp. 16-38.

Glaserfeld, E. (1985): Konstruktion der Wirklichkeit und des Begriffs der Objektivität. In: Carl von Siemens Stiftung (ed.): *Einführung in den Konstruktivismus*. Oldenbourg Verl., München, pp. 1-26.

Goedkoop, M. (1995): *The Eco-indicator 95; Final Report*. Vol. 9523, Netherlands Agency for Energy and the Environment (NOVEM), Ammersfoort.

Goedkoop, M. and R. Spriensma (2000): *The Eco-indicator 99; A Damage Oriented Method for Life Cycle Impact Assessment, Methodology Report*. 2nd ed., Pré Consultants B.V., Amersfoort.

Gordon, T.J. (1994): *Trend Impact Analysis*. United Nations University, Washington/D.C.

Grace, R., R.K. Turner and I. Walter (1978): Secondary Materials and International Trade. In: *Journal of Environmental Economics and Management*, **5**:172-186.

Graf, H.P. (2001): *Personal communication*. SBB AG, Abfallbewirtschaftung, Schaffhausen.

Grant, D. (1999): Recycling and Market Power; A More General Model and Re-evaluation of the Evidence. In: *International Journal of Industrial Organization*, **17**(1):59-80.

Greenley, G.E. and G.R. Foxall (1997): Multiple stakeholder orientation in UK Companies and the Implications For Company Performance. In: *Journal of Management Studies*, **34**:259-284.

Grober, U. (1999): Der Erfinder der Nachhaltigkeit. In: *Die Zeit*, **1999**(48): 98.

Guinée, J.B., Gorrée, M., Heijungs, R., Huppes, G., Kleijn, R., de Koning, A., van Oers, L., Wegener Sleesewijk, A., Suh, S., Udo de Haes, H.A., de Brujin, H., van Duin, R., Huijbregts, M.A.J., Lindeijer, E.W., Roorda, A.A.H., van der Ven, B.L. and Weidema, B.P. (2002): *Life cycle assessment; an operational guide to the ISO standards*. Eco-efficiency in industry and science: Volume 7, Kluwer Academic Publishers, Dordrecht.

Guinée, J.B. and R. Heijungs (1995): A Proposal for the Definition of Resource Equivalency Factors for Use in Product Life-Cycle Assessment. In: *Environmental Toxicology and Chemistry*, **14**:917-925.

Guinée, J.B., Heijungs, R. and Huppes, G. (2004): Economic Allocation: Examples and Derived Decision Tree. In: *International Journal for Life Cycle Assessment*, **9**(1):23-33.

Guinée, J.B., R. Heijungs, L. van Oers, D. van de Meent, T. Vermeire and M. Rikken (1996): *LCA Impact Assessment of Toxic Releases; Generic Modelling of Fate, Exposure and Effect for Ecosystems and Human Beings with Data for about 100 Chemicals*. Product Policy Report 1996/21, Ministry of Housing, Spatial Planning and Environment, The Hague (NL).

Guinée, J.B., H.A. Udo de Haes and G. Huppes (1993): Quantitative Life Cycle Assessment of Products: Goal Definition and Inventory. In: *Journal of Cleaner Production*, **1**(1):3ff.

Gungor, A. and S.M. Gupta (1999): Issues in Environmentally Conscious Manufacturing and Product Recovery: a Survey. In: *Computers & Industrial Engineering*, **36**(4):811-853.

Haag, D. and M. Kaupenjohann (2001): Parameters, Prediction, Post-normal Science and the Precautionary Principle–a Roadmap for Modelling for Decision-Making. In: *Ecological Modelling*, **144**:45-60.

Habermas, J. (1987): *Theorie des kommunikativen Handelns*; Bd. 1, Handlungsrationalität und gesellschaftliche Rationalisierung. 4th ed., Suhrkamp Verl., Frankfurt a.M.

Habersatter, K. (1991): *Ökobilanzen von Packstoffen, Stand 1990*. Schriftenreihe Umwelt SRU Nr. 132, Bundesamt für Umwelt, Wald und Landschaft (BUWAL), Bern.

Hallett, A.J.H. and P. Ramanujam (1990): The Role of Futures Markets as Stabilizers of Commodity Earnings. In: L.A. Winters and D. Sapsford (eds.): *Primary Commodity Prices*. Cambridge University Press, Cambridge, pp. 183-212.

Hannsen, O.J. (1995): Preventive Environmental Strategies for Product Systems. In: *Journal of Cleaner Production*, **3**(4):181-187.

Hannsen, O.J. (1999): Sustainable Product Systems – Experiences Based on Case Projects in Sustainable Product Development. In: *Journal of Cleaner Production*, **7**(1):27-41.

Harbeke, T. (1998): Gebrauchtholz in der Holzwerkstoffindustrie. In: *Umweltschutz in der Holzwerkstoffindustrie*; Tagungsband der Fachtagung am 24. und 25. Juni in Göttingen. Institut für Holzbiologie und Holztechnologie, Göttingen, pp. 66-83.

Hart, S.L. (1995): A Natural-resource-based View of the Firm. In: *Academy of Management Review*, **20**:986-1014.

Hartwick, J.M. (1977): Intergenerational Equity and the Investing of Rents from Exhaustible Resources. In: *American Economic Review*, **67**:972-974.

Hauschild, M. and J. Potting (2000): *Guideline on Spatial Differentiation in Life Cycle Impact Assessment – the EDIP2000 Methodology*. Guidelines from the Danish Environmental Protection Agency, Final Draft, Technical University of Denmark, Delft.

Heijungs, R. (1994): The Problem of Allocation: Some More Complications. In: G. Huppes and F. Schneider (eds.): *Proceedings of the European Workshop Allocation in LCA*. Centrum voor Milieukunde (CML), Leiden, Society of Environmental Toxicology and Chemistry (SETAC), Brussels, pp. 8-12.

Heijungs, R. (1997): *Economic Drama and the Environmental Stage; Formal Derivation of Algorithmic Tools for Environmental Analysis and Decision-Support from a Unified Epistemological Principle*. Diss, Centrum voor Milieukunde (CML), Rijksuniversiteit, Leiden.

Heijungs, R. and R. Frischknecht (1998): Special View on the Nature of the Allocation Problem. In: *International Journal for Life Cycle Assessment*, **3**(5):321-332.

Heijungs, R., J.B. Guinée, G. Huppes, R.M. Lankreijer, U.H.A. de Haes, A. Wegener Sleeswijk, A.M.M. Ansems, P.G. Eggels, R. van Duin and H.P. de Goede (1992): *Environmental Life Cycle Assessment of Products, Guide and Background*. Centre for Milieukunde, Leiden.

Heintz, B. and P.F. Baisnée (1991): System Boundaries. In: J.A. Fava and R. Denison (eds.): *Workshop Report: Life Cycle Assessment, Inventory, Classification, Valuation, Data Bases*. SETAC Europe, Brussels, pp. 35-52.

Heiskanen, E. (2002): The Institutional Logic of Life Cycle Thinking. In: *Journal of Cleaner Production*, **10**(2002):427-437.

Hekkert, M.P., L.A.J. Joosten and E. Worrell (2000a): Analysis of the Paper and Wood Flow in The Netherlands. In: *Resources, Conservation and Recycling*, **30**(1):29-48.

Hekkert, M.P., L.A.J. Joosten and E. Worrell (2000b): Reduction of CO_2 Emissions by Improved Management of Material and Product Use; The Case of Transport Packaging. In: *Resources, Conservation and Recycling*, **30**(1):1-27.

Henstock, M.E. (1980): Some Barriers of the Use of Materials Recovered from Municipal Solid Waste. In: *Resources Policy*, **9**(3):240-252.

Hermann, R., K. Burger and H.P. Smit (1990): Commodity Policy; Price Stabilization Versus Financing. In: L.A. Winters and D. Sapsford (eds.): *Primary Commodity Prices*. Cambridge University Press, Cambridge, pp. 240-302.

Hertwich, *E.G.*, J.K. Hammitt and W.S. Pease (2000): A Theoretical Foundation for Life-Cycle Assessment: Recognising the Role of Values in Environmental Decision making. In: *Journal of Industrial Ecology*, **4**(1):13-28.

Hertwich, *E.G.*, R. Heijungs and J.B. Guinée (1996): Resource Depletion in Life-Cycle Assessment; Letter to the Editor and The Authors' Reply. In: *Environmental Toxicology and Chemistry*, **15**(9):1442-1444.

Hettinga, S. (1996): Recycled Wood: the Ideal Filler for Plastic. In: R. H. Falk (ed.): *The Use of Recycled Wood and Paper in Building Applications*. The Forests Products Society, Madison/Wis., pp. 82-83.

Hicks, J.R. (1946): *Value and Capital*. 2nd ed., Oxford University Press, Oxford.

Hill, W. (1991): Basisperspektiven der Managementforschung. In: *Die Unternehmung*, **45**(1):2-15.

Hillier, W., R.J. Murphy, D.J. Dickinson and J.N.B. Bell (1997): Life Cycle Assessment (LCA) of Treated Timber Poles. In: *Proceedings of the Southeastern Pole Conference*, Forest Products Society, Madison/Wis., pp. 113-122.

Hobeika, L. (1990): On the Efficiency of the Aluminium Market. In: *Proceedings of the Metal Bulletin's 6th International Aluminium Conference*, November 11-13, 1990, Singapore.

Hofer, P. (2000a): Ansätze zu einer schweizerischen Altholzpolitik; vom Überfluss zum Mangel und zurück. In: *SAH bulletin CSRB: Altholz zwischen Verwertung und Beseitigung*, **5**: 16-23.

Hofer, P. (2000b): Ökologische Produkte mit ökologischer Rückführung? Altholz zwischen Verwertung und Beseitigung. In: *SAH bulletin CSRB: Altholz zwischen Verwertung und Beseitigung*, **5**:2-3.

Hofer, P., L. Morf, R. Taverna and K. Richter (2001): *Speicherung von CO_2 in Holzlagern im Zivilisationskreislauf um Emissionseffekte der Substitution bei gesteigerter Holzverwendung*. GEO Partner, Zürich.

Hofstetter, P. (1998): *Perspectives in Life Cycle Impact Assessment; A Structured Approach to Combine Models of the Technosphere, Ecosphere and Valuesphere*. Diss., Eidgenössische Technische Hochschule (ETH), Zürich.

Hofstetter, P., T. Mettier and O. Tietje (eds.) (1999): *Ansätze zum Vergleich von Umweltschäden; Nachbearbeitung des 9. Diskussionsforums Ökobilanzen vom 4. Dezember 1998, ETH Zürich*. Umweltnatur- und Umweltsozialwissenschaften (UNS), Eidgenössische Technische Hochschule (ETH), Zürich.

Hofstetter, P. and M. Scheringer (eds.) (1997): *Schutzgüter und ihre Abwägung aus der Sicht verschiedener Disziplinen; Vorbereitende Unterlagen zum 5. Diskussionsforum Ökobilanzen vom 17. Oktober 1997 an der ETH Zürich*. Umweltnatur- und Umweltsozialwissenschaften (UNS) und Laboratorium für Technische Chemie, Eidgenössische Technische Hochschule (ETH), Zürich.

Hofstetter, T., T. Baumgartner and R.W. Scholz (2000): Modelling the Valuesphere and the Ecosphere: Integrating the Decision Maker's Perspectives into LCA. In: *International Journal for Life Cycle Assessment*, **5**(3):161-175.

Hojman, D.E. (1980): The IBA and Cartel Problems, Prices, Policy Objectives and Elasticities. In: *Resources Policy*, **9**(4):290-302.

Hollander, A. and P. Lasserre (1988): Monopoly and the Preemption of Competitive Recycling. In: *International Journal of Industrial Organization*, **6**(4):489-497.

Holzkamp, K. (1972): *Kritische Psychologie*. Fischer Taschenbuch Verl., Frankfurt a.M.

Hopfenbeck, W. and C. Jasch (1995): *Öko-Design, Umweltorientierte Produktpolitik*. Verl. Moderne Industrie, Landsberg/Lech.

Horne-Brine, P. (1996): The Explosive Growth of Wood Recycling in the Pacific Northwest. In: R. H. Falk (ed.): *The Use of Recycled Wood and Paper in Building Applications*. The Forests Products Society, Madison/Wis., pp. 34-40.

Hotelling, H. (1931): The Economics of Exhaustible Resources. In: *Journal of Political Economy*, **39**:137-175.

Houghton, J.T., L.G. Meira Filho, B.A. Callander, N. Harris, A. Kattenberg and K. Maskell (eds.) (1996): *Climate Change 1995; The Science of Climate Change*. Cambridge University Press, published for the International Panel on Climate Change (IPCC), Cambridge.

Hull, C.J. (1997): *Options, Futures, and Other Derivates*. 3rd ed., Prentice-Hall International, London.

Huppes, G. (1992): Allocating Impacts of Multiple Economic Processes in LCA. In: J.A. Fava and R. Denison (eds.): *Workshop Report: Life Cycle Assessment, Inventory, Classification, Valuation, Data Bases*. SETAC Europe, Brussels, pp. 57-70.

Huppes, G. (1994): A General Method for Allocation in LCA. In: G. Huppes and F. Schneider (eds.): *Proceedings of the European Workshop Allocation in LCA*. Centrum voor Milieukunde (CML), Leiden, Society of Environmental Toxicology and Chemistry (SETAC), Brussels, pp. 74-90.

Huppes, G. (2000a): Economic Allocation and Value-Corrected Substitution; Comments to 'Economic Allocation in LCA' by Frank Werner and Klaus Richter. In: *International Journal for Life Cycle Assessment*, **5**(4):189.

Huppes, G. (2000b): *Personal communication*. Centre of Environmental Sciences (CML), Leiden.

Huppes, G. and F. Schneider (1994): *Proceedings of the European Workshop Allocation in LCA*. Centrum voor Milieukunde (CML), Leiden, Society of Environmental Toxicology and Chemistry (SETAC), Brussels.

Huttunen, A. and T. Pirttila (1998): Price Dynamics on the Recovered Paper Market. In: *International Journal of Production Economics*, **56/57**:261-273.

Jarnehammer, A. (2001): *Personal communication*. Swedish Institute for Wood Technology Research, Stockholm.

Johnson-Laird, P.N. (1983): *Mental Models*. Cambridge University Press, Cambridge.

Jönson, G. (1998): Paper Packaging Designed for Recycling. In: J.E.M. Klostermann and A. Tukker (eds.): *Product Innovation and Eco-efficiency, Twenty-three Industry Efforts to Reach the Factor 4*. Eco-Efficiency in Industry, Vol. 1, Kluwer Academic Publ., Dordrecht, Boston, London, pp. 273-282.

Jörgensen, A.M. (1996): *Stakeholders and the Weighting of Environmental Problems*. Desk Study, IMSA, Amsterdam.

Jungbluth, N. (2000): *Umweltfolgen des Nahrungsmittelkonsums: Beurteilung von Produktmerkmalen auf Grundlage einer modulare Ökobilanz*. Verl. Dissertation.de, Berlin.

Jungbluth, N., O. Tietje and R.W. Scholz (2000). Food purchases: Impacts from the consumers' point of view investigated with a modular LCA. *The International Journal of Life Cycle Asessment*, *5*(3):134-142.

Jungermann, H., Pfister, H.-R., & Fischer, K. (1998). *Die Psychologie der Entscheidung*. Spektrum Akademischer Verl., Heidelberg.

Jungermann, H. and P. Slovic (1993): Charakteristika individueller Risikowahrnehmung. In: Bayrische Rück (ed.): *Risiko ist ein Konstrukt*. Gesellschaft und Unsicherheit, Bd. 2, Knesebeck GmbH, München, pp. 89-107.

Jungmeier, G., A. Merl, F. McDarby, C. Gallis, C. Hohenthal, A.K. Petersen and K. Spanos (2001a): End of Use and End of Life Aspects in LCA of Wood Products – Selection of Waste Management Options and LCA Integration. In: G. Jungmeier (ed.): *Life Cycle Assessment of Forestry and Forest Products; Achievements of COST Action E9 Working Group 3 'End of Life': Recycling, Disposal and Energy Generation*. Joanneum, Institute of Energy Research, Graz, pp. 4/1-4/25.

Jungmeier, G., F. Werner, A. Jarnehammer, C. Hohenthal and K. Richter (2001b): Allocation in LCA of Forestry Products – Recommendations from COST E9. In: *International Journal for Life Cycle Assessment,* **7**(5):290-294.

Jungmeier, G., F. Werner, A. Jarnehammer, C. Hohenthal and K. Richter (2001c): Allocation in Multi Product Systems – Recommendations for LCA of Wood-based Products. In: G. Jungmeier (ed.): *Life Cycle Assessment of Forestry and Forest Products; Achievements of COST Action E9 Working Group 3 'End of Life: Recycling, Disposal and Energy Generation'*. Joanneum, Institute of Energy Research, Graz, pp. 6/1-6/24.

Kantor, B. (1979): Rational Expectations and Economic Thought. In: *American Economic Review*, **17**(12):1422-1441.

Karlsson, R. (1994): LCA as a Guide for the Improvement of Recycling. In: G. Huppes and F. Schneider (eds.): *Proceedings of the European Workshop Allocation in LCA*. Centrum voor Milieukunde (CML), Leiden, Society of Environmental Toxicology and Chemistry (SETAC), Brussels, pp. 18-28.

Karlsson, R. (1995): *Recycling in Life Cycle Assessments*. Technical Environmental Planning, Report 1995:6, Chalmers University of Technology, Gothenburg.

Keel, A. (2001): *Personal communication*. Schweizerische Vereinigung für Holzenergie (VHe), Zürich.

Keeney, R.L. (1992): *Value-focused Thinking; A Path to Creative Decisionmaking*. Harvard University Press, Cambridge (MA), London.

Keeney, R.L. and H. Raiffa (1976): *Decision with Multiple Objectives; Preferences and Value Tradeoffs*. John Wiley & Sons, New York, Santa Barbara, London, Sydney, Toronto.

Keller, C., M. Wegmüller and R. Winkler (1999): *Ecodesign in Swiss Industries – Overview of the Current Situation; An Empirical Research among Marketing and Research & Development Departments*. Verl. Eco-Performance, Zürich.

Kibat, K.-D. (1986): Interdependenzen zwischen den einzelnen Rohstoffen der Papierindustrie und zwischen dem Rohstoffmarkt und dem Markt der Papiererzeugnisse. In: K.J. Thomé-Kozmiensky (ed.): *Recycling von Holz, Zellstoff und Papier; Beiträge des 5. Internationalen Recycling Congresses*. EF-Verl. für Energie- und Umwelttechnik, Berlin, pp. 62-79.

Kim, S., T. Hwang and K.M. Lee (1997): Allocation for Cascade Recycling System. In: *International Journal for Life Cycle Assessment*, **2**:217-222.

Kimura, F. (1999): Life Cycle Design for Inverse Manufacturing. In: *Proceedings of the First International Symposium on Environmentally Conscious Design and Inverse Manufacturing*, 1999, Tokyo, pp. 995-999.

Kimura, F. and H. Suzuki (1996): Product Life Cycle Modelling for Inverse Manufacturing. In: F.-L. Klause and H. Jansen (eds.): *Life-Cycle Modelling for Innovative Products and Processes; Proceedings of the IFIP WG5.3 International Conference on Life-Cycle Modelling for Innovative Products and Processes*, Berlin, Germany. Chapman & Hall, London, pp. 80-89.

Kinkley, C.-C. (1984): Testing the Rational Expectations Hypothesis in a Secondary Materials Market. In: *Journal of Environmental Economics and Management*, **11**:282-291.

Kleindorfer, P., H.C. Kunreuther and P.J.H. Schoemaker (1993): *Decision Sciences; An Integrative Perspective*. Cambridge University Press, Cambridge/MA.

Klöpffer, W. (1996): Allocation Rule for Open-loop Recycling in Life Cycle Assessment. In: *International Journal for Life Cycle Assessment*, **1**(1):27-31.

Klöpffer, W. (1998): Subjective is not Arbitrary. In: *International Journal for Life Cycle Assessment*, **3**(2):61-61.

Kohler, M., T. Künniger, P. Schmid, E. Gujer, R. Crockett and M. Wolfensberger (2000): Inventory and Emission Factors of Creosote, Polycyclic Aromatic Hydrocarbons (PAH), and Phenols from Railroad Ties Treated with Creosote. In: *Environmental Science & Technology*, **34**:4766-4772.

Koopmans, T.C. (1951): Analysis of Production as an Efficient Combination of Activities. In: T.C. Koopmans, A. Alchian, G.B. Dantzig, N. Georgescu-Roegen, PA. Samuleson and Tucker A.W. (eds.): *Activity Analysis of Production and Allocation*. John Wiley & Sons, Chapman & Hall, New York, London, pp. 33-97.

Korhonen, J., M. Wihersaari and I. Savolainen (2001): Industrial Ecosystems in the Finnish Forest Industry: Using the Material and Energy Flow Model of a Forest Ecosystem in a Forest Industry System. In: *Ecological Economics*, **39**:145-161.

Krone, K. (2000): *Aluminium-Recycling; Vom Vorstoff bis zur fertigen Legierung*. Vereinigung Deutscher Schmelzhütten e.V., Düsseldorf.

Krooss, J., P. Stolz, U. Thurmann, W. Wosniok, R.-D. Peek and H. Giese (1998): Statistisch ermittelte Hintergrundkonzentrationen für Schwermetall- und Holzschutzmittel-Wirkstoffgehalte naturbelassener Althölzer. In: *Holz-Zentralblatt*, **124**(46):689-695.

Kühne, G. and U. Schwarz (1997): Möglichkeiten der Detoxifizierung von steinkohlenteerölimprägnierten Althölzern und deren stoffliche Nutzung. In: A. Kharazipour and E. Roeffael (eds.): *Recyclingkonzepte in der Holzwerkstoffindustrie*. Institut für Holzbiologie und Holztechnologie, Göttingen, pp. 75-89.

Künniger, T. and K. Richter (1998): *Ökologischer Vergleich von Eisenbahnschwellen in der Schweiz; Streckenschwellen aus vorgespanntem Beton, Profilstahl und teerölimprägniertem Buchenholz*. Forschungs- und Arbeitsbericht 115/38, Eidg. Materialprüfungs- und Forschungsanstalt (EMPA), Dübendorf.

Kusian, R. (2001): Schwarze Pumpe produziert Treibstoff aus Altholz. In: *Holz-Zentralblatt*, **127**(31): 455.

Lafleur, M.C.C. and P.J. Fraanje (1997): Towards Sustainable Use of the Renewable Resource Wood in the Netherlands – a Systematic Approach. In: *Resources, Conservation and Recycling*, **20**(1):19-29.

Lalive d'Epinay, A., H. Gugerli, D. Gilgen, A. Binz, G. Lehmann and R. Frischknecht (2000): *Ökologische Sanierung von Bürobauten*. Untersuchung im Rahmen des IEA BCS Annex 31: Energy Related Environmental Impact of Buildings, Bundesamt für Energiewirtschaft (BEW) und Zentrum für Energie und Nachhaltigkeit im Bauwesen (ZEN), Bern, Dübendorf.

Lane, D. (1994): Modelling as Learning: a Consultancy Methodology for Enhancing Learning in Management Teams. In: J.D.W. Morecroft and J.D. Sterman (eds.): *Modelling for Learning Organizations*. Productivity Press, Portland/OR, pp. 85-117.

Langer, K. (1986): Einsatz von Wirbelschichtfeuerungen in der Papier- und Zellstoffindustrie. In: K.J. Thomé-Kozmiensky (ed.): *Recycling von Holz, Zellstoff und Papier; Beiträge des 5. Internationalen Recycling Congresses*. EF-Verl. für Energie- und Umwelttechnik, Berlin, pp. 137-143.

Leach, M.A., A. Bauen and N.J.D. Lucas (1997): A Systems Approach to Materials Flows in Sustainable Cities; A Case Study of Paper. In: *Journal of Environmental Planning and Management*, **40**(6):705-723.

Lee, C.-H. (1998): Formulation of Resource Depletion Index. In: *Resources, Conservation and Recycling*, **24**(3-4):285-298.

Legrand, B. (1997): Prospects for the Aluminium Industry. In: *Materials Science Forum*, **242**:3-10.

Leontief, W. (1966): *Input-Output Economics*. Oxford University Press, New York.

Lewis, H. (1996): Data Quality for Life Cycle Assessment. In: *Proceedings of the Conference Life Cycle Assessment: Shaping Australia's Environmental Future*, 1996, Melbourne, http://www.cfd.rmit.edu.au/outcomes/papers/LCA-DataQ.html

Liao, T.W. (1996): A Fuzzy Multicriteria Decision-Making Method for Material Selection. In: *Journal of Manufacturing Systems*, **15**(1):1-12.

Lindeijer, E.W. (1994): Allocating Recycling for Integrated Chain Management: Taking Account of Quality Losses. In: G. Huppes and F. Schneider (eds.): *Proceedings of the European Workshop Allocation in LCA*. Centrum voor Milieukunde (CML), Leiden, Society of Environmental Toxicology and Chemistry (SETAC), Brussels, pp. 29-35.

Lindeijer, E.W. and G. Huppes (2001): Partitioning Economic In- and Outputs to Product Systems. In: J.B. Guinée, M. Gorrée, R. Heijungs, G. Huppes, R. Kleijn, A. de Koning, L. van Oers, A. Wegener Sleesewijk, S. Suh, H.A. Udo de Haes, H. de Brujin, R. van Duin, M.A.J. Huijbregts, E. Lindeijer, A.A.H. Roorda, B.L. van der Ven and B.P. Weidema (eds.): *Life Cycle Assessment; An Operational Guide to the ISO Standards; Final Report, May 2001*. Ministry of Housing, Spatial Planning and Environment (VROM) and Centrum voor Milieukunde (CML), Rijksuniversiteit, Leiden,
http://www.leidenuniv.nl/interfac/cml/lca2/index.html.

Lindfors, L.G., K. Christiansen, L. Hoffman, Y. Virtanen, V. Juntilla, O.J. Hanssen, A. Rønning, T. Ekvall and G. Finnveden (1995a): *Nordic Guidelines on Life-Cycle Assessment*. Nord 1995:502, Nordic Council of Ministers, Kopenhagen.

Lindfors, L.G., K. Christiansen, L. Hoffman, Y. Virtanen, V. Juntilla, A. Leskinen, O.J. Hanssen, A. Rønning, T. Ekvall and G. Finnveden (1995b): *LCA-Nordic Technical Reports No 1-9*. TemaNord 1995:502, Nordic Council of Ministers, Kopenhagen.

Lindquist, K.-T. (1995): The Existence of Factor Substitution in the Primary Aluminium Industry; A Multivariate Error-Correction Approach Using Norwegian Panel Data. In: *Empirical Economics*, **20**(3):361-381.

Lipsey, R.G. and P.O. Steiner (1972): *Economics*. 3rd ed., Harper & Row Publ., New York, Evanston, San Francisco, London.

Lovelock, J. (1995): *GAIA – a New Look at Life on Earth*. Oxford University Press, Oxford.

Lübkert, B. (1991): *Life-Cycle Analysis; Working Paper; IDEA, an International Database for Ecoprofile Analysis; a Tool for Decision Makers*. WP-91-30, International Institute for Applied Systems Analysis, Laxenburg.

Luhmann, N. (1994): *Die Wissenschaft der Gesellschaft*. Suhrkamp Verl., Frankfurt a.M.

Lundie, S. (1999): *Ökobilanzierung und Entscheidungstheorie: Praxisorientierte Produktbewertung auf der Basis gesellschaftlicher Werthaltungen*. Springer Verl., Berlin, Heidelberg.

Lysheim, A. (1999): *Personal communication*, Hydro Aluminium Metal Products, Oslo.

Mac, A. (2002): When Firms Make Sense of Environmental Agendas of Society. In: *Journal of Cleaner Production*, **2002**(10):259-269.

Maag, M. (2000): *Personal communication*, Maag Recycling AG, Winterthur.

Maibach, M., D. Peter and B. Seiler (1995): *Ökoinventar Transporte; Grundlagen für den ökologischen Vergleich von Transportystemen und den Einbezug von Transportystemen in*

Ökobilanzen; Technischer Schlussbericht. SPP Umwelt, Modul 5, Auftragsnummer 5001-34730, Verl. INFRAS, Zürich.

Maizels, A. (1984): A Conceptual Framework for Analysis of Primary Commodity Markets. In: *World Development*, **12**(1):25-41.

Manahan, S.E. (1999): *Industrial Ecology; Environmental Chemistry and Hazardous Waste*. Lewis Publ., Boca Raton, London, New York, Washington/D.C.

Martin, R.E. (1982): Monopoly Power and the Recycling of Raw Materials. In: *Journal of Industrial Economics*, **30**(4):405-419.

Marutzky, R. (1997): Entsorgung von Rest- und Gebrauchtholz: Eine Bestandesaufnahme der aktuellen Situation in Deutschland. In: *Holz-Zentralblatt*, **123**(146):2201/2204.

Matthews, R.W., G.-J. Nabuurs, V. Alexeyev, R.A. Birsey, A. Fischlin, J.P. Maclaren, G. Marland and D.T. Price (1996): WG3 Summary: Evaluation the Role of Forest Management and Forest Products in the Carbon Cycle. In: Apps, M.J. and Price, D.T. (eds.): *Forest Ecosystems, Forest Management and the Global Carbon Cycle*. NATO ASI Series, Vol. I 40, Springer-Verl., Berlin, Heidelberg, pp. 163-172.

Maturana, H. and F. Varela (1987): *Der Baum der Erkenntnis: die biologischen Wurzeln des menschlichen Erkennens*. Scherz Verl., Bern.

Mauch, S., T. von Stokar, D. Zürcher, J. Heldstab, C. Schreyer, B. Seiler, A. Scheidegger, V. Zanola and J. Tschirren (1995): *Quantitative Aspekte einer zukunftsfähigen Schweiz; Arbeitsbericht*. INFRAS, Zürich.

Meffert, H. and M. Kirchgeorg (1998): *Marktorientiertes Umweltmanagement: Konzeption Strategie und Implementierung mit Praxisfällen*. 3. ed., Schäffer Poeschel, Stuttgart.

Melo, M.T. (1999): Statistical Analysis of Metal Scrap Generation: the Case of Aluminium in Germany. In: *Resources, Conservation and Recycling*, **26**:91-113.

Mengeloglu, F. and D.J. Gardner (2000): Recycled CCA-treated Lumber in Flakeboards: Evaluation of Adhesives and Flakes. In: *Forest Products Journal*, **50**(2):41-45.

Mertz, T. (1999): Wege nach Ökotopie; Eine Gesprächsrunde mit Prof. Dr. S. Hofmeister, Prof. Dr. J. Huber und Prof. Dr. A.v. Gleich. In: *Politische Ökologie*, **62**:8-12.

Messner, F. (1999): *Nachhaltiges Wirtschaften mit nicht-erneuerbaren Ressourcen*. Verl. Peter Lang, Frankfurt a.M., Berlin, Bern, Brussels, New York, Wien.

Michanickl, A. and C. Boehme (1995): *Verfahren zur Wiedergewinnung von Spänen und Fasern aus Holzwerkstoffreststücken, Altmöbeln, Produktionsrückständen, Abfällen und anderen holzwerkstoffhaltigen Materialien*. DE-OS 195 09 152 A1.

Miettinen, P. and R.P. Hämäläinen (1997): How to Benefit from Decision Analysis in Environmental Life Cycle Assessment. In: *European Journal of Operational Research*, **102**(2):279-294.

Minsch, J. (2000): Nachhaltige Entwicklung – vom Schlagwort zu Innovationsperspektiven. In: *Proceedings of the 11. Schweizerisches Status-Seminar Energie- und Umweltforschung im Bauwesen*, 2000, Zürich, pp. 33-40.

Mintzberg, H. (1983): *Power In and Around Organisations*. Prentice-Hall Publ., Englewood Cliffs.

Mohamed, A. and T. Celik (1998): An Integrated Knowledge-based System for Alternative Design and Materials Selection and Cost Estimation. In: *Expert Systems With Applications*, **14**(3):329-339.

Möller, A. and S. Herrlich (1994): REHOLZ – ein neuer Werkstoff aus Altholz. In: E. Westkemper (ed.): *Marktorientiert, Schlank und Umweltgerecht; Recycling und ökologische Aspekte; 9. Holztechnisches Kolloquium*. Vulkan-Verl., Braunschweig, pp. 6.1-6.12.

Morecroft, J.D.W. (1994): Executive Knowledge, Models, and Learning. In: J.D.W. Morecroft and J.D. Sterman (eds.): *Modelling for Learning Organizations*. Productivity Press, Portland/OR, pp. 3-28.

Morecroft, J.D.W. and J.D. Sterman (eds.) (1994): *Modelling for Learning Organizations*. System Dynamics Series, Productivity Press, Portland/OR.

Müller, D. (1996): Szenarien zur nachhaltigen regionalen Holzbewirtschaftung. In: *Schweizerische Zeitschrift für Forstwesen*, **147**(11):873-885.

Müller-Wenk, R. (1996a): Examining the Shape of Damage Functions for Some Impact Categories. In: A. Braunschweig, R. Förster, P. Hofstetter and R. Müller-Wenk (eds.): *Developments in LCA Valuation*. IWÖ-Diskussionsbeitrag Nr. 32, Institut für Wirtschaft und Ökologie, Univ. St. Gallen, St. Gallen, pp. 66-83.

Müller-Wenk, R. (1996b): The Shape of the Damage Function and the Choice of Valuation Factors. In: A. Braunschweig, R. Förster, P. Hofstetter and R. Müller-Wenk (eds.): *Developments in LCA Valuation*. IWÖ-Diskussionsbeitrag Nr. 32, Institut für Wirtschaft und Ökologie, Univ. St. Gallen, St. Gallen, pp. 50-65.

Münster, H.P. (1992): Schrottanfall, Sammelwege, Mengen und Preise. In: W. Wuth and J. Zuchowski (eds.): *Recycling metallhaltiger Sekundärstoffe in der Metallindustrie*. Vorträge beim 26. Metallurgischen Seminar des Fachausschusses für Metallurgische Aus- und Weiterbildung der GDMB. Heft 63, GDMB Gesellschaft Deutscher Metallhütten- und Bergleute, Clausthal-Zellerfeld, pp. 1-26.

Neumann, J. (1986): Sammlung und Sortierung von Rohstoffen – besonders Altpapier – und deren Optimierung. In: K.J. Thomé-Kozmiensky (ed.): *Recycling von Holz, Zellstoff und Papier; Beiträge des 5. Internationalen Recycling Congresses*. EF-Verl. für Energie- und Umwelttechnik, Berlin, pp. 40-57.

Newell, S.A. and F.R. Field (1998): Explicit Accounting Methods for Recycling in LCI. In: *Resources, Conservation and Recycling*, **22**(1-2):31-45.

Newell, A. and H. A. Simon (1972): *Human Problem Solving*. Prentice-Hall, Englewood Cliffs/NJ.

Nijhof, G.H. and P.C. Rem (1999): Upgrading Nonferrous Metal Scrap for Recycling Purposes. In: *JOM-Journal of the Minerals, Metals and Materials Society*, **51**(8):20-23.

Ninck, M. (1997): *Zauberwort Nachhaltigkeit*. vdf Hochschulverlag AG an der ETH, Zürich.

Nussbaumer, T. (1994): Anforderungen bei der thermischen Verwertung von Holzreststoffen und -abfällen. In: E. Westkemper (ed.): *Marktorientiert, Schlank und Umweltgerecht; Recycling und ökologische Aspekte; 9. Holztechnisches Kolloquium*. Vulkan-Verl., Braunschweig, pp. 9.1-9.20.

Nussbaumer, T., P. Neuenschwander, P. Hasler, A. Jenni and R. Bühler (1997): *Energie aus Holz; Vergleich der Verfahren zur Produktion von Wärme, Strom und Treibstoff aus Holz*. Im Auftrag des Bundesamtes für Energiewirtschaft (BEW), ENET, Bern.

Odum, P.E. (1991): *Prinzipien der Ökologie; Lebensräume, Stoffkreisläufe, Wachstumsgrenzen*. Spektrum der Wissenschaften, Heidelberg.

Orpin, J. (1996): The Use of Salvaged and Resawn Wood in Fine Homebuilding. In: R. H. Falk (ed.): *The Use of Recycled Wood and Paper in Building Applications*. The Forests Products Society, Madison/Wis., pp. 101-103.

Östermark, U. and T. Rydberg (1995): Reuse Versus Recycling of PET-Bottles; A Case Study of Ambiguities in Life Cycle Assessment. In: *Proceedings of the Recovery, Recycling, Reintegration 1995* (*R'95*), 1995, Geneva, pp. I.249-I.253.

Ovens, W.G. (1997): Principles of Selection. In: G.T. Murray (ed.): *Handbook of Materials Selection for Engineering Applications*. Marcel Dekker Inc., New York, Basel, Hong Kong, pp. 1-70.

Partington, E.C. (1992): Strategic Inventory Management of Renewable Energy Resources; Aluminium as an Energy Storage Medium and Market Commodity. In: *International Journal of Production Economics*, **26**:211-216.

Patosaari, P. (2000): The Swiss Case. Main Results of the Sustainability Assessment of Swiss Forest Policy and its Merits in an International Context. In: *Schweizerische Zeitschrift für Forstwesen*, **151**(12):480-483.

Patterson, M. (1998): Commensuration and Theories of Value in Ecological Economics. In: *Ecological Economics*, **25**(1):105-125.

Pearce, D.W. (1993): *Economic Values and the Natural World*. Earthscan, London.

Pearce, D.W. and G. Atkinson (1995): Measuring Sustainable Development. In: D.W. Bromley (ed.): *Handbook of Environmental Economics*. Basil Blackwell, Oxford, Cambridge, pp. 166-181.

Peek, R.-D. (1998): Schadstoffe in Holzabfällen und ihre Bedeutung beim Recycling. In: *Umweltschutz in der Holzwerkstoffindustrie; Tagungsband der Fachtagung am 24. und 25. Juni in Göttingen*. Institut für Holzbiologie und Holztechnologie, Göttingen, pp. 107-119.

Pennington, N. and R. Hastie, R. (1993). Reasoning in Explanation-Based Decision Making. In: *Cognition*, **49**:123-163.

Peylo, A. (1998a): *Schnellerkennung von Holzkontaminationen – Ein Überblick*. Bundesforschungsanstalt für Forst- und Holzwirtschaft (BFH), Hamburg.

Peylo, A. (1998b): Schnellerkennung von Holzkontaminationen (1). In: *Holz-Zentralblatt*, **124**(46):690-692.

Pfeffer, J. and G.R. Salancik (1978): *The External Control of Organizations – A Resource Dependence Perspective*. Harper & Row Publ., New York.

Plätzer, E., U. Hamm and L. Göttsching (1996): Ökobilanzen: Lassen sich Umweltbelastungen gerecht auf die Papierkette verteilen? In: *Papier*, **50**(10A):V63-V70.

Plinke, E., H. Wolff and H. Meckel (1999): *PVC und Nachhaltigkeit; Systemqualität als Massstab; Ausgewählte Produktsysteme im Vergleich*. Arbeitsgemeinschaft PVC und Umwelt e.V. und Steuerungsgruppe Dialogprojekt PVC und Nachhaltigkeit, Deutscher Instituts-Verl., Köln.

Plume, G.D. (1996): Reclaimed Timber: A Modern Construction Material. In: R. H. Falk (ed.): *The Use of Recycled Wood and Paper in Building Applications*. The Forests Products Society, Madison/Wis., pp. 104-107.

Pohl, C., M. Ros, B. Waldeck and F. Dinkel (1996): Imprecision and Uncertainty in LCA. In: S. Schaltegger (ed.): *Life Cycle Assessment (LCA) – quo vadis?* Birkhäuser Verl., Basel, Boston, Berlin, pp. 51-67.

Poole, S., T.A. Bhamra, S. Evans, T.C. McAloone, M. Simon and A. Sweatman (1999): Integrating Environmental Decisions into the Product Development Process; The Later Stage. In: *Proceedings of the First International Symposium on Environmentally Conscious Design and Inverse Manufacturing*, 1999, Tokyo, pp. 334-337.

Portisch, W. (1996): *Überwachung und Berichterstattung des Aufsichtsrats im Stakeholder-Agency-Modell*. Diss., Univ. Lüneburg, Lüneburg.

Portisch, H.H. (1997): Life-Cycle Engineering and Design. In: G.E. Dieter, S.R. Lampman, G.M Davidson and B.R Sanders (eds.): *Materials Selection and Design*. ASM Handbook, Vol 20, ASM International, Materials Park/OH, pp. 93-100.

Pötzschke, M. (1993): Sekundärrohstoffe aus Gebrauchsgütern; Wiederverwertungstechnologien für Altprodukte. In: Anonymous (ed.): *Recyclinggerechte Produktentwicklung*. Bericht Nr. 1089, VDI-Verl., Düsseldorf, pp. 19-45.

Quinkertz, R., G. Rombach and D. Liebig (2001): A Scenario to Optimise the Energy Demand of Aluminium Production Depending on the Recycling Quota. In: *Resources, Conservation and Recycling*, **33**(2001):217-234.

Radke, V. (1999): *Nachhaltige Entwicklung; Konzept und Indikatoren aus wirtschaftstheoretischer Sicht*. Umwelt und Ökonomie, Vol. 30, Physica-Verl., Heidelberg.

Randall, A. and M.C. Farmer (1995): Benefits, Costs, and the Safe Minimum Standard of Conservation. In: D.W. Bromley (ed.): *Handbook of Environmental Economics*. Basil Blackwell, Oxford (UK), Cambridge/MA, pp. 26-44.

Rawls, J. (1971): *A Theory of Justice*. Harvard University Press, Cambridge/MA.

Rayner, S. (1993): Risikowahrnehmung, Technologieakzeptanz und institutionelle Kultur; Fallstudien für einige neue Definitionen. In: Bayrische Rück (ed.): *Risiko ist ein Konstrukt*. Gesellschaft und Unsicherheit, Bd. 2, Knesebeck GmbH, München, pp. 213-242.

Reagan-Cirincione, P. (1991): *Improving the Accuracy of Forecasts: A Process Intervention Combining Social Judgement Analysis and Group Facilitation*. Diss., State University of New York at Albany, Albany/NY.

Rechberger, H. (1999): *Entwicklung einer Methode zur Bewertung von Stoffbilanzen in der Abfallwirtschaft*. Diss., Technische Univ. Wien, Fakultät für Bauingenieurwesen, Wien.

Regitz, M. and J. Falbe (1995): *CD- Römpp Chemie Lexikon, Version 1.0*. 9. Aufl., Georg Thieme Verl., Stuttgart.

Remmler, N. and M. Fischer (1996): *Kosten und Leistung der Bereitstellung von Waldhackschnitzeln*. Bayerische Landesanstalt für Wald und Forstwirtschaft, Freising.

Rescher, N. (1993): *Rationalität: eine philosophische Untersuchung über das Wesen und die Rechtfertigung von Vernunft*. Königshausen & Neumann GmbH, Würzburg.

Richter, K. (1999): *Life-Sys Wood, FAIR CT95-0726; LCA of Standard Window Frames; Final Document*. Eidgenössische Materialprüfungs- und Forschungsanstalt (EMPA), Dübendorf.

Richter, K. (2000): Altholzverwertung als Teil des Produktlebenszyklus. In: *SAH bulletin CSRB: Altholz zwischen Verwertung und Beseitigung*, **5**:10-15.

Richter, K. (2001): *Personal communication*. Eidg. Materialprüfungs- und Forschungsanstalt (EMPA), Duebendorf.

Richter, K. and K. Brunner (1996): *Ökologische Bewertung von Wärmeschutzglas; Integraler Vergleich verschiedener Verglasungsvarianten*. Eidgenössiche Materialprüfungs- und Forschungsanstalt (EMPA), Duebendorf.

Richter, K., T. Künniger and K. Brunner (1996): *Ökologische Bewertung von Fensterkonstruktionen verschiedener Rahmenmaterialien (ohne Verglasung)*. EMPA-SZFF-Forschungsbericht, Schweizerische Zentralstelle für Fenster- und Fassadenbau (SZFF), Dietikon.

Rink, C. (1993): Aluminium, Automobil und Recycling. In: Anonymous (ed.): *Recyclinggerechte Produktentwicklung*. Bericht Nr. 1089, VDI-Verl., Düsseldorf, pp. 109-127.

Ritzén, S. and M. Norell (1999): Environmental Consciousness in Integrated Product Development. In: *Proceedings of the First International Symposium on Environmentally Conscious Design and Inverse Manufacturing*, 1999, Tokyo, pp. 346-350.

Robinson, T.J.C. (1980): Classical Foundations of the Contemporary Economic Theory of Non-Renewable Resources. In: *Resources Policy*, **9**(4):278-289.

Roeffael, E. (1997): Stand der Kenntnisse beim Recycling von Holzwerkstoffen. In: A. Kharazipour and E. Roeffael (eds.): *Recyclingkonzepte in der Holzwerkstoffindustrie*. Institut für Holzbiologie und Holztechnologie, Göttingen, pp. 4-14.

Rohrbaugh, J. (1979): Improving the Quality of Group Judgement: Social Judgement Analysis and the Delphi Technique. In: *Organizational Behavior and Human Performance*, **24**:73-92.

Rohrbaugh, J. (1981): Improving the Quality of Group Judgement: Social Judgement Analysis and the Nominal Group Technique. In: *Organizational Behavior and Human Performance*, **28**:272-288.

Rombach, G. (1998): Aluminium in offenen und geschlossenen Kreisläufen. In: *ALUMINIUM*, **74**(6):421-424.

Rombach, G. (1999): Komplexität und Machbarkeit von Werkstoffbilanzen. In: Anonymous (ed.): *Nutzen von Ökobilanzen: Vorträge beim 36. Metallurgischen Seminar des Fachausschusses für metallurgische Aus- und Weiterbildung der GDMB*. Schriftenreihe der GDMB Gesellschaft für Bergbau, Metallurgie, Rohstoff- und Umwelttechnik, Heft 85, GDMB-Informationsgesellschaft mbH, Hamburg, pp. 113-130.

Roth (1996): *Personal communication* by R. Roth, Alu Menzikon Guss AG, Menzikon.

Rotmans, J. and M. von Asselt (1996): Integrated Assessment: A Growing Child on Its Way to Maturity. In: J. Rotmans and M. von Asselt (eds.): *Climate Change*. Kluwer Academic Press, Bilthoven.

Rydberg, T. (1995): Cascade Accounting in Life Cycle Assessment Applied to Polymer Recycling. In: *Polymer Recycling*, **1**(4):233-241.

Sachs, L. (1974): *Angewandte Statistik; Planung und Auswertung, Methoden und Modelle*. Springer-Verl., Berlin, Heidelberg, New York.

Salthammer, T., H. Klipp, I. H.-M., R.-D. Peek and R. Marutzky (1993): Untersuchungen über Emissionen bei der thermischen Verwertung von holzschutzmittelhaltigen Holzresten. In: *Proceedings of the 15. Dreiländerholzschutztagung*, 1993, Garmisch-Partenkirchen, pp. 143-153.

Sattler (1992): Automatische Sortierung geshredderter NE-Metallschrotte. In: W. Wuth and J. Zuchowski (eds.): *Recycling metallhaltiger Sekundärstoffe in der Metallindustrie*. Vorträge beim 26. Metallurgischen Seminar des Fachausschusses für Metallurgische Aus- und Weiterbildung der GDMB. Heft 63, GDMB Gesellschaft Deutscher Metallhütten- und Bergleute, Clausthal-Zellerfeld, pp. 125-134.

Schaltegger, S. (1996): Eco-efficiency of LCA; The Necessity of a Site-Specific Approach. In: S. Schaltegger (ed.): *Life Cycle Assessment (LCA) – quo vadis?* Birkhäuser Verl., Basel, Boston, Berlin, pp. 131-149.

Schaltegger, S. (1997a): Economics of Life Cycle Assessment; Inefficiency of the Present Approach. In: *Business Strategy and the Environment*, **6**:1-8.

Schaltegger, S. (1997b): Information Cost, Quality of Information and Stakeholder Involvement – The Necessity of International Standards of Ecological Accounting. In: *Eco-Management and Auditing*, **4**:87-97.

Schaltegger, S. and F. Figge (2000): Shareholder Value dank Stakeholder-Beziehungen; Messung des Werts als strategisches Optimierungsinstrument. In: *Neue Zürcher Zeitung (NZZ)*, **264**(252):103.

Schaltegger, S. and A. Sturm (1992): *Ökologieorientierte Entscheidungen in Unternehmen; Ökologisches Rechnungswesen statt Ökobilanzierung: Notwendigkeit, Kriterien, Konzepte*. Schriftenreihe des Instituts für Betriebswirtschaft, Wirtschaftswissenschaftliches Zentrum der Univ. Basel, Band 27, Verl. Paul Haupt, Bern, Stuttgart, Wien.

Schärer, W. and C. Jakobi (2000): Assessment zur Nachhaltigkeit der Schweizer Forstpolitik – Die Sicht der Eidgenössischen Forstdirektion. In: *Schweizerische Zeitschrift für Forstwesen*, **151**(12):484-488.

Scheller, M. (2001): *Altholzmarkt in der Schweiz; Bericht zur Untersuchung*. Eidgenössische Materialprüfungs- und Forschungsanstalt (EMPA), Dübendorf.

Scheringer, M., A. Wegener Sleesewijk, M. Herrchen, G. Koller, M. Margni and P. Hofstetter (1999): *Chemikalienbewertung in Ökobilanzen; Vorbereitende Unterlagen zum 10. Diskussionsforum Ökobilanzen vom 28. April 1999 an der ETH Zürich*. Eidgenössische Technische Hochschule (ETH), Laboratorium für Technische Chemie, Zürich.

Scherm, E. and A. Frey (1999): *Planung I: Einordnung und Grundlagen der Planung*. Fern-Univ. Gesamtschule, Hagen.

Schey, J.A. (1997): Manufacturing Processes and Their Selection. In: G.E. Dieter, S.R. Lampman, G.M Davidson and B.R. Sanders (eds.): *Materials Selection and Design*. ASM Handbook, Vol. 20, ASM International, Materials Park/OH, pp. 687-704.

Schierge, F. (1996): *Die Bedeutung ökologischer Anspruchgruppen für den strategischen Planungsprozess eines Unternehmens – eine theoretische Abhandlung mit empirischer Validierung*. Diss., Univ. Lüneburg, Lüneburg.

Schneider, F. (1994): Allocation and Recycling: Enlarging to the Cascade System. In: G. Huppes and F. Schneider (eds.): *Proceedings of the European Workshop Allocation in LCA*. Centrum voor Milieukunde (CML), Leiden, Society of Environmental Toxicology and Chemistry (SETAC), Brussels, pp. 39-53.

Schneider, F. (1996): *Analyse des Reemplois, Recyclages, Valorisations de Déchets par l'Etude de Systèmes Cascades*. Diss., Institut National des Sciences Appliquées de Lyon (INSA), Université de Lyon, Lyon.

Scholz, R.W. (1987): *Cognitive Strategies in Stochastic Thinking*. D. Reidel Publishing Company, Dordrecht.

Scholz, R.W. and O. Tietje (2001): *Embedded Case Study Methods; Integrating Quantitative and Qualitative Methods*. SAGE Publications, Thousand Oaks/CA.

Scholz, R.W. and A. Weidenhaupt (1998): Fazit Ökobilanzen; Ökobilanzen auf dem Weg in die Praxis. In: *Bulletin ETH Zürich*, **268**:28-41.

Scholz, R. W. and A. C. Zimmer (1997). Qualitative aspects of decision making: A challenge for decision research. *Psychologische Beiträge* **39**:7-13.

Schrägle, R. (2001): Produktequalität gesichert; Qualitätskonzept für die Spanplattenproduktion des SIA Werks in Italien. In: *Schweizerische Holzzeitung*, **2001**(26):14.

Schuler, A. (2000): Von der Nachhaltigkeit als Beschränkung zur nachhaltigen Entwicklung als Programm. In: *Schweizerische Zeitschrift für Forstwesen*, **151**(12):497-501.

Schulz, H. (1972): Holz im Kreislauf der Wirtschaftsgüter. In: *Holz-Zentralblatt*, **98**(101):1415-1417.

Schulz, H. (1993): Entwicklung der Holzverwendung im 19., 20. und 21 Jahrhundert. In: *Holz als Roh- und Werkstoff*, **51**:78-82.

Schumpeter, J.A. (1993/1939): *Theorie der wirtschaftlichen Entwicklung*. 8. Aufl., Originalausgabe New York 1939, Duncker & Humbolt, Berlin.

Schwarz, M. and M. Thompson (1990): *Divided We Stand; Redefining Politics, Technology and Social Choice*. University of Pennsylvania Press, Philadelphia.

Schweinle, J. (2000): *Analyse und Bewertung der forstlichen Produktion als Grundlage für weiterführende forst- und holzwirtschaftliche Produktlinien-Analysen*. Mitteilungen der Bundesforschungsanstalt für Forst- und Holzwirtschaft (BFH), Nr. 184, 2nd ed., Komissionsverl. Max Wiedebusch, Hamburg.

Schweinle, J. (2001): *Personal communication*. Bundesforschungsanstalt für Forst- und Holzwirtschaft (BFH), Hamburg.

Sell, J. (1997): *Eigenschaften und Kenngrössen von Holzarten*. 4. überarbeitete und erweiterte Aufl., Lignum bei Baufachverl. Zürich, Dietikon.

Senge, P.M. and J.D. Sterman (1994): Systems Thinking and Organizational Learning: Acting Locally and Thinking Globally in the Organization of the Future. In: J.D.W. Morecroft and J.D. Sterman (eds.): *Modelling for Learning Organizations*. Productivity Press, Portland/OR, pp. 195-216.

Sieben, G. and T. Schildbach (1994): *Betriebswirtschaftliche Entscheidungstheorie*. 4th ed., Werner-Verl., Düsseldorf.

Siebert, H. (1983): *Ökonomische Theorie natürlicher Ressourcen*. Verl. J.C.B. Mohr, Tübingen.

Siebert, H. (1984): Allocation Policy in a Resource – Exporting Country. In: Siebert H (ed.): *The Resource Sector in Open Economy*. Lecture Notes in Economics and Mathematical Systems, 200, Springer-Verl., Berlin, pp. 1-18.

Sikkema, R. and G.-J. Nabuurs (1995): Forest and Forest Products: the Challenge for a Better Carbon Balance. In: A. Frühwald and B. Solberg (eds.): *Life-Cycle Analysis – A Challenge for Forestry and Forest Industry*. EFI Proceedings No 8, European Forest Institute, Hamburg, pp. 169-175.

Sillekens, W.H., T. Sano, M. Terasaki, K. Matsuno and J.A.G. Kals (1997): Formability of Recycled Aluminium Alloy 5017. In: *Journal of Materials Processing Technology*, **65**(1-3):252-260.

Sirkin, T. and M. ten Houten (1994): The Cascade Chain; A Theory and Tool for Achieving Resource Sustainability with Applications for Product Design. In: *Resources, Conservation and Recycling*, **10**(3):213-277.

Skog, K.E. and G.A. Nicholson (1998): Carbon Cycling Through Wood Products; The Role of Wood and Paper Products in Carbon Sequestration. In: *Forest Products Journal*, **48**(7/8):75-83.

Slade, M.E. (1980): An Econometric Model of the U.S. Secondary Copper Industry; Recycling Versus Disposal. In: *Journal of Environmental Economics and Management*, **7**:123-141.

Smith, R.L. and R.-J. Shiau (1998): An Industry Evaluation of the Reuse, Recycling, and Reduction of Spent CCA Wood Products. In: *Forest Products Journal*, **48**(2):44-48.

Smith, V.K. (1978): Measuring Natural Resource Scarcity; Theory and Practice. In: *Journal of Environmental Economics and Management*, **5**:150-171.

Solow, R.M. (1974): The Economics of Resources, or the Resources of Economics. In: *American Economic Review*, **64**(2):1-14.

Solow, R.M. (1993): An Almost Practical Step Toward Sustainability. In: *Resources Policy*, **19**:162-172.

Spencer-Brown, G. (1972): *Laws of Form*. Julian Publ., New York.

Staehle, H. and M.E. Nork (1992): Umweltschutz und Theorie der Unternehmung. In: U. Steger (ed.): *Handbuch des Umweltmanagements – Anforderungen und Leistungsprofile von Unternehmen und Gesellschaft*. C.H. Beck Verl., München.

Stahel, R., B.A. Schuler and E. Ledergerber (1987): *Altholz – eine vernachlässigte Ressource*. NFP 12 Holz, erneuerbare Rohstoff- und Energiequelle, Verl. Rüegger, Grüsch.

Steens, B. and S.O. Ryding (1992): *The EPS-Enviro-Accounting Method; An Application of Environmental Accounting Principles for Evaluation and Valuation of Environmental Impact in Product Design*. IVL Report No B1080, Swedish Environmental Research Institute (IVL), Gothenburg.

Stern, D.I. (1999): Use Value, Exchange Value and Resource Scarcity. In: *Energy Policy*, **27**(8):469-476.

Streissler, E. (1980): Die Knappheitsthese; Begründete Vermutung oder vermutete Fakten. In: H. Siebert (ed.): *Erschöpfbare Ressourcen; Verhandlungen auf der Arbeitstagung des Vereins für Socialpolitik der Gesellschaft für Wirtschafts- und Sozialwissenschaften in Mannheim 1979*. Neue Folge Band 108, Verl. Duncker & Humbolt, Berlin, pp. 9-36.

Stuip, J. (1998): Experiences with the Application of Secondary Material in the Building and Construction Industry. In: J.E.M. Klostermann and A. Tukker (eds.): *Product Innovation and Eco-efficiency, Twenty-three Industry Efforts to Reach the Factor 4*. Eco-Efficiency in Industry, Vol. 1, Kluwer Academic Publ., Dordrecht, Boston, London, pp. 199-212.

Sutton, W.R.J. (1993): *The World's Need for Wood. The Globalisation of Wood: Supply, Processes, Products, and Markets*. Forest Products Society, Madison/Wis.

Swan, P.L. (1980): Alcoa: The Influence of Recycling on Monopoly Power. In: *Journal of Law and Economics*, **88**(1):76-99.

Taguchi, G. (1986): *Quality Engineering; Minimierung von Verlusten durch Prozessbeherrschung*. gfmt-Gesellschaft für Management und Technologie-Verlags KG, München.

Thompson, J. (1967): *Organizations in Action*. McGraw Hill Publ., New York.

Thompson, M., R. Ellis and A. Wildavsky (1990): *Cultural Theory*. Westview Press, Boulder.

Tillman, A.M., T. Ekvall, H. Baumann and T. Rydberg (1994): Choice of System Boundaries in Life Cycle Assessment. In: *Journal of Cleaner Production*, **2**(1):21-29.

Turner, R.K., R. Grace and D.W. Pearce (1977): The Economics of Waste Paper Recycling. In: D.W. Pearce and I. Walter (eds.): *Resource Conservation: Social and Economic Dimensions of Recycling*. Longman Publ., London, pp. 296-343.

Uchida, H.H., Y. Ozu, S. Ohga and M. Ohga (1999): Possibility of Inverse Manufacturing of Coated Aluminium Product. In: *Proceedings of the First International Symposium on Environmentally Conscious Design and Inverse Manufacturing*, 1999, Tokyo, pp. 985-988.

Udry, H. (2001): *Personal communication*. Bundesamt für Statistik (BfS), Neuchâtel.

Ulli-Beer, S. (2002): The Role of Mental Models and Group Model Building in the Debate About Sustainability at the Local Level. In: M. Flury (ed.): *Local Environmental Management in a North-South Perspective; Issues of Participation and Knowledge Management*. Themenheft des Schwerpunktprogramms Umwelt (SPPU). Birkhäuser Verl., Basel.

Ulrich, H. (1981): Die BWL als anwendungsorientierte Sozialwissenschaft. In: M. Geist and R. Köhler (eds.): *Die Führung des Betriebes*. Poeschel, Stuttart, pp. 1-26.

Vaassen, B. (1996): *Die narrative Gestalt(ung) der Wirklichkeit; Grundlinien einer postmodern orientierten Epistemologie der Sozialwissenschaften*. Verl. Viehweg, Braunschweig, Wiesbaden.

van Asselt, M., J. Rotmans, M. den Elzen and H.B.B. Hilderink (1995): *Uncertainty in Integrated Assessment Modelling; a Cultural Perspective-based Approach*. GLOBO Report Series No 9, RIVM Report No. 461502009, Bilthoven.

van Weenen, J.C. (1995): Towards Sutainable Product Development. In: *Journal of Cleaner Production*, **3**(1-2):95-100.

Varian, H.R. (1995): *Grundzüge der Mikroökonomik*. 3. Aufl., R. Oldenbourg Verl., München, Wien.

Various (1992): *Materials Interactions Relevant to Recycling of Wood-Based Materials; Symposium held April 27-29, 1992, San Francisco/CA*. Materials Research Society Symposium Proceedings, Vol 266, Materials Research Society, Pittsburgh/PE.

Various (1997): Fourth International Conference on Woodfiber-Plastic Composites. In: *Proceedings of the Fourth International Conference on Woodfiber-Plastic Composites*, 1997, Madison/Wis., pp. 1-324.

Various (2000): *Land Use, Land-use Change, and Forestry; a Special Report of the IPCC*. Cambridge University Press, Cambridge.

Vásquez, M., M. Liz and J. Aracil (1996): Knowledge and Reality: Some Conceptual Issues in System Dynamics Modelling. In: *System Dynamics Review*, **12**(1):21-37.

Vennix, J.A.M. (1996): *Group Model Building; Facilitating Team Learning Using System Dynamics*. John Wiley & Sons, Chichester, New York, Brisbane, Toronto, Singapore.

Vennix, J.A.M. and J.W. Gubbels (1992): Knowlegde Elicitation in Conceptual Model Building: A Case Study In Modelling A Regional Dutch Health Care System. In: *European Journal of Operational Research*, **59**(1992):85-101.

Vennix, J.A.M., A. Henk, A. Akkermans and A.J.A. Rouwette (1996): Group Model-Building to Facilitate Organizational Change: An Exploratory Study. In: *System Dynamics Review*, **12**(1):39-58.

Vester, F. (1980): *Neuland des Denkens; vom technokratischen bis zum kybernetischen Zeitalter.* Deutsche Verlags-Anstalt GmbH, Stuttgart.

Vigon, B.W., D.A. Tolle, B.W. Cornaby, H.C. Latham, C.L. Harrison, T.K. Boguski, R.G. Hunt and J.D. Sellers (1993): *Life Cycle Assessment: Inventory Guidelines and Principles.* EPA/600/R-92/245, United States Environmental Protection Agency, Office of Research and Development, Washington/D.C., Cincinnati/OH.

Vock, W. (2000): Informationen zum Altholzfluss und Resultate aus neuen Untersuchungen. In: *SAH bulletin CSRB: Altholz zwischen Verwertung und Beseitigung*, **5**:4-6.

Vock, W., A. Jenni and P.A. Steinmann (1993): *Energie 2000-DIANE 8; Energie aus Altholz und Altpapier; Übersicht Schweiz: Situationsanalyse der energetischen Nutzung von Altholz und Altpapier.* Bundesamt für Energiewirtschaft (BEW), Bern.

Vogtländer, J.G., H.C. Brezet and C.F. Hendriks (2001): Allocation in Recycling Systems. In: *International Journal for Life Cycle Assessment*, **6**(6):344-355.

von Hayek, F.A. (1945): The Use of Knowledge in Society. In: *The American Economic Review*, **35**(4):519-530.

von Winterfeldt, V. and W. Edwards (1986): *Decision Analysis and Behavioral Research.* Cambridge University Press, New York.

Voss, A. (1998): *Aufkommen und Zusammensetzung schutzmittelbehandelter Althölzer und ihre Entsorgung.* Mitteilungen der Bundesforschungsanstalt für Forst- und Holzwirtschaft, Nr. 188, Bundesforschungsanstalt für Forst- und Holzwirtschaft (BFH), Hamburg.

Walter, R. (1998): *Personal communication*, INDRA Recycling GmbH. Hockenheim.

Watzlawick, P. (ed.) (1984): *Die erfundene Wirklichkeit.* Piper Verl., München.

Wegener, G., M. Beudert, A. Frühwald, K. Dreiner and M. Scharai-Rad (1994): *Bewertung von Holz im Vergleich mit anderen Werkstoffen unter dem Aspekt der CO_2-Bilanz.* Holzabsatzfonds, Bonn.

Weidema, B.P. (1998a): Application Typologies for Life Cycle Assessment – A Review. In: *International Journal for Life Cycle Assessment*, **3**(4):237-240.

Weidema, B.P. (1998b): New Developments in the Methodology for Life Cycle Assessment. In: *Proceedings of the 3rd International Conference on Ecobalance*, 1998, Tsukuba, http://www.ipt.dtu.dk/employ/bow/Developh.htm.

Weidema, B.P. (1999): Some Important Aspects of Market-based System Delimitation in LCA – with a Special View to Avoiding Allocation. In: *Report of a Danish-Dutch Workshop on LCA Methodologies*, September 1999. CML, Leiden, pp. 33-46.

Weidema, B.P., N. Frees and A.-M. Nielsen (1999): Marginal Production Technologies for Life Cycle Inventories. In: *International Journal for Life Cycle Assessment*, **4**(1):48-55.

Weis, N., U. Thurmann and W. Wosniok (1999): Praktikable Altholzanalyse im Versuch; Probenahmestrategie zur Erfassung von Belastungen inhomogener Altholzmischsortimente. In: *Holz-Zentralblatt*, **125**(145):1964.

Weiss, V. (1997): Computer-Aided Materials Selection. In: G.E. Dieter, S.R. Lampman, G.M Davidson and B.R. Sanders (eds.): *Materials Selection and Design.* ASM Handbook, Vol. 20, ASM International, Materials Park/OH, pp. 309-314.

Wellmer, F.-W. (1997): Factors Useful for Predicting Future Mineral-Commodity Supply Trends. In: *Geologische Rundschau*, **86**:311-321.

Wellmer, F.-W. (1998): Lebensdauer und Verfügbarkeit energetischer und mineralischer Ressource. In: *Erzmetall*, **51**(10):663-675.

Wenzel, H. (1998): Application Dependency of LCA Methodology: Key Variables and Their Mode of Influencing the Method. In: *International Journal for Life Cycle Assessment*, **3**(5):281-288.

Wenzel, H. and L. Alting (1999): Danish Experience with the EDIP Tool for Environmental Design of Industrial Products. In: *Proceedings of the First International Symposium on*

Environmentally Conscious Design and Inverse Manufacturing, 1999, Tokyo, pp. 370-379.

Wenzel, H., M. Hauschild and E. Rasmussen (1996): *Milijøvurdering af produkter*. Milijøstyrelsen and Dansk Industri, Copenhagen.

Werner, F. (1990): *Die Abfallwirtschaft als Teil der Rohstoffwirtschaft: eine Untersuchung über den ökonomischen und empirischen Einfluss auf die unternehmerischen Entscheidungen in der betrieblichen Rohstoffwirtschaft, im Recycling sowie in der Umwelterhaltung als Funktionen in der Materialwirtschaft*. VDI-Verl., Düsseldorf.

Werner, F. (1999): Nachhaltigkeit im Holzbau. In: *Schweizer Ingenieur und Architekt*, **117**(4):30-32.

Werner, F. (2000): *Treatment of Recycling of Aluminium in LCA; Development and Evaluation of the Value-Corrected Substitution Procedure Applied to Window Frames*. Research and Work Report 115/47, Eidg. Materialprüfungs- und Forschungsanstalt (EMPA), Duebendorf.

Werner, F., R. Taverna, P. Hofer and K.Richter (2004): Greenhouse Gas Dynamiks of an Increased Use of Wood in Buildings in Switzerland. In: *Climatic Change*, in press.

Werner, F. and K. Richter (1998a): *Ökobilanz eines Aluminiumfensters nach Massgabe gegebener Recycling-Stoffströme; Eine Ergänzung zur EMPA/SZFF-Fensterstudie*. Eidg. Materialprüfungs- und Forschungsanstalt (EMPA), Duebendorf.

Werner, F. and K. Richter (1998b): *Ökologische Untersuchung von WICONA-Aluminiumfenstern als Ergänzung zur EMPA/SZFF-Fensterstudie unter Verwendung eines neuen Allokationsansatzes für Alu-Recycling-Prozesse*. Eidg. Materialprüfungs- und Forschungsanstalt (EMPA), Duebendorf.

Werner, F. and K. Richter (2000a): Economic Allocation and Value-Corrected Substitution; Reply to the 'Letter to the Editor' by Gjalt Huppes. In: *International Journal for Life Cycle Assessment*, **5**(4):189-190.

Werner, F. and K. Richter (2000b): Economic Allocation in LCA: A Case Study About Aluminium Window Frames. In: *International Journal for Life Cycle Assessment*, **5**(2):79-83.

Wernick, I.K. and N.J. Themelis (1998): Recycling Metals for the Environment. In: *Annual Review of Energy and the Environment*, **23**:465-497.

Wiegand, U., M. Quetting and C. Mehlisch (1996): *Endverbrauch des Holzes in der Schweiz 1996*. Umwelt-Materialien Nr. 94, Bundesamt für Umwelt, Wald und Landschaft (BUWAL), Bern.

Wildavsky, A. (1993): Vergleichende Untersuchung zur Risikowahrnehmung: Ein Anfang. In: Bayrische Rück (ed.): *Risiko ist ein Konstrukt*. Gesellschaft und Unsicherheit, Bd. 2, Knesebeck GmbH, München, pp. 191-211.

Willeitner, H. and C. Bucki (1994): Marktorientiert, Schlank und Umweltgerecht; Recycling und ökologische Aspekte; 9. Holztechnisches Kolloquium. In: E. Westkemper (ed.): *Marktorientiert, Schlank und Umweltgerecht; Recycling und ökologische Aspekte; 9. Holztechnisches Kolloquium*. Vulkan-Verl., Braunschweig, pp. 5.1-5.10.

Winkler, H.D. (1996): *Problematik der Altholzentsorgung. byse-Symposium Altholzrecycling – ein Markt mit Perspektiven*. Umweltamt Nordrhein-Westfalen, Düsseldorf.

Withagen, C.A.A.M. (1996): Sustainability and Investment Rules. In: *Economics Letters*, **53**:1-6.

Wittke, B. (1998): Recycling von Holzwerkstoffen im industriellen Massstab. In: *Umweltschutz in der Holzwerkstoffindustrie; Tagungsband der Fachtagung am 24. und 25. Juni in Göttingen*. Institut für Holzbiologie und Holztechnologie, Göttingen, pp. 90-96.

Wöhe, G. (1996): *Einführung in die allgemeine Betriebswirtschaftslehre*. 19th ed., Verl. Franz Vahlen GmbH, Müchnen.

Woodward, R. (1995): Aluminium Alloys for Engineering Applications. In: *Proceedings of the Conference Materials Selection and Design*, 1995, London, pp. 161-168.
Zimbardo, P.G. (1992): *Psychologie*. 5. Aufl., Springer Verl., Berlin, Heidelberg, New York.
Zimmermann, P., G. Doka, F. Huber, A. Labhardt and M. Ménard (1996): *Ökoinventare für Entsorgungsprozesse; Grundlagen zur Integration der Entsorgung in Ökobilanzen*. ESU-Reihe Nr. 1/96, Eidg. Technische Hochschule (ETH), Zürich.
Züst, R. (1998): Ökologie-orientierte Beurteilung von Produkten und Dienstleistungen – Grundlagen und Beispiele. In: R. Züst and A. Schlatter (eds.): *Eco-Performance; Beiträge zum betrieblichen Umweltmanagement*. Verl. Eco-Performance, Zürich, pp. 67-83.

Standards

ISO/EN 14001 (1996): Environmental Management Systems – Specification with Guidance for Use (ISO/EN 14001:1996). European Committee for Standardization, Brussels.
ISO/TR 14025:2000: Environmental Labels and Declarations – Type III Environmental Declarations (ISO/TR 14025:2000(E)). International Organization for Standardization, Geneva.
ISO/EN 14040 (1997): Environmental Management – Life Cycle Assessment – Principles and Framework (ISO/EN 14040:1997). ISO/TC 207/SC5, CEN European Committee for Standardization, Brussels.
ISO/EN 14041 (1998): Environmental Management – Life Cycle Assessment – Goal and Scope Definition and Inventory Analysis (ISO/EN 14041:1998(E)). International Organization for Standardization, Geneva.
ISO/EN 14042 (1999): Environmental Management – Life Cycle Assessment – Life Cycle Impact Assessment (ISO/EN 14042:1999(E)). International Organization for Standardization, Geneva.
ISO/EN 14043 (1999): Environmental Management – Life Cycle Assessment – Life Cycle Interpretation (ISO/EN 14043:1999(E)). International Organization for Standardization, Geneva.
ISO/CD 14048 (2000): Environmental Management – Life Cycle Assessment – Data Documentation Format (ISO/CD 14048). International Organization for Standardization, Geneva.
ISO/TR 14049 (2000): Environmental Management – Life Cycle Assessment – Examples of Application of ISO 14041 to Goal and Scope Definition and Inventory Analysis (ISO/TR 14049:2000). International Standard Organisation (ISO), Geneva.
DIN 1725/5 (1986): Aluminiumlegierungen, Gusslegierungen, Blockmetall (Masseln) Flüssigmetall, Zusammensetzung (DIN 1725:1986, Teil 5). Deutsches Institut für Normung e.V. (DIN), Stuttgart.
ISO/EN 8402 (1995): Quality Management and Quality Assurance – Vocabulary (ISO/EN 8402:1994). European Committee for Standardization (CEN), Brussels.
ISO/DIS 9000 (1999): Quality Management Systems- Fundamentals and Vocabulary (ISO/DIS 9000:1999). Schweizerische Normenvereinigung (SNV), Zurich.

Swiss and German Legislation, International Agreements

Swiss legislation

Bundesgesetz über den Umweltschutz (Umweltschutzgesetz, USG) (SR 814.01), vom 7. Oktober 1983 (Stand am 21 Dezember 1999), http://www.admin.ch/ch/d/sr/8/814.01.de.pdf.

Bundesgesetz über die Reduktion der CO_2-Emissionen (SR 641.71), vom 8. Oktober 1999 (Stand am 18. April 2000), http://www.admin.ch/ch/d/sr/6/641.71.de.pdf.

Luftreinhalteverordnung (LVR) (SR 814.318.142.1), vom 16. Dezember 1985 (Stand am 28. März 2000), http://www.admin.ch/ch/d/sr/8/814.318.142.1.de.pdf.

Verordnung über den Verkehr mit Sonderabfällen (VVS) (SR 814.610), vom 12. November 1986 (Stand am 1. September 1998), http://www.admin.ch/ch/d/sr/8/814.610.de.pdf.

Technische Verordnung zu Abfällen (TVA) (SR 814.600.), vom 10. Dezember 1990 (Stand am 28. März 2000, http://www.admin.ch/ch/d/sr/8/814.600.de.pdf.

Verordnung über umweltgefährdende Stoffe (Stoffverordnung, StoV) (SR 814.013), vom 9. Juni 1996 (Stand am 6. März 2001), http://www.admin.ch/ch/d/sr/8/814.013.de.pdf.

German legislation

Gesetz zur Förderung der Kreislaufwirtschaft und Sicherung der umweltverträglichen Beseitigung von Abfällen (Kreislaufwirtschafts- und Abfallgesetz – KrW-(AbfG), vom 27. September 1994, BGBL. I 1994, S. 2705, http://www.deponie.de/gesetze/krwabfg.pdf.

International agreements

Basler Übereinkunft vom 22. März 1989 über die Kontrolle der grenzüberschreitenden Verbringung gefährlicher Abfälle und ihrer Entsorgung (mit Anlagen) (SR 0.814.05, vom 22. März 1989 (Stand am 1. Oktober 1997), http://www.admin.ch/ch/d/sr/i8/0.814.05.de.pdf.

Decision of the Council concerning the control of transfrontier movements of wastes destined for recovery opertations (OECD decision C(92)39/FINAL), adopted by the Council at its 778th Session on 30th March 1992, http://www.buwal.ch/abfall/docu/pdf/c_92_39_final.pdf.

Index

Eco-Efficiency in Industry and Science

1. J.E.M. Klostermann and A. Tukker (eds.): *Product Innovation and Eco-efficiency.* Twenty-three Industry Efforts to Reach the Factor 4. 1997 ISBN 0-7923-4761-7
2. K. van Dijken, Y. Prince, T. Wolters, M. Frey, G. Mussati, P. Kalff, O. Hansen, S. Kerndrup, B. Søndergård, E. Lopes Rodrigues and S. Meredith (eds.): *Adoption of Environmental Innovations.* The Dynamics of Innovation as Interplay Between Business Competence, Environmental Orientation and Network Involvement. 1999 ISBN 0-7923-5561-X
3. M. Bartolomeo, M. Bennett, J.J. Bouma, P. Heydkamp, P. James, F. de Walle and T. Wolters: *Eco-Management Accounting.* 1999 ISBN 0-7923-5562-8
4. P.P.A.A.H. Kandelaars: *Economic Models of Material-Product Chains for Environmental Policy Analysis.* 1999 ISBN 0-7923-5794-9
5. J. de Beer: *Potential for Industrial Energy-Efficiency Improvement in the Long Term.* 2000 ISBN 0-7923-6282-9
6. K. Green, P. Groenewegen and P.S. Hofman (eds.): *Ahead of the Curve.* Cases of Innovation in Environmental Management. 2001 ISBN 0-7923-6804-5
7. J.B. Guinée (ed.): *Handbook on Life Cycle Assessment.* Operational Guide to the ISO Standards. 2002 ISBN 1-4020-0228-9
8. T.J.N.M. de Bruijn and A. Tukker (eds.): *Partnership and Leadership.* Building Alliances for a Sustainable Future. 2002 ISBN 1-4020-0431-1
9. M. Bennett, J.-J. Bouma and T. Wolters (eds.): *Environmental Management Accounting.* Informal and Institutional Developments. 2002 ISBN 1-4020-0552-0; Pb: ISBN 1-4020-0553-9
10. N. Wrisberg and H.A. Udo de Haas (eds.): *Analytical Tools for Environmental Design and Management in a Systems Perspective.* 2002 ISBN 1-4020-0453-2
11. R. Heijungs and S. Suh: *The Computational Structure of Life Cycle Assessment.* 2002 ISBN 1-4020-0672-1
12. M. Bennett, P.M. Rikhardsson and S. Schaltegger (eds.): *Environmental Management Accounting - Purpose and Progress.* 2003 ISBN 1-4020-1365-5; Pb: ISBN 1-4020-1366-3
13. R.U. Ayres, L.W. Ayres and I. Råde: *The Life Cycle of Copper, Its Co-Products and Byproducts.* 2003 ISBN 1-4020-1552-6
14. R. Kuehr and E. Williams (eds.): *Computers and the Environment: Understanding and Managing their Impacts.* 2003 ISBN 1-4020-1679-4
15. K. Blok, H.L.F. de Groot, E.E.M. Luiten and M.G. Rietbergen: *The Effectiveness of Policy Instruments for Energy-Efficiency Improvement in Firms.* The Dutch Experience. 2004 ISBN 1-4020-1965-3
16. M.S. Wenk: *The European Union's Eco-Management and Audit Scheme (EMAS).* 2005 ISBN 1-4020-3212-9
17. F. Werner: *Ambiguities in Decision-oriented Life Cycle Inventories.* The Role of Mental Models and Values. 2005 ISBN 1-4020-3253-6